Methods in Molecular Biology

Series Editor
John M. Walker
School of Life and Medical Sciences
University of Hertfordshire
Hatfield, Hertfordshire, UK

For further volumes:
http://www.springer.com/series/7651

For over 35 years, biological scientists have come to rely on the research protocols and methodologies in the critically acclaimed *Methods in Molecular Biology* series. The series was the first to introduce the step-by-step protocols approach that has become the standard in all biomedical protocol publishing. Each protocol is provided in readily-reproducible step-by step fashion, opening with an introductory overview, a list of the materials and reagents needed to complete the experiment, and followed by a detailed procedure that is supported with a helpful notes section offering tips and tricks of the trade as well as troubleshooting advice. These hallmark features were introduced by series editor Dr. John Walker and constitute the key ingredient in each and every volume of the *Methods in Molecular Biology* series. Tested and trusted, comprehensive and reliable, all protocols from the series are indexed in PubMed.

Measles and Related Morbilliviruses

Methods and Protocols

Edited by

Dzwokai Z. Ma

Molecular, Cellular, and Developmental Biology, University of California Santa Barbara, Santa Barbara, CA, USA

Christian K. Pfaller

Department of Molecular Medicine, Mayo Clinic, Rochester, MN, USA

Humana Press

Editors
Dzwokai Z. Ma
Molecular, Cellular, and Developmental Biology
University of California Santa Barbara
Santa Barbara, CA, USA

Christian K. Pfaller
Department of Molecular Medicine
Mayo Clinic
Rochester, MN, USA

ISSN 1064-3745 ISSN 1940-6029 (electronic)
Methods in Molecular Biology
ISBN 978-1-0716-3869-9 ISBN 978-1-0716-3870-5 (eBook)
https://doi.org/10.1007/978-1-0716-3870-5

This Humana imprint is published by the registered company Springer Science+Business Media, LLC, part of Springer Nature.
The registered company address is: 1 New York Plaza, New York, NY 10004, U.S.A.

Paper in this product is recyclable.

Preface

Morbilliviruses form a genus in the family *Paramyxoviridae* within the order *Mononegavirales*. As such, they possess a non-segmented, single-stranded, encapsidated RNA genome of negative polarity, which is transcribed and replicated by a viral RNA-dependent RNA polymerase [1]. The viral particles have a lipid envelope that is decorated with two glycoproteins, the hemagglutinin and fusion proteins. These two proteins mediate receptor binding and membrane fusion, respectively [1]. Within the morbilliviruses, the prototype species is the human-pathogenic measles virus (MeV). It is one of the most contagious viruses known to date and has been a major cause of childhood morbidity and mortality throughout the history of mankind [1]. Since the development of a live-attenuated vaccine in the 1950s by John Franklin Enders [2], measles cases and fatalities have dropped worldwide, and the World Health Organization had set the goal for measles elimination in the 1980s [3]. However, this goal has faced several setbacks in the last 30 years due to increasing vaccine hesitancy in all parts of the world. The COVID-19 pandemic further intensified the global resurgence of MeV infection. This was even seen in countries that previously maintained the status of measles elimination for nearly two decades, such as the United States [3]. Today, measles and its causative MeV remain a major risk to human health due to not only its highly infectious nature but also its pathogenic capability.

The substantial pathogenicity of morbilliviruses arises from their molecular biology. They enter a host through droplet transmission in the upper respiratory tract [1]. However, in contrast to most other respiratory viruses, morbilliviruses first infect sentinel immune cells, such as dendritic cells and macrophages, present in the lung tissue [1]. Infection of both immune cells requires the expression of the surface receptor protein CD150, also known as signaling lymphocytic activation molecule (SLAM) [4]. This initial infection is followed by the prodromal phase, during which the virus affects the myeloid and lymphocytic immune cells throughout the lymphatic system. This subsequently leads to severe leukopenia and prolonged immunosuppression and can result in serious complications from opportunistic secondary infections [1]. Moreover, virus-mediated killing of memory B cells can induce the permanent loss of preexisting immunity, a condition recently termed immune amnesia [5, 6]. Finally, the virus returns to the upper respiratory tract, where infected immune cells transmit it to epithelial cells expressing the nectin-4 receptor protein [7, 8]. After replicating in these epithelial cells, viral particles are eventually shed into the airway lumen and are transmitted to the next host [9]. In rare cases, about 1 in 10,000 MeV can enter the brain and establishes a persistent infection that always results in fatal neurodegenerative diseases such as subacute sclerosing panencephalitis (SSPE) or measles inclusion body encephalitis (MIBE) [10]. Some other morbilliviruses exhibit increased neurotropism compared to MeV.

While MeV naturally infects only humans and non-human primates, other morbilliviruses infect different host species. Canine distemper virus (CDV) infects carnivores such as domestic dogs, raccoons, and ferrets [11]. The recently identified feline morbilliviruses (FeMV) infect feline species [12]. The extensive range of morbillivirus hosts is further emphasized by the viruses' ability to infect a wide range of species (e.g., cattle by rinderpest virus, and sheep and goats by small ruminant morbillivirus) [1]. Since pathogenesis of all morbilliviruses is very similar in their respective hosts, investigation of CDV in ferrets can serve as a surrogate model for the pathogenesis of MeV infection in humans, as well as for the development of novel antiviral treatments for morbillivirus-related diseases. A notable exemption may be FeMV, since it is associated with atypical severe kidney disease in cats, in addition to the hallmarks of morbillivirus infection.

This book provides a source for protocols to study MeV and related morbilliviruses and will also be relevant to researchers studying other members of *Paramyxoviridae* and *Mononegavirales*. We have collected state-of-the-art protocols to investigate the molecular biology of morbilliviruses, including cell entry, replication, and virus-host interactions. Additionally, this volume contains methods for studying morbillivirus infection by employing novel tissue culture and in vivo models and for analyzing the biological implications of infection by using primary tissues isolated from infected animals. Lastly, we present cutting-edge diagnostic tools to evaluate humoral immunity against MeV.

We would like to extend our sincere gratitude to those who contributed to individual chapters and those who provided assistance throughout the publication process.

Santa Barbara, CA, USA *Dzwokai Z. Ma*
Rochester, MN, USA *Christian K. Pfaller*

References

1. Griffin DE (2022) Measles virus. In: Howley PM, Knipe DM, Whelan S, Freed EO (eds) Fields virology: RNA viruses, vol 3, 7 edn. Wolters Kluwer Health, Philadelphia, pp 228–266
2. Enders JF, Katz SL, Milovanovic MV, Holloway A (1960) Studies on an attenuated measles-virus vaccine. I. Development and preparations of the vaccine: technics for assay of effects of vaccination. N Engl J Med 263:153–159. https://doi.org/10.1056/nejm196007282630401
3. WHO Team IVB (2017) Measles vaccines: WHO position paper. Wkly Epidemiol Rec 92(17): 205–228
4. Tatsuo H, Ono N, Tanaka K, Yanagi Y (2000) SLAM (CDw150) is a cellular receptor for measles virus. Nature 406(6798):893–897. https://doi.org/10.1038/35022579
5. Mina MJ, Kula T, Leng Y, Li M, de Vries RD, Knip M, Siljander H, Rewers M, Choy DF, Wilson MS, Larman HB, Nelson AN, Griffin DE, de Swart RL, Elledge SJ (2019) Measles virus infection diminishes preexisting antibodies that offer protection from other pathogens. Science 366(6465): 599–606. https://doi.org/10.1126/science.aay6485
6. Petrova VN, Sawatsky B, Han AX, Laksono BM, Walz L, Parker E, Pieper K, Anderson CA, de Vries RD, Lanzavecchia A, Kellam P, von Messling V, de Swart RL, Russell CA (2019) Incomplete genetic reconstitution of B cell pools contributes to prolonged immunosuppression after measles. Sci Immunol 4(41). https://doi.org/10.1126/sciimmunol.aay6125
7. Muhlebach MD, Mateo M, Sinn PL, Prufer S, Uhlig KM, Leonard VH, Navaratnarajah CK, Frenzke M, Wong XX, Sawatsky B, Ramachandran S, McCray PB, Jr., Cichutek K, von Messling V, Lopez M, Cattaneo R (2011) Adherens junction protein nectin-4 is the epithelial receptor for measles virus. Nature 480 (7378):530–533. https://doi.org/10.1038/nature10639
8. Noyce RS, Bondre DG, Ha MN, Lin LT, Sisson G, Tsao MS, Richardson CD (2011) Tumor cell marker PVRL4 (nectin 4) is an epithelial cell receptor for measles virus. PLoS Pathog 7(8):e1002240. https://doi.org/10.1371/journal.ppat.1002240
9. Hippee CE, Singh BK, Thurman AL, Cooney AL, Pezzulo AA, Cattaneo R, Sinn PL (2021) Measles virus exits human airway epithelia within dislodged metabolically active infectious centers. PLoS Pathog 17(8):e1009458. https://doi.org/10.1371/journal.ppat.1009458

10. Cattaneo R, Schmid A, Eschle D, Baczko K, ter Meulen V, Billeter MA (1988) Biased hypermutation and other genetic changes in defective measles viruses in human brain infections. Cell 55(2):255–265. https://doi.org/10.1016/0092-8674(88)90048-7
11. Beineke A, Puff C, Seehusen F, Baumgärtner W (2009) Pathogenesis and immunopathology of systemic and nervous canine distemper. Vet Immunol Immunopathol 127(1–2):1–18. https://doi.org/10.1016/j.vetimm.2008.09.023
12. Woo PC, Lau SK, Wong BH, Fan RY, Wong AY, Zhang AJ, Wu Y, Choi GK, Li KS, Hui J, Wang M, Zheng BJ, Chan KH, Yuen KY (2012) Feline morbillivirus, a previously undescribed paramyxovirus associated with tubulointerstitial nephritis in domestic cats. Proc Natl Acad Sci USA 109(14):5435–5440. https://doi.org/10.1073/pnas.1119972109

Contents

Contributors

DANIELLE E. ANDERSON • *Victorian Infectious Disease Reference Laboratory, The Peter Doherty Institute for Infection and Immunity, Melbourne, VIC, Australia*

FELIX G. M. ANDRES • *Division of Veterinary Medicine, Paul-Ehrlich-Institute, Langen, Germany*

MATTHEW B. T. ASUELIME-SMITH • *The Department of Molecular, Cell, and Developmental Biology, University of California Santa Barbara, Santa Barbara, CA, USA*

BETTINA BANKAMP • *Centers for Disease Control and Prevention, National Center for Respiratory Diseases, Division of Viral Diseases, Viral Vaccine Preventable Diseases Branch, Atlanta, GA, USA*

LOUIS-MARIE BLOYET • *Centre International de Recherche en Infectiologie (CIRI), Inserm, CNRS, Université Claude Bernard Lyon 1, ENS Lyon, Lyon, France; Department of Molecular Microbiology, Washington University School of Medicine in St. Louis, St. Louis, MO, USA*

DENNY BÖTTCHER • *Faculty of Veterinary Medicine, Institute of Veterinary Pathology, Leipzig University, Leipzig, Germany*

JOANNA BRUNEL • *Centre International de Recherche en Infectiologie (CIRI), Inserm, CNRS, Université Claude Bernard Lyon 1, ENS Lyon, Lyon, France*

JOHANNES BUSCH • *Faculty of Veterinary Medicine, Institute of Virology, Leipzig University, Leipzig, Germany*

CHANTAL COMBREDET • *Institut Pasteur, Université Paris Cité, Innovation Lab: Vaccines, Paris, France*

MELISSA M. COUGHLIN • *Centers for Disease Control and Prevention, National Center for Respiratory Diseases, Division of Viral Diseases, Viral Vaccine Preventable Diseases Branch, Atlanta, GA, USA*

ROBERT M. COX • *Center for Translational Antiviral Research, Georgia State University Institute for Biomedical Sciences, Atlanta, GA, USA*

STEPHEN CROOKE • *Division of Viral Diseases, National Center for Immunization and Respiratory Diseases, Centers for Disease Control and Prevention (CDC), Atlanta, GA, USA*

THIBAUT DOUCHÉ • *Institut Pasteur, Université Paris Cité, Proteomics Platform, Mass Spectrometry for Biology, CNRS, Paris, France*

W. PAUL DUPREX • *Center for Vaccine Research, School of Medicine, University of Pittsburgh, Pittsburgh, PA, USA; Department of Microbiology and Molecular Genetics, School of Medicine, University of Pittsburgh, Pittsburgh, PA, USA*

GIMENA FEIJÓO • *Faculty of Veterinary, Veterinary Clinics & Hospital, Universidad de la República, Montevideo, Uruguay*

DENIS GERLIER • *Centre International de Recherche en Infectiologie (CIRI), Inserm, CNRS, Université Claude Bernard Lyon 1, ENS Lyon, Lyon, France*

QUENTIN GIAI GIANETTO • *Institut Pasteur, Université Paris Cité, Proteomics Platform, Mass Spectrometry for Biology, CNRS, Paris, France; Institut Pasteur, Université Paris Cité, Bioinformatics Hub, Paris, France*

Claudia Gonzales • *Centre International de Recherche en Infectiologie (CIRI), INSERM, CNRS, Université de Lyon, Université Claude Bernard Lyon 1, École Normale Supérieure de Lyon, Lyon, France*

Branka Horvat • *Centre International de Recherche en Infectiologie (CIRI), INSERM, CNRS, Université de Lyon, Université Claude Bernard Lyon 1, École Normale Supérieure de Lyon, Lyon, France*

Mathieu Iampietro • *Centre International de Recherche en Infectiologie (CIRI), INSERM, CNRS, Université de Lyon, Université Claude Bernard Lyon 1, École Normale Supérieure de Lyon, Lyon, France*

Victoria Iribarnegaray • *Faculty of Veterinary, Department of Pathobiology, Universidad de la República, Montevideo, Uruguay*

Timokratis Karamitros • *Bioinformatics and Applied Genomics Unit, Department of Microbiology, Hellenic Pasteur Institute, Athens, Greece*

Justin W. Kaufman • *Stead Family Department of Pediatrics, Carver College of Medicine, The University of Iowa, Iowa City, IA, USA*

Anastassia V. Komarova • *Institut Pasteur, Université Paris Cité, Interactomics, RNA and Immunity Laboratory, Paris, France*

Camila Larrañaga • *Faculty of Veterinary, Department of Pathobiology, Universidad de la República, Montevideo, Uruguay*

Sandrine Le Guellec • *DTF-Aerodrug, R&D Aerosoltherapy, Department of DTF Medical (Saint Etienne, France), Faculté de Médecine, Université de Tours, Tours, France*

Ni Li • *Scripps Research Institute, University of Florida, Jupiter, FL, USA*

Dzwokai Z. Ma • *Molecular, Cellular, and Developmental Biology, University of California Santa Barbara, Santa Barbara, CA, USA*

Annette Mankertz • *National Reference Center Measles, Mumps, Rubella, Robert Koch-Institute, Berlin, Germany*

Mariette Matondo • *Institut Pasteur, Université Paris Cité, Proteomics Platform, Mass Spectrometry for Biology, CNRS, Paris, France*

Zachary Matson • *Centers for Disease Control and Prevention, National Center for Respiratory Diseases, Division of Viral Diseases, Viral Vaccine Preventable Diseases Branch, Atlanta, GA, USA*

Andreas Mentis • *Central Laboratories, Iaso General Hospital, Athens, Greece*

Sara Mercader • *Division of Viral Diseases, National Center for Immunization and Respiratory Diseases, Centers for Disease Control and Prevention (CDC), Atlanta, GA, USA*

Jérôme Montharu • *PST-A, University of Tours, Tours, France*

Marie Mura • *Institut Pasteur, Université Paris Cité, Innovation Lab: Vaccines, Paris, France; Institut de recherche biomédicale des armées, Immunopathologie, Brétigny-sur-Orge, France*

Valerie Najburg • *Institut Pasteur, Université Paris Cité, Innovation Lab: Vaccines, Paris, France*

Chanakha K. Navaratnarajah • *Department of Molecular Medicine, Mayo Clinic, Rochester, MN, USA*

Takashi Okura • *Department of Virology 3, National Institute of Infectious Diseases, Tokyo, Japan*

Yaw Shin Ooi • *Programme in Emerging Infectious Diseases, Duke-National University of Singapore Medical School, Singapore, Singapore*

CHRISTIAN K. PFALLER • *Division of Veterinary Medicine, Paul-Ehrlich-Institute, Langen, Germany; Department of Molecular Medicine, Mayo Clinic, Rochester, MN, USA*

RICHARD K. PLEMPER • *Center for Translational Antiviral Research, Georgia State University Institute for Biomedical Sciences, Atlanta, GA, USA*

VASILIKI POGKA • *Bioinformatics and Applied Genomics Unit, Department of Microbiology, Hellenic Pasteur Institute, Athens, Greece*

LINDA J. RENNICK • *Center for Vaccine Research, School of Medicine, University of Pittsburgh, Pittsburgh, PA, USA; Department of Microbiology and Molecular Genetics, School of Medicine, University of Pittsburgh, Pittsburgh, PA, USA*

OLIVIER REYNARD • *Centre International de Recherche en Infectiologie (CIRI), INSERM, CNRS, Université de Lyon, Université Claude Bernard Lyon 1, École Normale Supérieure de Lyon, Lyon, France*

SABINE SANTIBANEZ • *National Reference Center Measles, Mumps, Rubella, Robert Koch-Institute, Berlin, Germany*

MORITOSHI SATO • *Graduate School of Arts and Sciences, The University of Tokyo, Tokyo, Japan*

BEVAN SAWATSKY • *Division of Veterinary Medicine, Paul-Ehrlich-Institute, Langen, Germany*

MICHAEL SIEG • *Faculty of Veterinary Medicine, Institute of Virology, Leipzig University, Leipzig, Germany*

OLIVER SIERING • *Division of Veterinary Medicine, Paul-Ehrlich-Institute, Langen, Germany*

KRISTMUNDUR SIGMUNDSSON • *Programme in Cardiovascular & Metabolic Disorders, Duke-National University of Singapore Medical School, Singapore, Singapore*

BRAJESH K. SINGH • *Stead Family Department of Pediatrics, Carver College of Medicine, The University of Iowa, Iowa City, IA, USA*

PATRICK L. SINN • *Stead Family Department of Pediatrics, Carver College of Medicine, The University of Iowa, Iowa City, IA, USA*

GABY SMITS • *Center for Immunology and Infectious Diseases and Vaccines, National Institute for Public Health and Environment, Bilthoven, The Netherlands*

MAINO TAHARA • *Department of Virology 3, National Institute of Infectious Diseases, Tokyo, Japan*

MAKOTO TAKEDA • *Graduate School of Medicine and Faculty of Medicine, Department of Microbiology, The University of Tokyo, Tokyo, Japan*

FRÉDÉRIC TANGY • *Oncovita, Paris, France*

NATASHA L. TILSTON-LUNEL • *Department of Microbiology and Immunology, Indiana University School of Medicine, Indianapolis, IN, USA*

ÉRICA URZUA • *Centre International de Recherche en Infectiologie (CIRI), Inserm, CNRS, Université Claude Bernard Lyon 1, ENS Lyon, Lyon, France*

THOMAS W. VAHLENKAMP • *Faculty of Veterinary Medicine, Institute of Virology, Leipzig University, Leipzig, Germany*

ROB VAN BINNENDIJK • *Center for Immunology and Infectious Diseases and Vaccines, National Institute for Public Health and Environment, Bilthoven, The Netherlands*

BELÉN VARELA • *Faculty of Veterinary, Department of Pathobiology, Universidad de la República, Montevideo, Uruguay*

LAURENT VECELLIO • *PST-A, University of Tours, Tours, France*

HEIDY VERA-PERALTA • *Institut Pasteur, Université Paris Cité, Innovation Lab: Vaccines, Paris, France; Institut de recherche biomédicale des armées, Immunopathologie, Brétigny-sur-Orge, France*
JOSÉ MANUEL VERDES • *Faculty of Veterinary, Department of Pathobiology, Universidad de la República, Montevideo, Uruguay*
KANJI YAMASAKI • *Faculty of Veterinary, Department of Pathobiology, Universidad de la República, Montevideo, Uruguay*
BIAOGUO YAN • *Programme in Emerging Infectious Diseases, Duke-National University of Singapore Medical School, Singapore, Singapore*
VICTORIA YOZZI • *Faculty of Veterinary, Department of Pathobiology, Universidad de la República, Montevideo, Uruguay*

Chapter 1

A Quantitative Fusion Assay to Study Measles Virus Entry

Chanakha K. Navaratnarajah

Abstract

We have adopted a real-time assay based on a dual-split reporter to assess cell-cell fusion mediated by the measles virus (MeV) membrane fusion machinery. This reporter system is comprised of two expression vectors, each encoding a segment of Renilla luciferase fused to a segment of GFP. To regain function, the two segments need to associate, which is dependent on cell-cell fusion between effector cells expressing the MeV fusion machinery and target cells expressing the corresponding MeV receptor. By measuring reconstituted luciferase activity, we can follow the kinetics of cell-cell fusion and quantify the extent of fusion. This assay lends itself to the study of the MeV fusion machinery comprised of the attachment and fusion glycoproteins, the matrix protein, and the MeV receptors. Moreover, entry inhibitors targeting attachment or fusion can be readily screened using this assay. Finally, this assay can be easily adopted to study the entry of other members of the *Paramyxoviridae*, as we have demonstrated for the henipaviruses.

Key words Virus entry, *Morbillivirus*, *Henipavirus*, Hemagglutinin, Fusion protein, Glycoprotein, Membrane, Inhibitor screening assay, In vitro

1 Introduction

Measles virus (MeV) entry is mediated by the concerted action of the receptor binding protein hemagglutinin (H) tetramers and fusion (F) protein trimers [1]. H-protein tetramers interact with the MeV receptors, signaling lymphocytic activation molecule [SLAM, CD150], nectin-4, or CD46 [2–5] at the cell surface, leading to conformational changes in the H-protein [6, 7]. This triggers F-trimers to undergo their own irreversible conformational changes leading to membrane fusion [8–10]. Cell-cell fusion mediated by the measles virus glycoproteins has been long used as a proxy to study virus entry. Moreover, MeV is a predominantly cell-associated virus, and as such, cell-to-cell spread is an important method of MeV spread and dissemination [11–14]. Thus, a quantitative fusion assay to study the kinetics of cell-cell fusion is an important tool to study MeV biology.

Dzwokai Z. Ma and Christian K. Pfaller (eds.), *Measles and Related Morbilliviruses: Methods and Protocols*, Methods in Molecular Biology, vol. 2808, https://doi.org/10.1007/978-1-0716-3870-5_1,

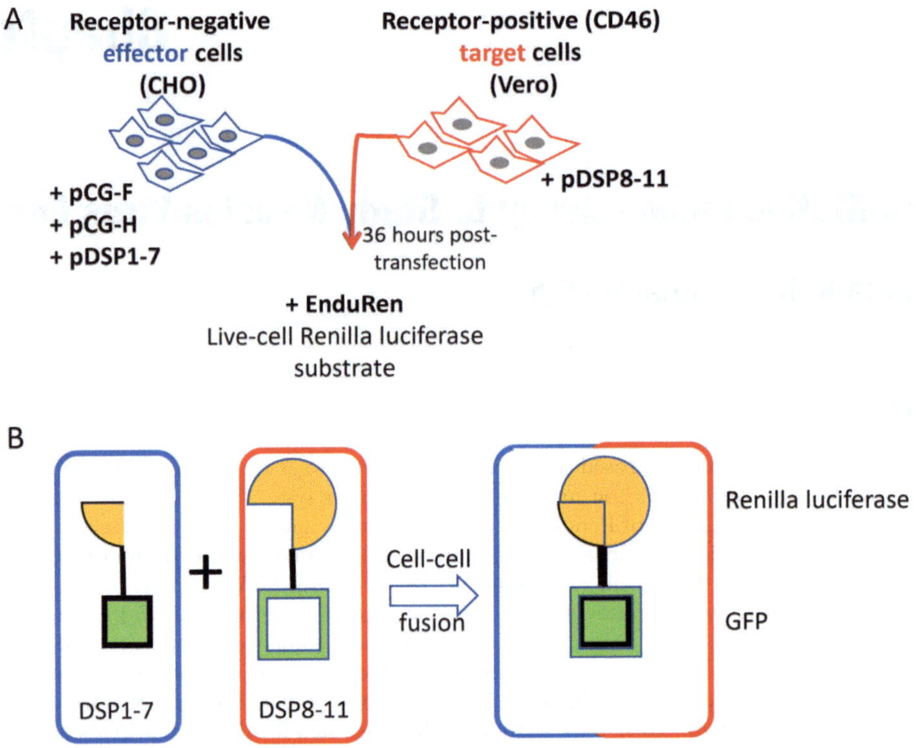

Fig. 1 Schematic outline of the dual-split reporter-based cell-cell fusion assay. (**a**) Receptor-negative effector cells (blue) are prepared by transfecting the expression plasmids encoding the MeV glycoproteins (F and H) and one segment of the dual-split reporter (DSP1-7) into CHO cells. Receptor-positive target cells (red) are prepared by transfecting the plasmid encoding the other segment of the dual-split reporter (DSP8-11) into Vero cells. Thirty-six hours post-transfection, the effector and target cells are mixed and luciferase activity measured using the cell-permeable Renilla luciferase substrate, EnduRen. (**b**) Cell mixing leads to cell-cell fusion mediated by the MeV glycoproteins expressed on the surface of the effector cells interacting with the corresponding receptors on the target cells. Content mixing between effector and target cells allows the two segments of the dual-split reporter to assemble reconstituting luciferase enzymatic activity and GFP fluorescence. Note: syncytia formation as depicted here is a not a prerequisite for reconstitution of function. Fusion pore formation allowing mixing of cell content is sufficient for detection of luciferase activity

We have adopted a real-time assay, previously used to study the HIV-1 envelope protein [15, 16], to monitor MeV mediated cell-cell fusion over time [7]. The assay is based on a dual-split protein (DSP) reporter developed by Kondo et al. [15]. The DSP constructs consist of two expression plasmids, one encoding the N-terminal fragment of *Renilla* luciferase fused to a fragment of green fluorescent protein (GFP) comprising beta-sheets 1 through 7 (DSP1-7) (Fig. 1). The other encodes the C-terminal fragment of *Renilla* luciferase fused to a fragment of GFP comprising beta-sheets 8 through 11 (DSP8-11).

In this assay, cell-cell fusion between an effector cell expressing the MeV glycoproteins and a target cell expressing one or more MeV receptors allows content mixing and the two halves of *Renilla* luciferase/GFP to functionally interact (Fig. 1a, b). This dual-functional reporter construct allows real-time monitoring of fusion by microscopy using the GFP reporter or by the enzymatic activity of *Renilla* luciferase on a cell-permeable live cell substrate, EduRen. The luciferase-based assay is especially suited for a kinetic analysis of cell-cell fusion. The high dynamic range of the luciferase assay allows us to discern even small differences in fusion kinetics over a broad range. Thus, this bifunctional cell fusion reporter is ideal for the real-time no invasive analysis of cell fusion kinetics. This assay is amenable to characterizing the role of different components of the MeV fusion machinery including the H- and F-proteins. For instance, by assessing the fusion function of H-protein mutants, we have gained a mechanistic understanding of the role of the attachment protein in triggering the F-protein [7]. We can also extend this characterization to other components of the fusion machinery. For example, co-expression of the MeV matrix (M) protein allows one to study how the M-protein modulates the fusion function of the glycoproteins. Moreover, we can compare the ability of different MeV receptors to support cell fusion. This assay is also compatible for screening MeV entry inhibitors such as small molecules or antibodies.

While we describe the use of this assay to study MeV mediated cell-cell fusion, the DSP assay can be easily adopted to study the entry of other viruses. The critical requirement is that fusion should occur at the plasma membrane. In fact, we have adopted this assay to study another member of the paramyxoviruses, namely, the henipaviruses which include the BSL4 pathogens Hendra and Nipah viruses [17], while others have adopted this assay to study coronavirus entry.

2 Materials

1. Effector cells (receptor negative): Chinese hamster ovary (CHO) cells.
2. Complete medium (CHO cells): F-12 K medium containing 10% fetal bovine serum (FBS).
3. Target cells (receptor positive): Vero cells (CD46 receptor), Vero/hSLAM (human SLAM receptor) [3], or Vero/N4 (nectin-4 receptor) [5].
4. Complete medium (Vero cells): Dulbecco's modified eagle medium (DMEM) containing 10% FBS.
5. pCG-H: MeV H-protein expression plasmid [7].
6. pCG-F: MeV F-protein expression plasmid [7].
7. pDSP1-7: dual-split protein 1–7 expression vector (a kind gift of Z. Matsuda; [15]).

8. pDSP8-11: dual-split protein 8–11 expression vector (a kind gift of Z. Matsuda; [15]).
9. GeneJuice transfection reagent (Millipore Sigma).
10. EnduRen Live Cell Substrate (Promega).
11. Versene: 0.48 mM EDTA in PBS.
12. 96-well black plate with clear bottom.
13. 37 °C, 5% CO_2 incubator.
14. 96-well plate compatible luminometer.
15. Fluorescent microscope.

3 Methods

3.1 Transfection of Effector and Target Cells

1. Seed the receptor-negative effector (CHO; *see* **Note 1**) cells in a six-well plate (5 × 10^5 cells/well, 2 mL). Similarly, seed the receptor-positive target cells (Vero, Vero/hSLAM, or Vero/N4; *see* **Note 2**) in a six-well plate (5 × 10^5 cells/well, 2 mL). Incubate the cells at 37 °C (5% CO_2) until 50–60% confluent (usually 24 h).
2. Transfect the effector and target cells when they reach 50–60% confluency (*see* **Note 3**). Transfect the effector cells with the plasmids expressing the MeV glycoproteins (pCG-F and pCG-H, 400 ng each/per well; *see* **Note 4**)) [7] and the plasmid expressing one-half of the fused *Renilla* luciferase/GFP subunit (pDSP1-7, 200 ng/well). Transfect the target cells with 1000 ng of pDSP8–11 encoding the other half of the fused *Renilla* luciferase/GFP subunit.

3.2 Quantify Fusion by Measuring Renilla Luciferase Activity

1. At 36 h post-transfection, wash the effector and target cells with PBS. Add 1 mL of Versene per well, and incubate the cells for about 10–15 min until they start to detach (*see* **Notes 5** and **6**).
2. Use a 1 mL pipette to gently resuspend the effector and target cells, and transfer to a 10 mL centrifuge tube. Be very gentle, as cell membrane damage may lead to high background. Centrifuge the cells at 100 *g* for 5 min. Discard the Versene, and then resuspend the cells in 500 μL of complete media.
3. Count the cells and add 1 × 10^4 cells (50 μL) each, of the effector and target cells per well of the clear bottom, black-wall 96-well plate (2 × 10^4 cells/well, 100 μL). *See* **Note 7**.
4. Add 100 μL of complete media supplemented with EnduRen, the *Renilla* luciferase cell-permeable substrate, to each well of the 96-well plate, according to the manufacturer's instructions. We diluted the stock EnduRen 1:1000 in the cell culture media (DMEM, 10% FBS) for a final concentration of 60 μM.

5. Take the time zero reading by measuring Renilla luciferase activity with a luminometer, and then incubate the 96-well plate at 37 °C (5% CO_2).
6. At regular intervals, measure the Renilla luciferase activity (*see* **Notes 8** and **9**). A plot of luciferase activity vs. time will reveal the kinetics of fusion for each condition tested. Additionally, GFP can also be used to monitor cell-cell fusion (*see* **Note 10**).

4 Notes

1. The choice of effector and target cells will depend on your experimental system. Ideally, the effector cells should be receptor negative to prevent cell-cell fusion prior to overlay with the target cells. If this is not possible, consider blocking fusion using a reversible inhibitor such as the fusion inhibitory peptide (FIP) [18].
2. The target cells will need to have one or more MeV receptors. You can also use a receptor-negative cell as a target cell by providing the appropriate receptor(s) in trans.
3. While we use GeneJuice (Millipore Sigma) as the transfection reagent, you may use your transfection reagent of choice. GeneJuice transfection reagent is composed of a nontoxic cellular protein and a small amount of a polyamine that we have found to be relatively nontoxic to cells and give a low background in this assay. We used a 1:3 ratio of DNA to GeneJuice reagent. For one well of a six-well plate, we mixed 3 μL of GeneJuice reagent to 100 uL of OptiMEM in a 1.5 mL Eppendorf tube. After 5 minutes, we added 1 μg of DNA (400 ng each of the pCG-H and pCG-F plasmids and 200 ng of pDSP1-7 for the effector cells and 1 μg of pDSP8-11 for the target cells). After 20 min of incubation, the transfection mix was added to the cells.
4. A control with effector cells lacking one or both MeV glycoproteins should be used as the negative control.
5. It is important to be very gentle when resuspending the cells by pipetting as damage to the cell membrane can lead to high background readings. We have found Versene works well for detaching and resuspending a variety of cell lines with minimal damage to the cell membrane.
6. If your cells of choice detach poorly by this method or are sensitive to Versene, you may consider using UpCell dishes (Thermo Fisher Scientific; Catalog No., 174,901) which allow the cells to dissociate from the plate by a temperature change from 37 °C to room temperature.

7. We recommend three to six replicates for each condition tested, which allows for statistical analysis.
8. For fusion mediated by the MeV vaccine-strain glycoproteins, we recommend 1-h intervals for a total duration of 10–12 h. The luciferase signal peaked between 9 and 10 h post cell mixing [7].
9. Avoid excessive temperature changes during Renilla luciferase activity measurements, and immediately place the plate back in the incubator after luminometer readings.
10. The reconstituted GFP will fluoresce as fusion progresses. However, the GFP fluorescence takes longer to appear than the signal from luciferase activity and is not readily quantifiable.

References

1. Navaratnarajah CK, Leonard VH, Cattaneo R (2009) Measles virus glycoprotein complex assembly, receptor attachment, and cell entry. Curr Top Microbiol Immunol 329:59–76. https://doi.org/10.1007/978-3-540-70523-9_4
2. Dörig RE, Marcil A, Chopra A, Richardson CD (1993) The human CD46 molecule is a receptor for measles virus (Edmonston strain). Cell 75(2):295–305. https://doi.org/10.1016/0092-8674(93)80071-L
3. Tatsuo H, Ono N, Tanaka K, Yanagi Y (2000) SLAM (CDw150) is a cellular receptor for measles virus. Nature 406(6798):893–897. https://doi.org/10.1038/35022579
4. Mühlebach MD, Mateo M, Sinn PL, Prüfer S, Uhlig KM, Leonard VH, Navaratnarajah CK, Frenzke M, Wong XX, Sawatsky B (2011) Adherens junction protein nectin-4 is the epithelial receptor for measles virus. Nature. 480 (7378):530–533. https://doi.org/10.1038/nature10639
5. Noyce RS, Bondre DG, Ha MN, Lin LT, Sisson G, Tsao MS, Richardson CD (2011) Tumor cell marker PVRL4 (Nectin 4) is an epithelial cell receptor for measles virus. PLoS Pathog 7(8):e1002240. https://doi.org/10.1371/journal.ppat.1002240
6. Navaratnarajah CK, Oezguen N, Rupp L, Kay L, Leonard VH, Braun W, Cattaneo R (2011) The heads of the measles virus attachment protein move to transmit the fusion-triggering signal. Nat Struct Mol Biol 18(2): 128–134. https://doi.org/10.1038/nsmb.1967
7. Navaratnarajah CK, Rosemarie Q, Cattaneo R (2016) A structurally unresolved head segment of defined length favors proper measles virus hemagglutinin tetramerization and efficient membrane fusion triggering. J Virol 90(1): 68–75. https://doi.org/10.1128/JVI.02253-15
8. Navaratnarajah CK, Negi S, Braun W, Cattaneo R (2012) Membrane fusion triggering: three modules with different structure and function in the upper half of the measles virus attachment protein stalk. J Biol Chem 287(46): 38543–38551. https://doi.org/10.1074/jbc.M112.410563
9. Apte-Sengupta S, Negi S, Leonard VH, Oezguen N, Navaratnarajah CK, Braun W, Cattaneo R (2012) Base of the measles virus fusion trimer head receives the signal that triggers membrane fusion. J Biol Chem 287(39): 33026–33035. https://doi.org/10.1074/jbc.M112.373308
10. Apte-Sengupta S, Navaratnarajah CK, Cattaneo R (2013) Hydrophobic and charged residues in the central segment of the measles virus hemagglutinin stalk mediate transmission of the fusion-triggering signal. J Virol 87(18): 10401–10404. https://doi.org/10.1128/JVI.01547-13
11. Mateo M, Sinn PL, Navaratnarajah CK, Mendoza C, Donohue R, Cattaneo R (2014) Measles virus rapid and discreet epithelial spread through Nectin-4: implications for oncolytic Virotherapy. In: Molecular therapy. Nature Publishing, New York, p S10. https://doi.org/10.1016/j.tim.2012.05.006
12. Singh BK, Hornick AL, Krishnamurthy S, Locke AC, Mendoza CA, Mateo M, Miller-Hunt CL, Cattaneo R, Sinn PL (2015) The Nectin-4/Afadin protein complex and intercellular membrane pores contribute to rapid spread of measles virus in primary human airway epithelia. J Virol 89(14):7089–7096. https://doi.org/10.1128/JVI.00821-15

13. Cattaneo R, Donohue RC, Generous AR, Navaratnarajah CK, Pfaller CK (2019) Stronger together: multi-genome transmission of measles virus. Virus Res 265:74–79. https://doi.org/10.1016/j.virusres.2019.03.007
14. Hippee CE, Singh BK, Thurman AL, Cooney AL, Pezzulo AA, Cattaneo R, Sinn PL (2021) Measles virus exits human airway epithelia within dislodged metabolically active infectious centers. PLoS Pathog 17(8):e1009458. https://doi.org/10.1371/journal.ppat.1009458
15. Kondo N, Miyauchi K, Meng F, Iwamoto A, Matsuda Z (2010) Conformational changes of the HIV-1 envelope protein during membrane fusion are inhibited by the replacement of its membrane-spanning domain. J Biol Chem 285(19):14681–14688. https://doi.org/10.1074/jbc.M109.067090
16. Kondo N, Miyauchi K, Matsuda Z (2011) Monitoring viral-mediated membrane fusion using fluorescent reporter methods. Curr Protoc Cell Biol Chapter 26:Unit 26. https://doi.org/10.1002/0471143030.cb2609s50
17. Laing ED, Navaratnarajah CK, Cheliout Da Silva S, Petzing SR, Xu Y, Sterling SL, Marsh GA, Wang LF, Amaya M, Nikolov DB, Cattaneo R, Broder CC, Xu K (2019) Structural and functional analyses reveal promiscuous and species specific use of ephrin receptors by cedar virus. Proc Natl Acad Sci USA 116(41):20707–20715. https://doi.org/10.1073/pnas.1911773116
18. Richardson CD, Choppin PW (1983) Oligopeptides that specifically inhibit membrane fusion by paramyxoviruses: studies on the site of action. Virology 131(2):518–532. https://doi.org/10.1016/0042-6822(83)90517-2

Chapter 2

A Semiquantitative Protein-Fragment Complementation Assay to Study Protein-Protein Interactions of the Polymerase Complex in *Cellula*

Joanna Brunel, Érica Urzua, Denis Gerlier, and Louis-Marie Bloyet

Abstract

Protein-fragment complementation assays (PCAs) are powerful tools to investigate protein-protein interactions in a cellular context. These are especially useful to study unstable proteins and weak interactions that may not resist protein isolation or purification. The PCA based on the reconstitution of the *Gaussia princeps* luciferase (split-luc) is a sensitive approach allowing the mapping of protein-protein interactions and the semiquantitative measurement of binding affinity. Here, we describe the split-luc protocol we used to map the viral interactome of measles virus polymerase complex.

Key words Protein-protein interaction, Protein-fragment complementation assay (PCA), *Gaussia princeps* luciferase, Split-luc reassembly assay, Measles virus (MeV), Polymerase complex, Nucleoprotein, Phosphoprotein, RNA-dependent RNA polymerase

1 Introduction

Protein-fragment complementation assays (PCA) are rapid and efficient techniques allowing the detection, and sometimes the semiquantification, of protein-protein interactions both in cells and in vitro. The general principle is to link two putative protein partners to two complementary fragments of a reporter protein to be reconstituted (Fig. 1). Upon the interaction between the protein candidates, the two fragments get close enough to reassemble the reporter and reconstitute its activity. When performed in cells, these assays do not require the purification of the protein complexes and are therefore particularly useful for unstable proteins and weak interactions.

Measles virus RNA synthesis machinery is made of two major components: (1) the polymerase complex, composed of the large protein (L) and its cofactor the phosphoprotein (P), and (2) its template, the nucleocapsid, made of the viral genome entirely

Dzwokai Z. Ma and Christian K. Pfaller (eds.), *Measles and Related Morbilliviruses: Methods and Protocols*, Methods in Molecular Biology, vol. 2808, https://doi.org/10.1007/978-1-0716-3870-5_2,

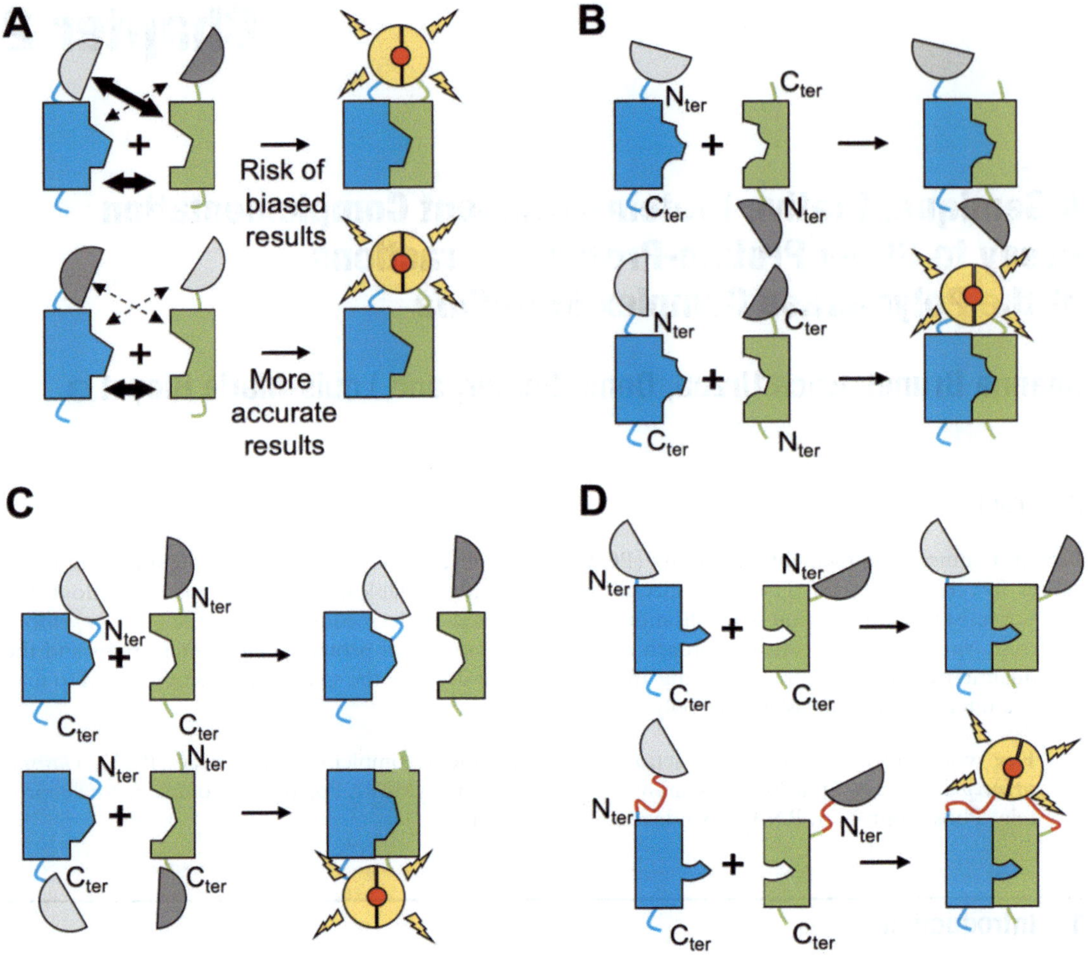

Fig. 1 Design of the fusion proteins. Schematic representation of the parameters to take into consideration: (**a**) the background affinity between the fragments of the reporter and the protein candidates, according to the assignment of the fragment to each candidate, (**b**) the orientation of the proteins in the complex, (**c**) the accessibility of the protein extremities, and (**d**) the length of the linker

encased into a homopolymer of nucleoproteins (N). Investigating the molecular partnership up to the detailed mapping of homo- and heterophilic interactions sites within the N-P-L complex has been rather challenging and has required several decades of tedious work by several groups, and it is yet to be completed. Several features of these proteins have hindered the understanding of the molecular architecture of the ribonucleoprotein complex. For instance, measles virus (MeV) L protein is a large multi-domain protein requiring the HSP90 chaperoning system and the binding to P to be properly folded and stable [1]. Also, when expressed alone, N binds and oligomerizes on cellular RNA, thus forming large and heterogeneous nucleocapsid-like structures.

Several PCAs have been used to investigate protein-protein interactions involved in MeV life cycle, including interactions between the viral proteins supporting RNA synthesis. The yeast two-hybrid system and its equivalent in mammalian cells, which are based on the reconstitution of the yeast GAL4 transcription activator [2], were used to map the interactions between MeV N, P, and L proteins [3, 4]. Although it allowed the detection of some essential interactions, the two-hybrid system has intrinsic limitations. For example, the reconstituted complex needs to enter the nucleus to activate transcription, and the rather large size of the components of the viral RNA synthesis machinery calls into question the efficiency of this translocation. Moreover, the N-terminal acidic region of P was found to constitutively activate the yeast reporter gene. Others have used PCAs based on fluorescent reporters (or bimolecular fluorescence complementation (BiFC) assays), such as the split-GFP approach which was used to visualize the homo- and hetero-oligomerization of the two surface glycoproteins of MeV [5] and the interaction between the α-MoRE of N ($N_{\alpha\text{-MoRE}}$) and the X domain of P (P_{XD}) [6]. However, the rather low signal-to-noise ratio and the irreversibility of the reconstituted fluorescent protein are important limitations [7, 8].

In 2006, Remy I. and Michnick S.W. developed a reversible, robust, and sensitive reassembly assay based on the *Gaussia princeps* luciferase (split-luc assay) [9]. It was further developed into a high-throughput assay to screen for protein-protein interactions in mammalian cells and showed excellent sensitivity and specificity [10].

Together with careful designs of protein domains based on algorithm predictors and/or available 3D protein structures, the split-luc assay allowed us to decipher an improved map of the binding sites between MeV P and L proteins [11] and between P and C proteins [12]. Likewise, this assay allowed us to demonstrate a correlation between the efficiency of the reinitiation of the transcription at intergenic junctions and the binding strength between MeV $N_{\alpha\text{-MoRE}}$ and P_{XD} [13, 14]. In this work, from a series of point mutants, the binding affinities measured with either the split-luc assay or the isothermal titration calorimetry of purified protein domains were found to nicely correlate, thus indicating that this assay is sensitive enough to detect modifications of the binding affinity between two proteins interacting with a K_D in the 10 μM range [14].

The experimental procedure takes 3 days with day 1, seeding the cells; day 2, transfection; and day 3, cell lysis and measurement of luminescence (Fig. 2). The results are expressed as normalized luminescence ratios (NLR). The NLR corresponds to the signal detected in the presence of the two fusion proteins divided by the sum of the signals measured when one fusion protein is co-expressed with the opposite fragment alone (Fig. 3c). For each

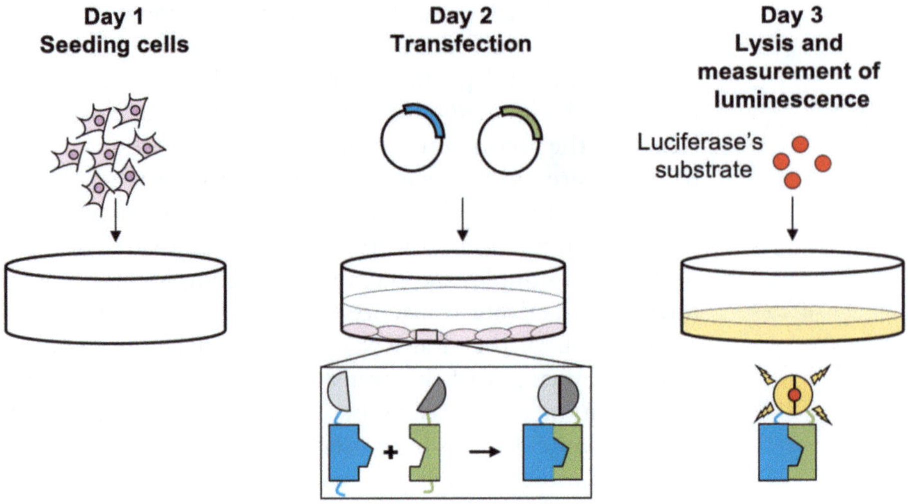

Fig. 2 Protein complementation assay. Schematic representation of the experiment. Plasmids encode fusion proteins between the protein candidates and the fragments of the *Gaussia* luciferase

interaction, these three measurements are made in three distinct wells (Fig. 3a, b). Thus, when performed in a 96-well plate, the split-luc assay allows the monitoring of the interaction between up to 15 protein pairs in 3 days (Figs. 2 and 3).

2 Materials

1. 96-well white-sided, clear bottom plates.
2. HEK-293 T cells (human embryonic kidney cells).
3. Dulbecco's Modified Eagle's Medium (DMEM) supplemented with 10% heat-inactivated fetal bovine serum (FBS).
4. Poly-D-lysine solution at 50 μg/mL.
5. jetPRIME transfection reagent (Polyplus Transfection).
6. Phosphate-buffered saline (PBS), pH 7.4.
7. *Renilla* Luciferase Assay System (Promega, E2810).

3 Methods

3.1 Considerations for the Design of the Fusion Proteins

The sensitivity and specificity of the PCAs are greatly influenced by the design of the chimeric proteins. For each interaction, three main parameters can be tuned:

Some protein candidates may have a significant affinity for a GLF. These unspecific interactions can interfere with the assay and

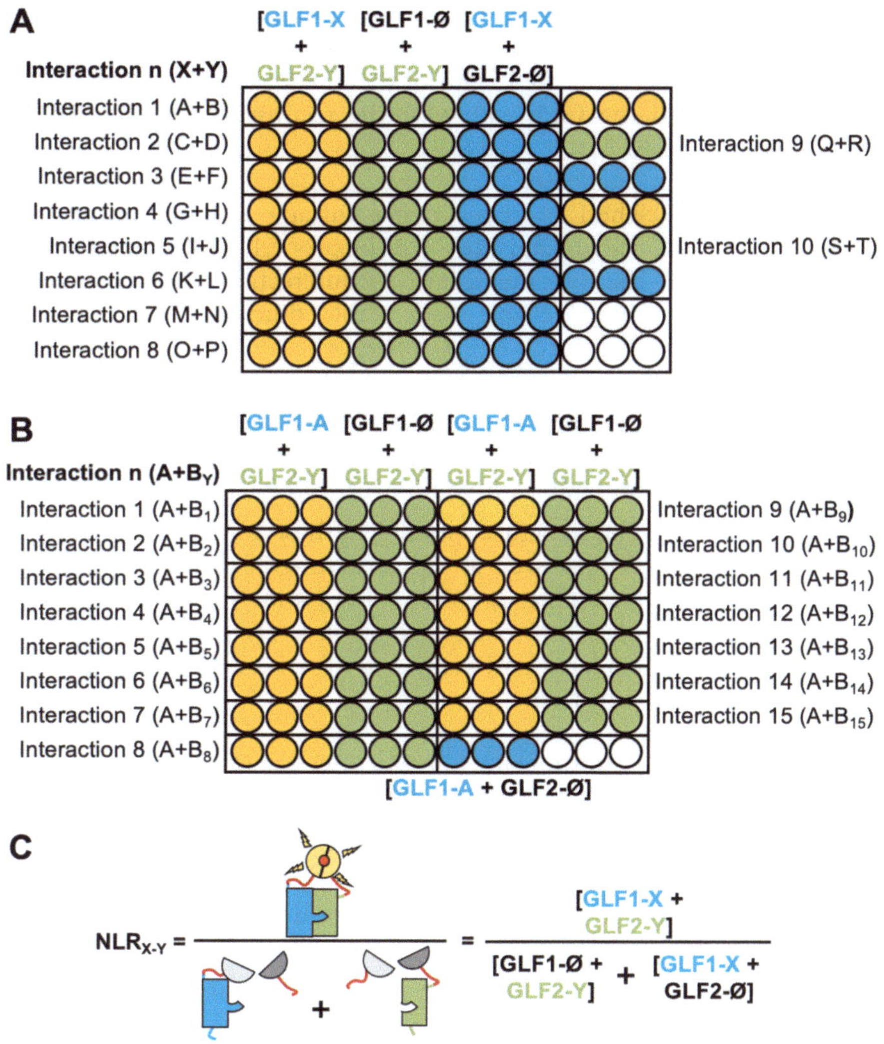

Fig. 3 Examples of experimental designs. (**a**) Measurement of ten independent interactions in a 96-well plate. (**b**) Measurement of 15 interactions between a protein A and 15 mutants of a protein B. (**c**) Schematic representation of the calculation of the normalized luminescence ratio (NLR)

3.1.1 The Assignment of the Gaussia Luciferase Fragments (GLF) to each Candidate

should be avoided by reversing the assignment of the GLF to each candidate (Fig. 1a) (see **Note 1**).

3.1.2 The Extremity to which the GLF Are Fused

The orientation of the proteins in the complex may position their extremities on the opposite side of the complex. The extremity of a protein candidate may also be buried in the protein's structure or may directly take part in the interaction. In these cases, different fusion combinations should be tested (Fig. 1b, c).

3.1.3 The Length of the Linker

The extremities of the protein candidates may be too far apart in the protein complex to allow the reconstitution of the luciferase. Increasing the length of the linker between the protein of interest and the GLF will favor the reconstitution (Fig. 1d) (*see* **Notes 2** and **3**).

3.1.4 Special Considerations Regarding Measles Virus and Related Viruses

1. The L protein is rather unstable and requires both the HSP90 activity and MeV P to be properly folded and stable [1]. Thus, the investigation of the interaction between L and P may be more reliable with stable domains of L rather than full-length L.
2. The interaction between $N_{\alpha\text{-MoRE}}$ and P_{XD} is quite weak and dynamic. The addition of oligomerization domains to the constructs increases the signal and allows better detection of mutants with even lower affinity [14].
3. The structures of full length or domains of MeV N and P [15–20], as well as the structure of the polymerase complex of parainfluenza virus 5 [21] and Newcastle virus [22] and of the C protein of Sendai virus [23] and Tupaia virus [24], are available and can be used to guide the design of the fusion constructs.

3.2 Seeding Cells

1. In a 96-well white-sided, clear bottom plate, add 40 μL poly-D-lysine per well, and make sure the bottom of each well is fully covered (*see* **Notes 4** and **5**).
2. Incubate for 1 h at 37 °C.
3. Remove the poly-D-lysine and wash with 200 μL of sterile water (*see* **Note 6**).
4. Remove the water and let the plate open to dry under the laminar hood (about 30 min).
5. For a 75 cm^2 flask, detach the HEK-293 T cells with 1 mL of trypsin, suspend cells with 9 mL of DMEM-10% FBS, and measure their concentration (*see* **Note 7**).
6. Add 2×10^4 cells per well in 100 μL DMEM-10% FBS medium (*see* **Note 8**).

3.3 Transfection

Twenty to twenty-four hours after seeding the cells, transfect the wells with 100 ng of each plasmid (200 ng total) using jetPRIME transfection reagent (*see* **Note 9**).

1. Per well, in 5 μL of jetPRIME buffer, add 100 ng of each plasmid.
2. Mix by vortexing for 10 s.
3. And 0.4 μL of jetPRIME reagent and vortex for 1 s.

4. Incubate for 10 min at room temperature.
5. Add 5 μL of the DNA solution to the medium.
6. Incubate the cells at 37 °C for 24 h (*see* **Note 10**).

3.4 Cell Lysis and Measurement

1. Remove the medium and wash the cells with 100 μL PBS.
2. Add 40 μL of lysis buffer per well.
3. Incubate for 30 min at room temperature on a plate shaker.
4. In the meantime, prepare the luminometer and set the injector to dispense 100 μL per well. Set the integration time to 10 s after the injection (*see* **Note 11**).
5. Place the plate in the luminometer and start the program.

4 Notes

1. For instance, we noticed that measles virus P protein has a significant affinity for the N-terminal fragment of the luciferase (GLF1). That is why we only used P fused to GLF1.
2. Linkers are usually rich in glycine and serine residues. A peptide tag (i.e., FLAG or HA) can also be inserted in the linker to allow easy and efficient detection by immunoblotting, especially useful to check the relative expression levels of the fusion proteins.
3. Ideally, the control plasmids with the GLF alone must be as similar as possible as the plasmids with the candidates, i.e., same promoter, Kozak sequence, linker, tag, etc.
4. Any 96-well plate with a flat bottom can be used. However, white plates give higher signals. White plates with a transparent bottom allow visual inspection of the cells, but plates with a white bottom will maximize the signal. It is also possible to seed and transfect cells in a standard plate and then transfer the cell lysates to a white plate before the addition of the substrate and measurement of the luminescence.
5. Coating with poly-D-lysine is important for cells with poor adhesion like HEK-293 T cells. Cells with good adherence may not need coating.
6. It is important to remove the excess of poly-D-lysine as it can be toxic to some cells.
7. Any cells can be used. We used HEK-293 T cells because they are human cells with a high transfection efficiency.
8. To get a homogeneous layer of cells in each well, we use all the materials at room temperature (cells, medium, plate) and keep the seeded plate at room temperature for 15 min or until the cells have fallen on the bottom of the plate.

9. The choice of the transfection reagent depends on the cells. HEK-293 T cells are efficiently transfected by a wide range of transfection reagents.
10. Some cells may require to change the medium 4 h post-transfection. In some cases, waiting longer than 24 h may be useful to increase the signal.
11. This assay is a luciferase "flash" assay that provides a strong but short signal. That is why using an injector is strongly recommended.

Acknowledgments

We thank all our past collaborators and in particular Carine Lazert for her crucial technical contributions to investigating the molecular partnership of MeV polymerase. We are indebted to Y. Jacob who kindly provide us with the molecular constructs encoding the N- and C-terminal domains of the *Gaussia* luciferase.

References

1. Bloyet LM, Welsch J, Enchery F, Mathieu C, de Breyne S, Horvat B, Grigorov B, Gerlier D (2016) HSP90 chaperoning in addition to phosphoprotein required for folding but not for supporting enzymatic activities of measles and nipah virus L polymerases. J Virol 90(15): 6642–6656. https://doi.org/10.1128/JVI.00602-16
2. Fields S, Song O (1989) A novel genetic system to detect protein-protein interactions. Nature 340(6230):245–246. https://doi.org/10.1038/340245a0
3. Harty RN, Palese P (1995) Measles virus phosphoprotein (P) requires the NH2- and COOH-terminal domains for interactions with the nucleoprotein (N) but only the COOH terminus for interactions with itself. J Gen Virol 76(Pt 11):2863–2867. https://doi.org/10.1099/0022-1317-76-11-2863
4. Chen M, Cortay JC, Gerlier D (2003) Measles virus protein interactions in yeast: new findings and caveats. Virus Res 98(2):123–129. https://doi.org/10.1016/j.virusres.2003.09.003
5. Quinn DJ, McFerran NV, Nelson J, Duprex WP (2012) Live-cell visualization of transmembrane protein oligomerization and membrane fusion using two-fragment haptoEGFP methodology. Biosci Rep 32(3):333–343. https://doi.org/10.1042/BSR20110100
6. Gruet A, Dosnon M, Vassena A, Lombard V, Gerlier D, Bignon C, Longhi S (2013) Dissecting partner recognition by an intrinsically disordered protein using descriptive random mutagenesis. J Mol Biol 425(18):3495–3509. https://doi.org/10.1016/j.jmb.2013.06.025
7. Kodama Y, Hu CD (2012) Bimolecular fluorescence complementation (BiFC): a 5-year update and future perspectives. BioTechniques 53(5):285–298. https://doi.org/10.2144/000113943
8. Bignon C, Gruet A, Longhi S (2022) Split-GFP reassembly assay: strengths and caveats from a multiparametric analysis. Int J Mol Sci 23(21):13167. https://doi.org/10.3390/ijms232113167
9. Remy I, Michnick SW (2006) A highly sensitive protein-protein interaction assay based on Gaussia luciferase. Nat Methods 3(12): 977–979. https://doi.org/10.1038/nmeth979
10. Cassonnet P, Rolloy C, Neveu G, Vidalain PO, Chantier T, Pellet J, Jones L, Muller M, Demeret C, Gaud G, Vuillier F, Lotteau V, Tangy F, Favre M, Jacob Y (2011) Benchmarking a luciferase complementation assay for detecting protein complexes. Nat Methods 8(12):990–992. https://doi.org/10.1038/nmeth.1773

11. Bloyet LM, Schramm A, Lazert C, Raynal B, Hologne M, Walker O, Longhi S, Gerlier D (2019) Regulation of measles virus gene expression by P protein coiled-coil properties. Sci Adv 5 (5):eaaw3702. https://doi.org/10.1126/sciadv.aaw3702
12. Pfaller CK, Bloyet LM, Donohue RC, Huff AL, Bartemes WP, Yousaf I, Urzua E, Claviere M, Zachary M, de Masson d'Autume V, Carson S, Schieferecke AJ, Meyer AJ, Gerlier D, Cattaneo R (2020) The C protein is recruited to measles virus ribonucleocapsids by the phosphoprotein. J Virol 94 (4):e01733-19. https://doi.org/10.1128/JVI.01733-19
13. Brunel J, Chopy D, Dosnon M, Bloyet LM, Devaux P, Urzua E, Cattaneo R, Longhi S, Gerlier D (2014) Sequence of events in measles virus replication: role of phosphoprotein-nucleocapsid interactions. J Virol 88(18): 10851–10863. https://doi.org/10.1128/jvi.00664-14
14. Bloyet LM, Brunel J, Dosnon M, Hamon V, Erales J, Gruet A, Lazert C, Bignon C, Roche P, Longhi S, Gerlier D (2016) Modulation of re-initiation of measles virus transcription at intergenic regions by PXD to NTAIL binding streNgth. PLoS Pathog 12(12): e1006058. https://doi.org/10.1371/journal.ppat.1006058
15. Johansson K, Bourhis JM, Campanacci V, Cambillau C, Canard B, Longhi S (2003) Crystal structure of the measles virus phosphoprotein domain responsible for the induced folding of the C-terminal domain of the nucleoprotein. J Biol Chem 278(45):44567–44573. https://doi.org/10.1074/jbc.M308745200
16. Gutsche I, Desfosses A, Effantin G, Ling WL, Haupt M, Ruigrok RW, Sachse C, Schoehn G (2015) Structural virology. Near-atomic cryo-EM structure of the helical measles virus nucleocapsid. Science 348(6235):704. https://doi.org/10.1126/science.aaa5137
17. Guryanov SG, Liljeroos L, Kasaragod P, Kajander T, Butcher SJ (2015) Crystal structure of the measles virus nucleoprotein core in complex with an N-terminal region of phosphoprotein. J Virol 90(6):2849–2857. https://doi.org/10.1128/jvi.02865-15
18. Kingston RL, Hamel DJ, Gay LS, Dahlquist FW, Matthews BW (2004) Structural basis for the attachment of a paramyxoviral polymerase to its template. Proc Natl Acad Sci USA 101(22):8301–8306. https://doi.org/10.1073/pnas.0402690101
19. Blocquel D, Habchi J, Durand E, Sevajol M, Ferron F, Erales J, Papageorgiou N, Longhi S (2014) Coiled-coil deformations in crystal structures: the measles virus phosphoprotein multimerization domain as an illustrative example. Acta Crystallogr D Biol Crystallogr 70(Pt 6):1589–1603. https://doi.org/10.1107/S139900471400234X
20. Communie G, Crepin T, Maurin D, Jensen MR, Blackledge M, Ruigrok RW (2013) Structure of the tetramerization domain of measles virus phosphoprotein. J Virol 87(12): 7166–7169. https://doi.org/10.1128/JVI.00487-13
21. Abdella R, Aggarwal M, Okura T, Lamb RA, He Y (2020) Structure of a paramyxovirus polymerase complex reveals a unique methyltransferase-CTD conformation. Proc Natl Acad Sci USA 117(9):4931–4941. https://doi.org/10.1073/pnas.1919837117
22. Cong J, Feng X, Kang H, Fu W, Wang L, Wang C, Li X, Chen Y, Rao Z (2023) Structure of the Newcastle disease virus L protein in complex with tetrameric phosphoprotein. Nat Commun 14(1):1324. https://doi.org/10.1038/s41467-023-37012-y
23. Oda K, Matoba Y, Irie T, Kawabata R, Fukushi M, Sugiyama M, Sakaguchi T (2015) Structural basis of the inhibition of STAT1 activity by Sendai virus C protein. J Virol 89(22):11487–11499. https://doi.org/10.1128/JVI.01887-15
24. Roy A, Chan Mine E, Gaifas L, Leyrat C, Volchkova VA, Baudin F, Martinez-Gil L, Volchkov VE, Karlin DG, Bourhis JM, Jamin M (2023) Orthoparamyxovirinae C proteins have a common origin and a common structural organization. Biomol Ther 13(3):455. https://doi.org/10.3390/biom13030455

Chapter 3

Design and Execution of In Vitro Polymerase Assays for Measles Virus and Related Mononegaviruses

Robert M. Cox and Richard K. Plemper

Abstract

Morbilliviruses such as measles virus (MeV) are responsible for major morbidity and mortality worldwide, despite the availability of an effective vaccine and global vaccination campaigns. MeV belongs to the mononegavirus order of viral pathogens that store their genetic information in non-segmented negative polarity RNA genomes. Genome replication and viral gene expression are carried out by a virus-encoded RNA-dependent RNA polymerase (RdRP) complex that has no immediate host cell analog. To better understand the organization and regulation of the viral RdRP and mechanistically characterize antiviral candidates, biochemical RdRP assays have been developed that employ purified recombinant polymerase complexes and synthetic RNA templates to monitor the initiation of RNA synthesis and RNA elongation in vitro. In this article, we will discuss strategies for the efficient expression and preparation of mononegavirus polymerase complexes, provide detailed protocols for the execution and optimization of RdRP assays, evaluate alternative options for the choice of template and detection system, and describe the application of the assay for the characterization of inhibitor candidates. Although MeV RdRP assays are the focus of this article, the general strategies and experimental approaches are readily transferable to related viruses in the mononegavirus order.

Key words Baculovirus expression, Insect cells, RNA-dependent RNA polymerase, In vitro RNA synthesis, Primer extension assay

1 Introduction

MeV is a highly contagious member of the morbillivirus genus in the paramyxovirus family, which is part of the mononegavirus order [1]. Two canonical morbillivirus receptors have been identified, signaling activation molecular (SLAM)/CD150, which is expressed predominantly on activated B and T cells, macrophages, and dendritic cells [2], and Nectin4, which is expressed on basolateral membranes of epithelial cells [3, 4]. Although spreading through the respiratory route, morbilliviruses are, accordingly, unable to directly infect respiratory epithelial cells from the apical site. Rather, CD150-positive alveolar macrophages (AMs) and

Dzwokai Z. Ma and Christian K. Pfaller (eds.), *Measles and Related Morbilliviruses: Methods and Protocols*, Methods in Molecular Biology, vol. 2808, https://doi.org/10.1007/978-1-0716-3870-5_3,

dendritic cells (DCs) are the first target of invading morbilliviruses, followed by viral replication in B and T cells after homing of infected AMs and DCs to mediastinal lymph nodes and peripheral blood mononuclear cells (PBMC)-driven primary viremia [5, 6]. The viruses become airborne again after Nectin4-mediated basolateral entry into the respiratory epithelium.

Consistent with the lymphotropic nature of morbilliviruses, MeV infection results in severe lymphocytopenia, causing acute immunosuppression and long-term destruction of preexisting, unrelated immune memory [7, 8]. Although the case-fatality rate of measles is with approximately 1% in low- and middle-income countries relatively low [9], most morbidity and mortality arises in the years after recovery as a consequence of this loss of immunity against unrelated infectious agents [10]. Over 3 years after the onset of the COVID-19 pandemic, over 40 million children worldwide are without immune protection against MeV infection due to interrupted vaccination campaigns in large geographical areas and temporary SARS-CoV-2 viral dominance [11–13]. The major health threat posed by reemerging MeV and newly emerging zoonotic morbilliviruses [14] urgently calls for the development of a new generation of morbillivirus inhibitors that can synergize with vaccines for a comprehensive anti-morbillivirus platform through improved disease management, rapid outbreak control, and protection of the immunocompromised [15, 16].

Successful morbillivirus replication critically depends on the viral RNA-dependent RNA polymerase (RdRP) complex, which mediates both replication of the viral RNA genome and synthesis of viral mRNAs [1]. Because host cells do not possess comparable enzymatic activity and the viral polymerase complexes are structurally unique, RNA virus RdRPs have emerged as a druggable target for the development of antiviral therapeutics [17], and several orally efficacious inhibitors of the paramyxovirus RdRP complex, specifically, have been developed [18–20]. For a detailed understanding of the mechanism of activity (MOA) of these and related viral RdRP inhibitors [21–24], biochemical RdRP assays have been employed.

All mononegavirus polymerase complexes are composed of a large (L) polymerase protein that contains all enzymatic activities required for phosphodiester bond formation, mRNA capping, and cap methylation and its obligatory molecular chaperone, the phosphor (P)-protein [1]. Only P-L complexes are folding competent, and only the complexes, through protein-protein interfaces involving the P protein, interact with the viral nucleo (N)-protein encapsidated genomic and antigenomic RNA [25]. Accordingly, P and L must be co-expressed to prepare bioactive polymerase complexes for in vitro RdRP assays that rely on purified RdRP complexes, synthetic RNA templates, and, typically, a ^{32}P-labeled nucleotide tracer to monitor RNA elongation [26–28].

Baculovirus-driven expression in insect cells has emerged as the leading system for P-L production, since the polymerase complexes exceed the capacity of the prokaryotic protein folding machinery and are cytotoxic, when overexpressed, in mammalian cells. Due to mononegavirus genome replication and gene expression in the cytosol [1], viral mRNAs have not evolved under the selective pressure of the eukaryotic splicing machinery and frequently harbor hidden splice sites that interfere with mRNA integrity and nuclear export competence. Codon optimization for mammalian or insect cells is, therefore, imperative for efficient expression.

In contrast to bona fide genomic and antigenomic viral RNA, synthetic templates for RdRP assays are not N-encapsidated, excluding in vitro monitoring of P interactions with the N proteins as the polymerase complex advances along the template. Two different template types have been used successfully in mononegavirus RdRP assays: (i) base-paired primer/template combinations that trigger primer extension and are thought to represent a committed polymerase complex in RNA elongation mode and (ii) a single RNA molecule harboring terminally an abbreviated promoter sequence that supports de novo initiation of a P-L polymerase complex at the promoter and thus allows appreciation of distinct RdRP initiation steps. In both assay types, RNA sequences have been identified that are capable of stimulating bioactivity of RdRP complexes of different viral origin [29, 30], but templates are not universal and the optimal sequence must be explored individually for each P-L combination.

Biochemical RdRP assays have been applied to the molecular characterization of both nucleoside analog and non-nucleoside antivirals. In the case of substrate analog inhibitors, the assay allowed differentiation between compounds mediating immediate chain termination and delayed chain termination, which often reflects secondary structure changes in the nascent RNA strand after incorporation of the analog, stalling or pausing the polymerase complex a few nucleotides downstream of the incorporation site [29]. The assay furthermore provided a platform to assess relative selectivity of the polymerase complex for the analog in comparison with the corresponding natural nucleotide. By contrast, several non-nucleoside polymerase inhibitors were identified that prevent structural reorganization of the polymerase complex itself, blocking, for instance, progression from pre-initiation to initiation or from initiation to elongation conformation [23, 25]. Among morbillivirus inhibitors specifically, compound ERDRP-0519 was demonstrated in RdRP assays to block all phosphodiester bond formation [21], whereas the structurally unrelated compound GHP-88309 prevented de novo initiation after incorporation of the first few nucleotides [18]. Resistance profiling and photoaffinity-based target mapping revealed that ERDRP-0519 interfered with proper positioning of the priming loop relative to the catalytic site for phosphodiester bond formation, whereas GHP-88309 targets trigger residues required for opening of an exit channel for the nascent RNA strand. In both cases, only the combination of RdRP assays, resistance profiling, and photoaffinity target mapping allowed appreciation of molecular MOA and inhibitor docking pose.

2 Materials

2.1 Cells

1. Competent *E. coli* – DH10bac.
2. Spodoptera frugiperda (SF9) cells.

2.2 Media

1. Luria-Bertani (LB) agar.
2. Luria-Bertani (LB) broth.
3. SOC broth.
4. Hink's TNM-FH media.
5. Grace's Insect Medium.

2.3 Enzymes

1. T4 Polynucleotide Kinase.

2.4 Chemicals and Reagents

1. Gentamycin.
2. Kanamycin.
3. Tetracycline.
4. Isopropyl ß-D-1-thiogalactopyranoside (IPTG).
5. 5-Bromo-4-chloro-3-indolyl β-D-galactopyranoside (X-gal).
6. Cellfectin.
7. Sodium phosphate monobasic (NaH_2PO_4) pH 7.5.
8. Tris-HCl pH 7.4.
9. Dithiothreitol (DTT).
10. Glycerol.
11. Sodium chloride (NaCl).
12. Imidazole.
13. NP40.
14. Manganese chloride ($MnCl_2$).
15. Pierce protease inhibitor.
16. Pierce universal nuclease.
17. NTPs (ATP, GTP, UTP, CTP).
18. α-^{32}P labeled GTP.
19. 20% polyacrylamide gels with 7 M urea in Tris-borate-EDTA buffer.

2.5 Consumables

1. T25 flasks.
2. 1.5 mL microcentrifuge tubes.
3. 2.0 mL microcentrifuge tubes.
4. 15 mL and 50 mL conical tubes.
5. T25 flasks.
6. Erlenmeyer shaker flasks.
7. Slide-a-Lyzer cassette.
8. X-ray films.

2.6 Equipment

1. BAS Storage Phosphor Screen MS 22040.
2. Typhoon FLA7000 imager.
3. Tabletop microcentrifuge.
4. RNA electrophoresis apparatus.
5. Heat block (30–95 °C).
6. Bacteria shaking incubator.
7. Insect cell incubator with shaker.
8. Water bath (37 °C and 42 °C).

3 Method

3.1 In Vitro RdRP Assay Design

1. In vitro RdRP assays have greatly contributed to the mechanistic understanding of viral polymerase function including interaction with the promoter, RNA synthesis initiation, and RNA elongation. Applied to antiviral drug development, the biochemical assays have fostered the mechanistic characterization of nucleoside analog and non-nucleoside inhibitor candidates, provided guidance for the synthetic optimization of target affinity, and established a platform for the evaluation of the molecular basis of viral resistance. Efficient baculovirus-based polymerase expression systems have been established for several pathogens in the mononegavirus order including the morbilliviruses and highly pathogenic henipaviruses, and the experience gained with these different origin proteins supports that established expression strategies may be transferable to newly emerging or reemerging mononegaviruses. Key decision points when establishing RdRP assays comprise the choice of template (monitoring of RNA elongation versus de novo polymerase initiation at the promoter), selection of the tracer (radioactive versus nonradioactive), detection method (film versus phosphorimager for densitometry), and, in antiviral assays involving nucleoside analog inhibitors, administration of the analog in competition with, or replacement of, the corresponding natural nucleotide.

3.2 Choice of Expression Vectors and Codon Optimization

1. Insect cells (*Spodoptera frugiperda*, SF9) were selected to express the MeV L and P proteins (strain IC-B).
2. Prior to construction of the RdRP expression plasmid, both the MeV L and MeV P genes were codon optimized for expression in eukaryotic cells. A C-terminal His-tag was added to the MeV P gene, which did not hinder productive interactions between MeV L and P.
3. The MeV L and P proteins were expressed using the Fastbac dual baculovirus expression vector (ThermoFisher Scientific), under control of the polyhedron (late) and P10 (early) promoters, for L and P, respectively. All plasmids were sequenced prior to generation of recombinant baculovirus stocks.

3.3 Generation of Bacmid DNA

1. To generate bacmid DNA for recombinant baculovirus stocks, competent DH10BAC *E. coli* (MAX Efficiency DH10Bac; ThermoFisher Scientific) were transformed with the MeV L-P Fastbac dual vector (Fig. 1).
2. Fastbac dual plasmid (5 ng) was mixed with competent bacteria (25 μL) and incubated on ice for 30–45 min. Bacteria were then heat shocked at 42 °C for 45 s and then placed back on ice.
3. After incubating on ice for 5 min, 225 μL SOC medium was added, and bacteria were incubated at 37 °C with shaking for at least 4 h.
4. Bacteria were then plated on to LB agar supplemented with gentamycin (100 μg/mL), kanamycin (50 μg/mL), IPTG (40 μg/mL), and tetracycline (10 μg/mL). At least 1 h prior

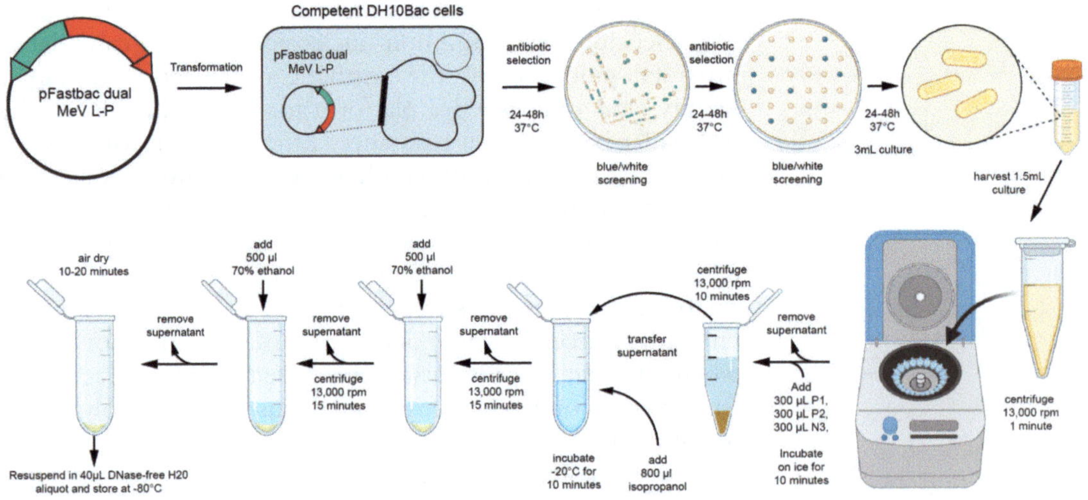

Fig. 1 Overview of the protocol discussed for pFastbac-MeV L-P bacmid purification

to plating bacteria, 120 μL X-Gal (20 mg/mL) was added to the surface of LB agar plates and spread evenly over the entire surface.

5. After adding bacteria (50–150 μL), plates were incubated at 37 °C for 24–48 h, until blue and white colonies could be clearly distinguished.
6. Bacmid containing colonies (white) were then selected and plated on new agar plates containing gentamycin, kanamycin, tetracycline, IPTG, and X-gal and incubated at 37 °C for 24–48 h to confirm colonies remained lacZ$^-$ (white colonies).
7. Colonies confirmed to contain bacmid DNA were selected and grown in 3 mL LB broth supplemented with gentamycin, kanamycin, tetracycline, and IPTG for 24–48 h.
8. After 24–48 h, 1.5 mL of culture were collected in a microcentrifuge tube and centrifuged for 1 min at 13,000 rpm.
9. The supernatant was then removed, and the pellet was resuspended in 300 μL of P1 buffer (Qiagen MiniPrep).
10. Cells were then lysed by adding 300 μL P2 buffer (Qiagen MiniPrep) and inverting the tube six to eight times.
11. After incubating for 3–5 min at room temperature, 300 μL of N3 buffer (Qiagen MiniPrep) was added, mixed by inverting the tube six to eight times, and incubated on ice for 10 min.
12. The solution was then clarified by centrifugation for 10 min at 13,000 rpm.
13. The supernatant was then removed and placed into a new 1.5 mL microcentrifuge tube and centrifuged for 10 min at 13,000 rpm to completely remove any debris.
14. Following this, the supernatant was carefully removed and placed into a 2 mL microcentrifuge tube containing 800 μL isopropanol and mixed gently by inverting the tube six to eight times.
15. The solution was then incubated at −20 °C for 15 min.
16. DNA was then pelleted by centrifugation at 13,000 rpm for 15 min.
17. The supernatant was then discarded, and the pellet was washed by adding 500 μL 70% ethanol and centrifuging the pellet for 5 min at 13,000 rpm. The wash step was repeated once.
18. Any remaining 70% ethanol was then removed, and the pellet was allowed to air-dry for 10–20 min.
19. Bacmid DNA was then resuspended by adding 40 μL DNase-free water and incubating at room temperature for 10–15 min. Bacmid DNA was then aliquoted into eight 5 μL aliquots and stored at −80 °C until further use.

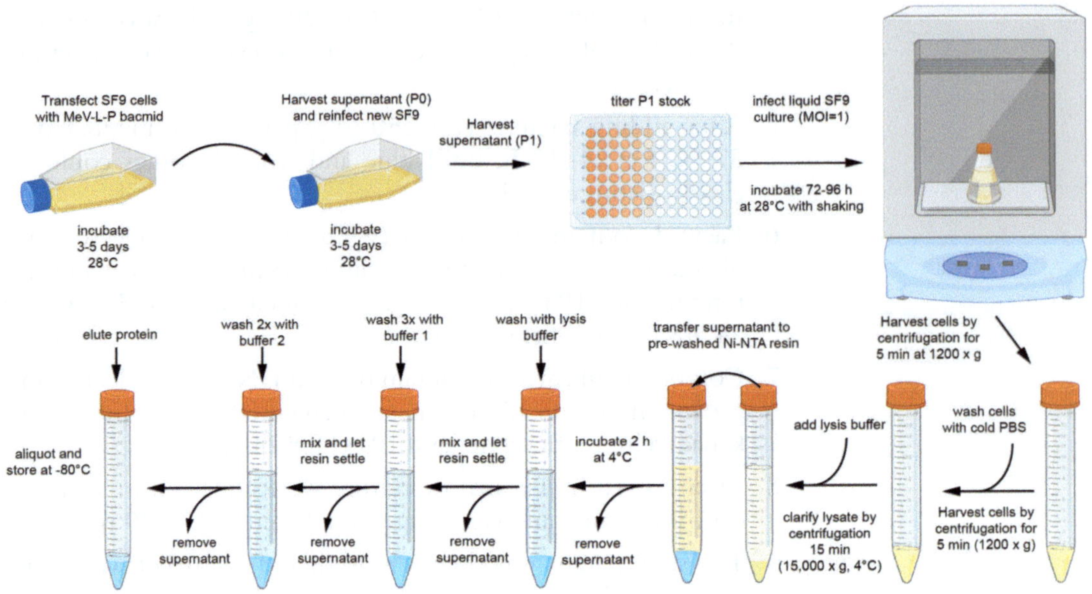

Fig. 2 Overview of baculovirus production and protein purification for MeV L-P complexes used in RdRP assays

3.4 Production of Recombinant Baculovirus

1. To generate recombinant baculovirus stocks, 8×10^6 SF9 cells were first added to a T25 flask and allowed to settle for 1–2 h (Fig. 2, top).
2. The transfection mixture was then prepared by preparing two solutions: Solution 1 containing 5 μL bacmid with 100 μL Hink's TNM-FH media (unsupplemented, Corning) and Solution 2 containing 6 μL Cellfectin in 100 μL Grace's Insect Medium (unsupplemented, Gibco).
3. Solutions 1 and 2 were then mixed gently by pipetting and incubated for 45 min at room temperature. Media was then removed from SF9 cells, and cells were washed once with unsupplemented culture media (Grace's Insect Medium, Gibco).
4. For each transfection, 800 μL culture media (unsupplemented) was added to the transfection mixture (Solutions 1 + 2) and added to SF9 cells.
5. The cells were then allowed to incubate for 4 h at 28 °C.
6. The transfection mixture was then removed, and 2 mL supplemented media was added to cells (Grace's supplemented with 10% FBS and penicillin-streptomycin (Gibco)).
7. After 3–5 days, the media was then removed and clarified by centrifugation for 10 min at 3000 rpm.
8. Supernatant containing P0 baculovirus stocks was stored at 4 °C.

9. To amplify virus stocks, SF9 cells (1.6×10^7 cells in a T75 flask plated) were infected by replacing the media with 2 mL P0 baculovirus stock.
10. Virus was allowed to adsorb for 1 h at 28 °C, and then 10 mL media was added to cells.
11. SF9 cells were incubated for 3–5 days until 100% CPE was observed.
12. Media (P1 stock) was then removed, clarified by centrifugation at 3000 rpm, and stored at 4 °C.
13. P1 baculovirus stocks were then titered by TCID50 titration [31].

3.5 Protein Expression

1. SF9 cells (100 mL of SF9 cells ($1–2 \times 10^6$ cells/mL)) in logarithmic growth phase were infected with P1 baculovirus (MOI = 1) (Fig. 2, right).
2. After 72–96 h, cells were equally divided into two 50 mL conical tubes and harvested by centrifugation ($1200 \times g$ for 5 min at 4 °C).
3. Each cell pellet was washed once with cold sterile PBS (10 mL) and centrifuged again for 5 min at $1200 \times g$ (4 °C).
4. Cells were then lysed by adding lysis buffer (50 mM NaH_2PO_4 pH 7.5, 150 mM NaCl, 20 mM imidazole, 0.5% NP40; 4 mL per pellet) with Pierce Protease Inhibitor and Pierce universal nuclease (ThermoFisher Scientific).
5. Lysates were then clarified by centrifugation ($15{,}000 \times g$ at 4 °C) for 15 min.
6. Supernatant was then removed and stored on ice prior to purifying MeV L-P complexes by Ni-NTA affinity chromatography. An aliquot of total cell lysate was taken for downstream analysis by SDS-PAGE (90 μL sample + 10 μL 1 M DTT + 100 μL 2× loading buffer).

3.6 Protein Purification

1. Prior to use, Ni-NTA agarose resin (800 μL; 50% slurry) was added to a 15 mL conical tube.
2. Resin was washed once with 5 mL sterile water and mixed gently by inversion four to six times.
3. The resin was then pelleted by centrifugation (2 min at $700 \times g$, 4 °C). The water was carefully removed.
4. The resin was then resuspended in 5 mL of lysis buffer (50 mM NaH_2PO_4 pH 7.5, 150 mM NaCl, 20 mM imidazole, 0.5% NP40) and pelleted by centrifugation (2 min at $700 \times g$, 4 °C). The lysis buffer was then removed, and the pellet was washed once more in 5 ml of lysis buffer.

5. After the second wash with lysis buffer, approximately 400 μL of lysis buffer was not discarded to prevent the resin from drying out.
6. The total soluble lysate (approximately 8 mL) was added to the 15 mL conical tube containing the washed resin and incubated for 2 h at 4 °C with rocking (Fig. 2, bottom).
7. After 2 h, the resin was allowed to settle for 5–7 min on ice, followed by centrifugation for 5 s at 300 × g (4 °C). An aliquot (90 μL) of unbound protein was taken for downstream analysis by SDS-PAGE (90 μL sample + 10 μL 1 M DTT + 100 μL 2× loading buffer). The remaining unbound protein was discarded.
8. The bound Ni-NTA resin was then washed three times with buffer 1 (50 mM NaH_2PO_4 pH 7.5, 150 mM NaCl, 60 mM imidazole, 0.5% NP40). In short, 4 mL of buffer 1 was used to resuspend the resin by gently inverting the tube three to four times.
9. The resin was allowed to settle for 5–7 min on ice, followed by centrifugation for 5 s at 300 × g (4 °C). An aliquot of each wash buffer 1 wash was taken for downstream analysis by SDS-PAGE (90 μL sample + 10 μL 1 M DTT + 100 μL 2× loading buffer).
10. The remaining buffer 1 wash was discarded, and the process was repeated twice with buffer 1.
11. The resin was then washed twice with 4 mL buffer 2 (50 mM NaH_2PO_4 pH 7.5, 150 mM NaCl, 60 mM imidazole, 0.5% NP40). After each wash with buffer 2, aliquots were taken for downstream analysis by SDS-PAGE (45 μL sample + 5 μL 1 M DTT + 50 μL 2× loading buffer).
12. Protein was then eluted by adding 2 mL of elution buffer (50 mM NaH_2PO_4 pH 7.5, 150 mM NaCl, 250 mM imidazole, 0.5% NP40).
13. The resin was gently resuspended by gently inverting the tube three to four times and allowed to settle on ice for 10 min.
14. The resin was then pelleted by centrifuging the tube for 2 min at 700 × g (4 °C).
15. The eluate was carefully harvested using a pipette and incubated on ice. An aliquot of eluate was taken for downstream analysis by SDS-PAGE (11.25 μL eluate +33.75 μL elution buffer +5 μL 1 M DTT + 50 μL 2× loading buffer).
16. Following this, 2 mL of elution buffer was used to resuspend the beads. An aliquot of the beads (containing a mixture of eluate and beads) was then taken for downstream analysis by SDS-PAGE (45 μL beads +5 μL 1 M DTT + 50 μL 2× loading buffer).

17. The eluate was then dialyzed overnight at 4 °C into buffer containing 20 mM Tris-HCl pH 7.4, 150 mM NaCl, 10% glycerol, using a Slide-a-Lyzer cassette (ThermoFisher; 10 K MWCO).
18. The following day, the dialysis buffer was switched to a buffer containing 20 mM Tris-HCl pH 7.4, 150 mM NaCl, 10% glycerol, 1 mM DTT, and allowed to dialyze overnight at 4 °C. Once dialysis was complete, an aliquot was taken for downstream analysis by SDS-PAGE (11.25 μL eluate +33.75 μL elution buffer +5 μl 1 M DTT + 50 μL 2× loading buffer).
19. Protein concentration was then determined, and the remaining eluate was aliquoted into 1.5 mL microcentrifuge tubes (low protein binding; Eppendorf) and stored at −80 °C.

3.7 *De Novo MeV RdRP* Assay

1. Purified MeV L-P complexes (50–150 ng) were diluted in 50 μL reaction buffer (50 mM Tris-HCL pH 7.4, 1 mM DTT, 10% glycerol, 3-4 mM MnCl2, 1 μM of RNA template (16-mer RNA 3′-UGGUCUUUUUGUUUC-5′ for nonbackpriming assays; 25-mer RNA 3′-UGCUCUU UUUUUCACAGUUUUUGAU-5′ for backpriming), 1 mM GTP (16-mer template) or 1 mM ATP (25-mer template), 1 mM UTP, 1 mM CTP, 50 μM ATP (16-mer template) or 50 μM GTP (25-mer template), 10 μCi of α-^{32}P labeled ATP (16-mer template) or 10 μCi of α-^{32}P labeled GTP (25-mer template) (PerkinElmer), and up to 5% DMSO containing small molecule antivirals (Fig. 3, bottom).
2. Assays were performed at 30 °C and allowed to continue for 5 h.
3. RNA was then ethanol precipitated overnight at −20 °C.
4. The following day, the RNA was resuspended in 50% deionized formamide with 10 mM EDTA, heat denatured for 5 min at 95 °C, and then separated through a 20% polyacrylamide gel containing 7 mM urea in Tris-borate-EDTA buffer.
5. De novo polymerization products were visualized by autoradiography using X-ray films or quantified by exposing a BAS Storage Phosphor Screen MS 22040 and scanning for signal on a Typhoon FLA7000 imager (GE Healthcare). Experimental controls included vehicle treated MeV L-P complexes and MeV L_{N774A}-P complexes. MeV L_{N774A} has a mutation in the RdRP active site and is enzymatically nonfunctional.
6. A 4-nt and 16-nt or 25-nt radiolabeled RNA molecular weight marker was included as reference on each gel. Molecular weight markers were created by labeling a 4-nt primer (5′-ACCA-3′) or the RNA template with γ-32P-ATP, using T4 Polynucleotide Kinase (New England Biolabs).

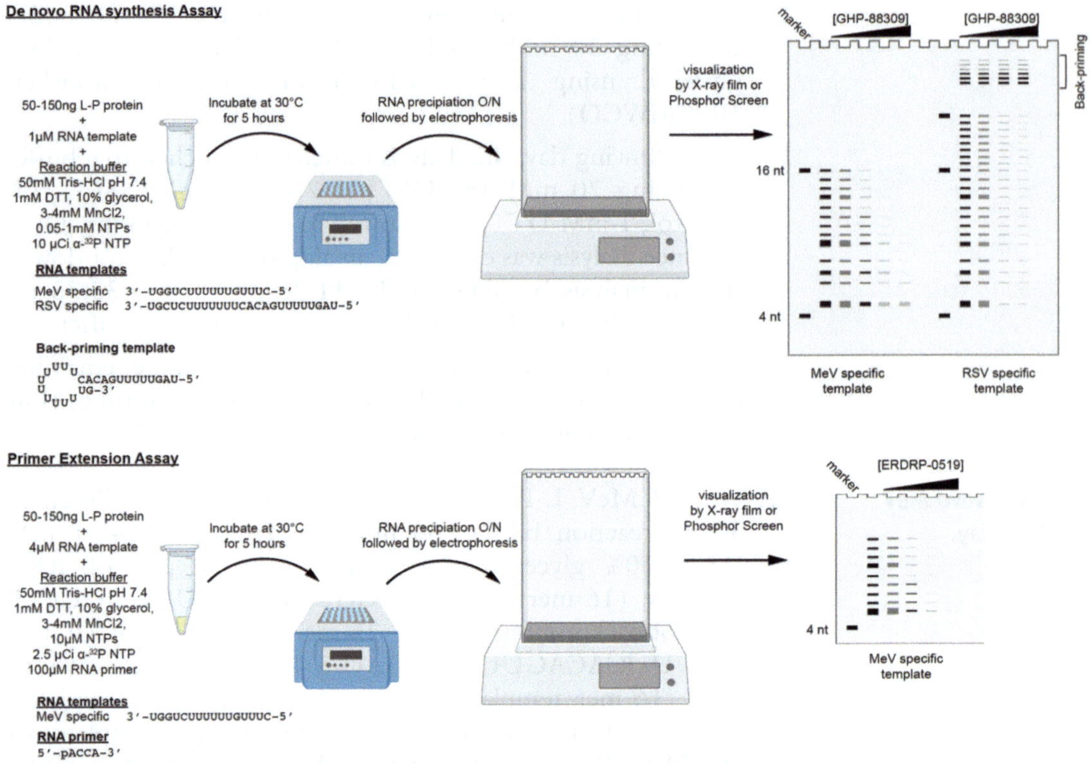

Fig. 3 Overview of RdRP assays discussed. (top) De novo RNA synthesis assay detailing the use of two distinct templates for RNA synthesis. Use of the 25-nt RSV specific template can result in backpriming, which is hypothesized to mimic RNA synthesis after initiation. Hypothetical results for RdRP assays performed in the presence of GHP-883089 are shown. (bottom) Primer extension assay that can be utilized to study the effects of polymerase inhibitors on elongation using the 16-nt MeV specific RNA template. Hypothetical results for primer extension assays performed in the presence of ERDRP-0519 are shown

3.8 Primer Extension Assays

1. Purified MeV L-P complexes (50–150 ng) were diluted in reaction buffer (50 mM Tris-HCL pH 7.4, 1 mM DTT, 10% glycerol, 3 mM MnCl2, 10 μM GTP and ATP, or 10 μM GTP only (when measuring single base extensions), 2.5 μCi of α-^{32}P labeled GTP (PerkinElmer) 4 μM of 16-nt RNA template (3-′-UGGUCUUUUUUGUUUC-5′, 100 μM of 5′ phosphorylated 4-nt RNA primer (5′-pACCA-3′; Perkin Elmer) and small molecule antivirals in up to 5% DMSO final concentration.
2. Primer extension reactions were allowed to continue for 5 h at 30 °C, after which the reaction was stopped with 1 volume of deionized formamide with 20 mM EDTA.
3. RNA was then ethanol precipitated overnight at −20 °C.
4. The following day, the RNA was resuspended in 50% deionized formamide with 10 mM EDTA, heat denatured for 5 min at 95 °C, and then separated through a 20% polyacrylamide gel containing 7 mM urea in Tris-borate-EDTA buffer.

5. Primer extension products were visualized by autoradiography using X-ray films or quantified by exposing a BAS Storage Phosphor Screen MS 22040 and scanning for signal on a Typhoon FLA7000 imager (GE Healthcare). Experimental controls included vehicle treated MeV L-P complexes, reactions with no template, and reactions with no primer. A 4-nt radiolabeled molecular weight marker was included as reference on each gel.
6. Molecular weight markers were created by labeling the 4-nt primer (5′-ACCA-3′) with γ-^{32}P-ATP, using T4 Polynucleotide Kinase (New England Biolabs).

4 Notes

1. Optimization of $MnCl_2$ concentration has been critical to achieve the best efficiency in assay performance. $MnCl_2$ concentration also effects the ratio of backpriming products to de novo elongation products. In addition to $MnCl_2$, $MgCl_2$ can be substituted, although optimization of salt concentrations is critical for assay performance.

Acknowledgments

We thank R. Fearns and J. Sourimant for helpful discussions. Figures were created with BioRender.com.

References

1. Plemper RK, Lamb RA (2020) Paramyxoviridae: the viruses and their replication. In: Knipe DM, Howley PM, Whelan S (eds) Fields virology, vol 1, 7th edn. Wolters Kluwer/Lippincott Williams & Wilkins, Philadelphia, pp 504–558
2. Tatsuo H, Ono N, Yanagi Y (2001) Morbilliviruses use signaling lymphocyte activation molecules (CD150) as cellular receptors. J Virol 75(13):5842–5850. https://doi.org/10.1128/jvi.75.13.5842-5850.2001
3. Muhlebach MD, Mateo M, Sinn PL, Prufer S, Uhlig KM, Leonard VH, Navaratnarajah CK, Frenzke M, Wong XX, Sawatsky B, Ramachandran S, McCray PB Jr, Cichutek K, von Messling V, Lopez M, Cattaneo R (2011) Adherens junction protein nectin-4 is the epithelial receptor for measles virus. Nature 480(7378):530–533. https://doi.org/10.1038/nature10639
4. Noyce RS, Bondre DG, Ha MN, Lin LT, Sisson G, Tsao MS, Richardson CD (2011) Tumor cell marker PVRL4 (nectin 4) is an epithelial cell receptor for measles virus. PLoS Pathog 7(8):e1002240. https://doi.org/10.1371/journal.ppat.1002240
5. de Swart RL (2017) Measles: what we have learned from non-human primate models. Drug Discov Today Dis Model 23:31–34. https://doi.org/10.1289/ehp.1408257
6. Laksono BM, de Vries RD, McQuaid S, Duprex WP, de Swart RL (2016) Measles virus host invasion and pathogenesis. Viruses 8(8):210. https://doi.org/10.3390/v8080210
7. Petrova VN, Sawatsky B, Han AX, Laksono BM, Walz L, Parker E, Pieper K, Anderson CA, de Vries RD, Lanzavecchia A, Kellam P, von Messling V, de Swart RL, Russell CA (2019) Incomplete genetic reconstitution of

B cell pools contributes to prolonged immunosuppression after measles. Sci Immunol 4(41): eaay6125. https://doi.org/10.1126/sciimmunol.aay6125

8. Mina MJ, Kula T, Leng Y, Li M, de Vries RD, Knip M, Siljander H, Rewers M, Choy DF, Wilson MS, Larman HB, Nelson AN, Griffin DE, de Swart RL, Elledge SJ (2019) Measles virus infection diminishes preexisting antibodies that offer protection from other pathogens. Science 366(6465):599–606. https://doi.org/10.1126/science.aay6485
9. Portnoy A, Jit M, Ferrari M, Hanson M, Brenzel L, Verguet S (2019) Estimates of case-fatality ratios of measles in low-income and middle-income countries: a systematic review and modelling analysis. Lancet Glob Health 7(4):e472–e481. https://doi.org/10.1016/S2214-109X(18)30537-0
10. Mina MJ, Metcalf CJ, de Swart RL, Osterhaus AD, Grenfell BT (2015) Long-term measles-induced immunomodulation increases overall childhood infectious disease mortality. Science 348(6235):694–699. https://doi.org/10.1126/science.aaa3662
11. Guha-Sapir D, Moitinho de Almeida M, Keita M, Greenough G, Bendavid E (2020) COVID-19 policies: remember measles. Science 369(6501):261. https://doi.org/10.1126/science.abc8637
12. Roberts L (2020) Why measles deaths are surging – and coronavirus could make it worse. Nature 580(7804):446–447. https://doi.org/10.1038/d41586-020-01011-6
13. Teasdale CA, Borrell LN, Shen Y, Kimball S, Zimba R, Kulkarni S, Rane M, Rinke ML, Fleary SA, Nash D (2022) Missed routine pediatric care and vaccinations in US children during the first year of the COVID-19 pandemic. Prev Med 158:107025. https://doi.org/10.1016/j.ypmed.2022.107025
14. Lee B, Ikegame S, Carmichael J, Wells H, Furler R, Acklin J, Chiu HP, Oguntuyo K, Cox R, Patel A, Kowdle S, Stevens C, Eckley M, Zhan S, Lim J, Hashiguchi T, Durigon EL, Schountz T, Epstein J, Plemper R, Daszak P, Anthony S (2021) Zoonotic potential of a novel bat morbillivirus. Res Sq. rs.3.rs-926789. https://doi.org/10.21203/rs.3.rs-926789/v1
15. Plemper RK, Hammond AL (2014) Synergizing vaccinations with therapeutics for measles eradication. Expert Opin Drug Discov 9(2): 201–214. https://doi.org/10.1517/17460441.2014.867324
16. Plemper RK (2020) Measles resurgence and drug development. Curr Opin Virol 41:8–17. https://doi.org/10.1016/j.coviro.2020.02.007
17. Cox R, Plemper RK (2015) The paramyxovirus polymerase complex as a target for next-generation anti-paramyxovirus therapeutics. Front Microbiol 6:459. https://doi.org/10.3389/fmicb.2015.00459
18. Cox RM, Sourimant J, Toots M, Yoon JJ, Ikegame S, Govindarajan M, Watkinson RE, Thibault P, Makhsous N, Lin MJ, Marengo JR, Sticher Z, Kolykhalov AA, Natchus MG, Greninger AL, Lee B, Plemper RK (2020) Orally efficacious broad-spectrum allosteric inhibitor of paramyxovirus polymerase. Nat Microbiol 5(10):1232–1246. https://doi.org/10.1038/s41564-020-0752-7
19. Wittwer K, Anderson DE, Pfeffermann K, Cox RM, Wolf JD, Santibanez S, Mankertz A, Plesker R, Sticher ZM, Kolkykhalov AA, Natchus MG, Pfaller CK, Plemper RK, von Messling V (2021) Small-molecule polymerase inhibitor protects non-human primates from measles and reduces shedding. Nat Commun 12(1):5233. https://doi.org/10.1038/s41467-021-25497-4
20. Krumm SA, Yan D, Hovingh ES, Evers TJ, Enkirch T, Reddy GP, Sun A, Saindane MT, Arrendale RF, Painter G, Liotta DC, Natchus MG, von Messling V, Plemper RK (2014) An orally available, small-molecule polymerase inhibitor shows efficacy against a lethal morbillivirus infection in a large animal model. Sci Transl Med 6 (232):232ra252. https://doi.org/10.1126/scitranslmed.3008517
21. Cox RM, Sourimant J, Govindarajan M, Natchus MG, Plemper RK (2021) Therapeutic targeting of measles virus polymerase with ERDRP-0519 suppresses all RNA synthesis activity. PLoS Pathog 17(2):e1009371. https://doi.org/10.1371/journal.ppat.1009371
22. Sourimant J, Lieber CM, Yoon JJ, Toots M, Govindarajan M, Udumula V, Sakamoto K, Natchus MG, Patti J, Vernachio J, Plemper RK (2022) Orally efficacious lead of the AVG inhibitor series targeting a dynamic interface in the respiratory syncytial virus polymerase. Sci Adv 8(25):eabo2236. https://doi.org/10.1126/sciadv.abo2236
23. Noton SL, Nagendra K, Dunn EF, Mawhorter ME, Yu Q, Fearns R (2015) Respiratory syncytial virus inhibitor AZ-27 differentially inhibits different polymerase activities at the promoter. J Virol 89(15):7786–7798. https://doi.org/10.1128/JVI.00530-15
24. Gotte M (2021) Remdesivir for the treatment of Covid-19: the value of biochemical studies.

Curr Opin Virol 49:81–85. https://doi.org/10.1016/j.coviro.2021.04.014

25. Fearns R, Plemper RK (2017) Polymerases of paramyxoviruses and pneumoviruses. Virus Res 234:87–102. https://doi.org/10.1016/j.virusres.2017.01.008
26. Deval J, Hong J, Wang G, Taylor J, Smith LK, Fung A, Stevens SK, Liu H, Jin Z, Dyatkina N, Prhavc M, Stoycheva AD, Serebryany V, Liu J, Smith DB, Tam Y, Zhang Q, Moore ML, Fearns R, Chanda SM, Blatt LM, Symons JA, Beigelman L (2015) Molecular basis for the selective inhibition of respiratory syncytial virus RNA polymerase by 2′-Fluoro-4′-Chloromethyl-cytidine triphosphate. PLoS Pathog 11(6): e1004995. https://doi.org/10.1371/journal.ppat.1004995
27. Noton SL, Aljabr W, Hiscox JA, Matthews DA, Fearns R (2014) Factors affecting de novo RNA synthesis and back-priming by the respiratory syncytial virus polymerase. Virology 462–463:318–327. https://doi.org/10.1016/j.virol.2014.05.032
28. Noton SL, Deflube LR, Tremaglio CZ, Fearns R (2012) The respiratory syncytial virus polymerase has multiple RNA synthesis activities at the promoter. PLoS Pathog 8(10): e1002980. https://doi.org/10.1371/journal.ppat.1002980
29. Gordon CJ, Lee HW, Tchesnokov EP, Perry JK, Feng JY, Bilello JP, Porter DP, Gotte M (2022) Efficient incorporation and template-dependent polymerase inhibition are major determinants for the broad-spectrum antiviral activity of remdesivir. J Biol Chem 298(2): 101529. https://doi.org/10.1016/j.jbc.2021.101529
30. Shareef AM, Ludeke B, Jordan P, Deval J, Fearns R (2021) Comparison of RNA synthesis initiation properties of non-segmented negative strand RNA virus polymerases. PLoS Pathog 17(12):e1010151. https://doi.org/10.1371/journal.ppat.1010151
31. LaBarre DD, Lowy RJ (2001) Improvements in methods for calculating virus titer estimates from TCID50 and plaque assays. J Virol Methods 96(2):107–126. https://doi.org/10.1016/s0166-0934(01)00316-0

Chapter 4

Optical Control of Mononegavirus Gene Expression and Replication

Maino Tahara, Takashi Okura, Moritoshi Sato, and Makoto Takeda

Abstract

Mononegaviruses are promising tools as oncolytic and transgene vectors for gene therapy and regenerative medicine. However, when mononegaviruses are used for therapeutic applications, the viral activity must be strictly controlled due to concerns about toxicity and severe side effects. With this technology, mononegavirus vectors can be grown where they are intended and can be easily removed when they are no longer needed. In particular, a photoswitch protein called Magnet (consisting of two magnet domains) is incorporated into the hinge region between the connector and methyltransferase domains of the mononegavirus polymerase protein (L protein) to disrupt the L protein functions. Blue light (470 ± 20 nm) irradiation causes the dimerization of the two magnet domains, and the L protein is restored to activity, allowing viral gene expression and virus replication. Since the magnet domains' dimerization is reversible, viral gene expression and replication cease when blue light irradiation is stopped.

Key words Measles virus, Rabies virus, Parainfluenza virus, Mononegavirus, Viral vectors, Optogenetics, Magnet

1 Introduction

The order of mononegaviruses includes the *Paramyxoviridae*, *Pneumoviridae*, *Bornaviridae*, *Filoviridae*, and *Rhabdoviridae* families. The utility of mononegaviruses in cancer therapy, gene therapy, regenerative medicine, and vaccine development is well known [1–3]. For clinical applications, controlling vector replication can reduce the risk of side effects caused by excessive vector growth and gene expression [4]. However, spatiotemporal control of viral vectors, including intentional control of gene expression and silencing and precisely scheduled vector removal, remains challenging.

Maino Tahara and Takashi Okura are equally contributed to this work.

Dzwokai Z. Ma and Christian K. Pfaller (eds.), *Measles and Related Morbilliviruses: Methods and Protocols*, Methods in Molecular Biology, vol. 2808, https://doi.org/10.1007/978-1-0716-3870-5_4,

The polymerase protein of mononegaviruses is a large protein (L protein) with five major domains (RNA-dependent RNA polymerase [RdRp] domain, capping domain [PRNTase], connector domain [CD], methyltransferase domain [MTase], and C-terminal domain [CTD]) and two hinge regions that functionally connect these domains (hinge 1 [H1] connects PRNTase and CT, and hinge 2 [H2] connects CD and MTase) [5] (Fig. 1). The fundamental functions and structure are conserved among all mononegaviruses [5–7]. The L protein is multifunctional and performs all of the processes in replication and transcription of the viral genome. Therefore, the L protein must have the appropriate arrangement of the above domains to exhibit these diverse functions [6].

We hypothesized that the activity of the mononegavirus L protein could be regulated by controlling this domain structure. The N-terminal and C-terminal regions can fold independently even when the L protein is separated at the H2 region. Still, they have no affinity for each other and do not exhibit polymerase activity [8]. However, polymerase activity is restored when a dimerization domain is artificially added to them [8]. We focused on an optogenetics approach using photoswitch proteins to achieve intentional control of L protein activity. Photoswitch proteins change their structure depending on the presence or absence of light irradiation, and this property can be used to manipulate various phenomena with light. However, most natural photoreceptors are unsuitable for high-precision control of protein functions because of their considerable molecular weight and slow response to light irradiation or blockage. Among photoswitch proteins, Vivid, a photoreceptor of *Neurospora crassa*, is small in size and has a simple property. It associates when exposed to blue light (dimer formation) and separates when blue light is stopped (monomerization). Furthermore, protein engineering has enabled Vivid to respond to blue light exposure and blockage as fast as a few seconds [9, 10]. This improved Vivid was named Magnet (consisting of two magnet domains, nMag and pMagHigh) (Fig. 1) because of its magnet-like properties.

The H2 region of the mononegavirus L protein was found to be able to insert peptides and proteins (such as green fluorescent protein [GFP]) of a certain size [11, 12]. By inserting Magnet at various positions in H2 (Fig. 1), we have tested the possibility of regulating the L protein and have so far produced viruses that can be precisely controllable for gene expression and growth with blue light in three different mononegaviruses. The detailed gene structure of the optically controllable mononegavirus L protein and the details of how it is produced are described.

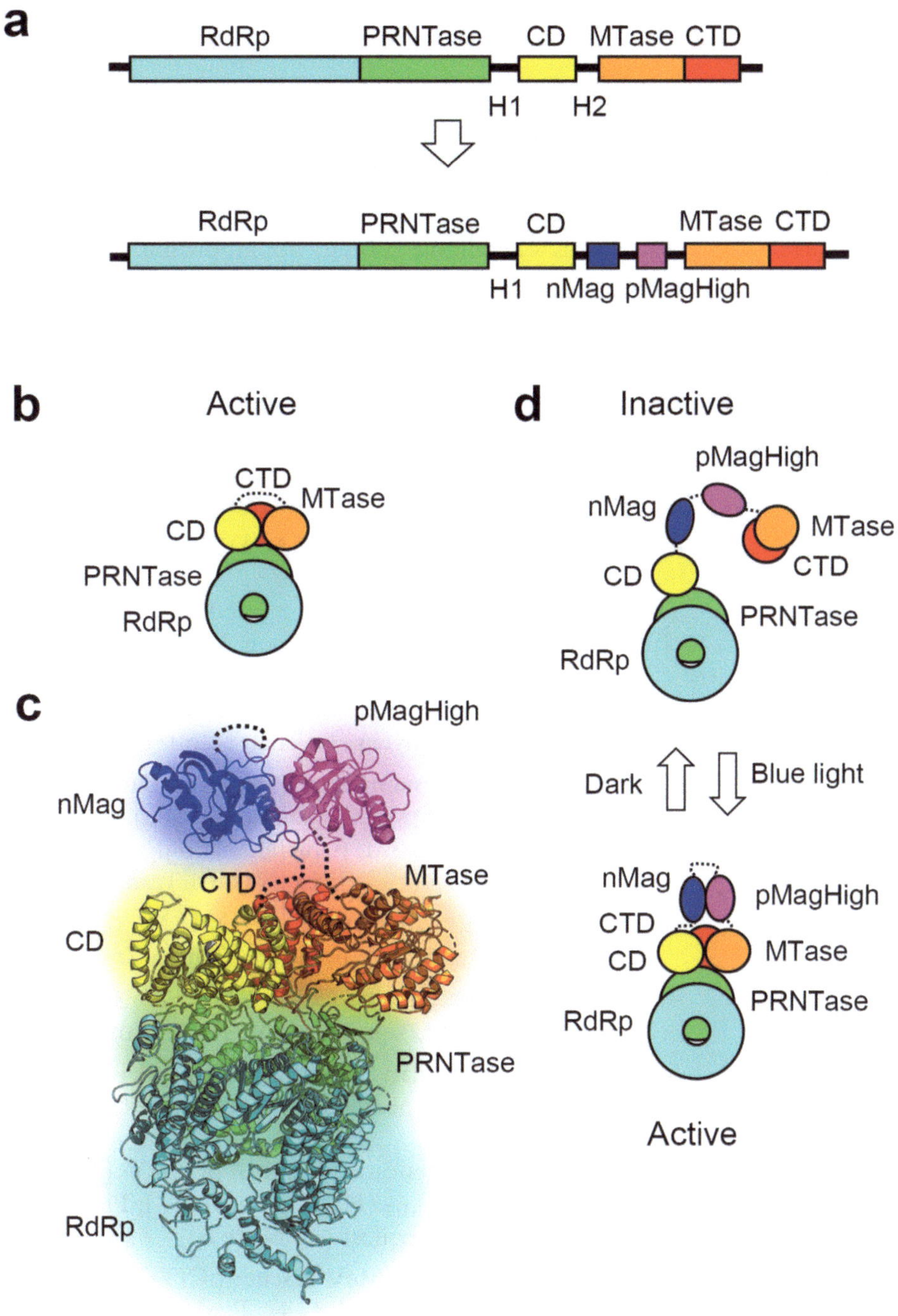

Fig. 1 Schematic diagram of optically controllable polymerase L protein. (**a**) Coding regions of L protein domains and insertion position of Magnet protein genes. RdRp, RNA-dependent RNA polymerase; PRNTase, capping domain; CD, connector domain; MTase, methyltransferase domain; CTD, C-terminal domain; H1, hinge 1; H2, hinge 2; nMag and pMagHigh, Magnet domains. (**b**) Schematic diagram of the domain structure of L protein. (**c**) The 3D structure of parainfluenza virus 5 L protein (PDB 6V85) and Vivid (3RH8). The location of the L protein and Vivid is just an imaginary diagram. (**d**) Conceptual diagram of optical control of the L protein

2 Materials

2.1 Preparation of Support Plasmids for MeV, RABV, and BPIV3 Rescue

1. Competent cells: *Escherichia coli* DH5α (TOYOBO).
2. Luria-Bertani (LB) medium: 10 g tryptone, 5 g yeast extract, 10 g NaCl per liter. Biological buffers have been added for proper pH adjustment. Autoclaved at 121 °C for 15 min.
3. LB plate containing 100 μg/mL ampicillin.
4. NucleoBond Xtra Midi kit (MACHEREY-NAGEL).

2.2 Preparation of Full-Length MeV, RABV, and BPIV3 Genome Plasmids

1. Competent cells: *Escherichia coli* DH5α (TOYOBO).
2. LB medium: 10 g tryptone, 5 g yeast extract, 10 g NaCl per liter. Biological buffers have been added for proper pH adjustment. Autoclaved at 121 °C for 15 min.
3. LB plate containing 100 μg/mL ampicillin.
4. NucleoBond Xtra Midi kit (MACHEREY-NAGEL).

2.3 Cell Culture

1. BHK/T7–9 cell maintenance medium: Eagle's Minimum Essential Medium (MEM, Nissui Pharmaceutical), 0.3% tri-phosphate broth (TPB), 2 mM L-glutamine, 0.15% sodium bicarbonate ($NaHCO_3$), 5% fetal bovine serum (FBS), and 600 μg/mL hygromycin (Nakalai-Tesque) (*see* **Note 1**).
2. Vero/hSLAM cell maintenance medium: Dulbecco's Modified Eagle's Medium (DMEM, Nacalai-Tesque), 0.15% $NaHCO_3$, 7.5% FBS, 100 unit/mL penicillin, 100 μg/mL streptomycin, 0.4 mg/mL G418 (Geneticin, Nakalai-Tesque) (*see* **Note 2**).
3. Vero cell maintenance medium: DMEM, 0.15% $NaHCO_3$, 7.5% FBS, 100 unit/mL penicillin, 100 μg/mL streptomycin.
4. 293T cell maintenance medium: DMEM, 0.15% $NaHCO_3$, 10% FBS, 100 unit/mL penicillin, 100 μg/mL streptomycin.
5. MDBK cell maintenance medium: MEM, 0.15% $NaHCO_3$, 2% FBS, 100 unit/mL penicillin, 100 μg/mL streptomycin.

2.4 Rescue of Optically Controllable MeV from Plasmids

1. Full-length MeV genome plasmids: p(+)MV-EGFP-L_{LMH} and p(+)MV-EGFP-L_{DMH} (Figs. 2 and 3) [13]. Plasmid sequences are available upon request (*see* **Note 3**).
2. Support plasmids for MeV rescue: pCITE-IC-N, pCITE-IC-PΔC, pCITE-ko-9301B-L [14]. Plasmid sequences are available upon request.
3. BHK-T7/9 cells (RIKEN BioResource Research Center; RCB4942) [15].
4. Vero/hSLAM cells (JCRB Cell Bank; JCRB1809) [16].
5. Dulbecco's phosphate-buffered saline (PBS).

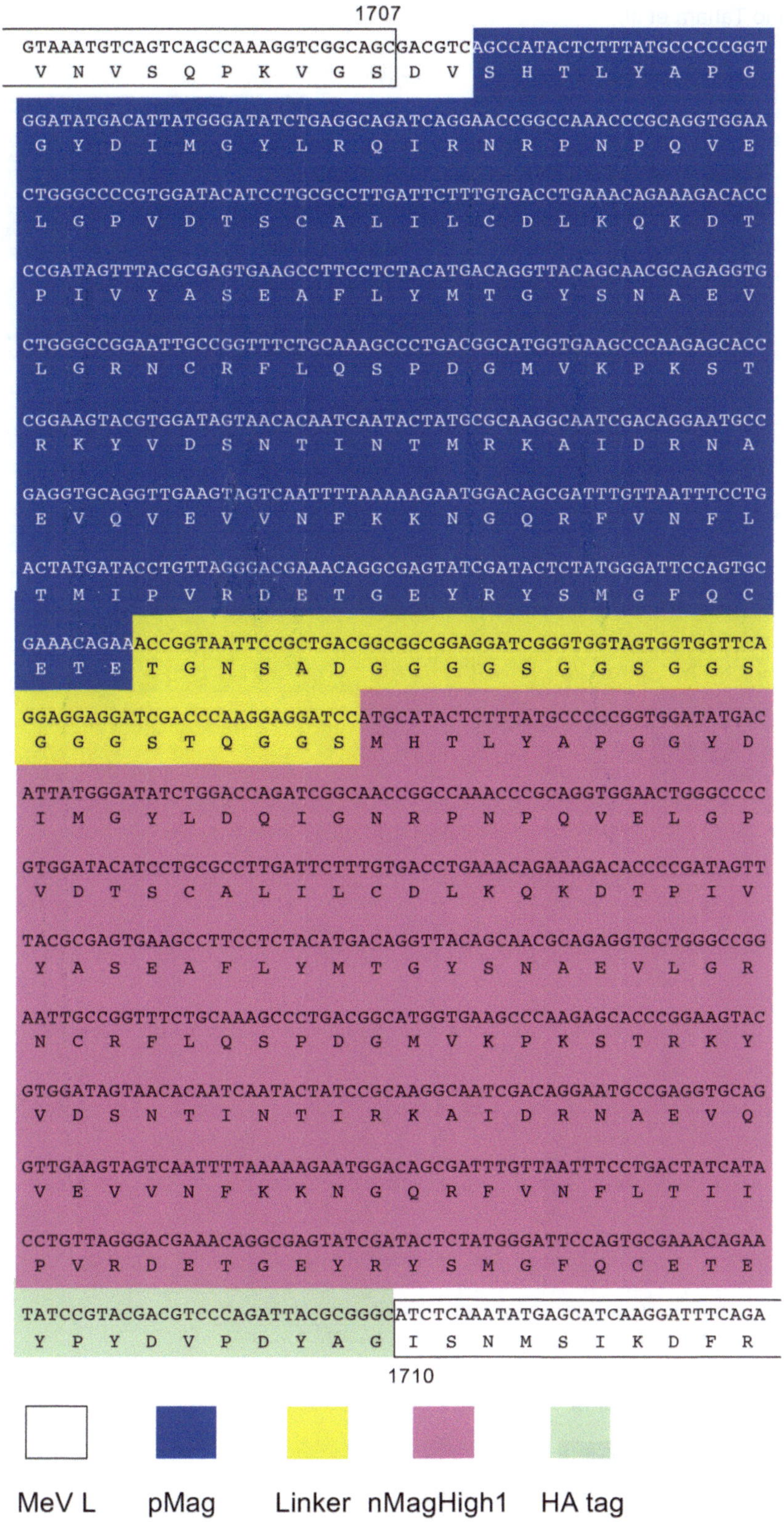

Fig. 2 Nucleotide and amino acid sequences of the Magnet domain insertion region of p(+)MV-EGFP-L_{LMH}. Measles virus L gene (MeV L), pMag, Linker, nMagHigh1, and HA tag regions are indicated in white, blue, yellow, pink, and light green, respectively

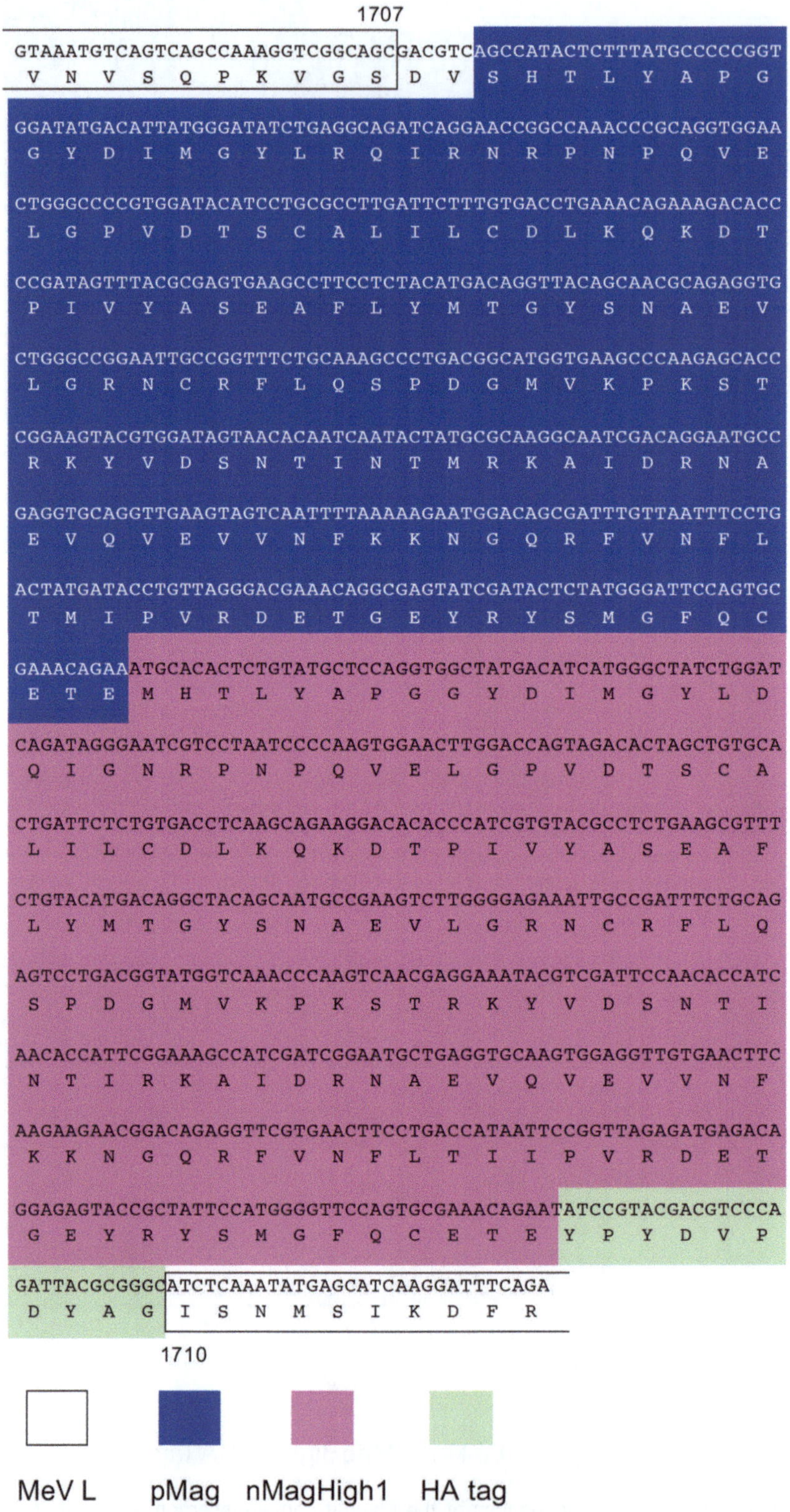

Fig. 3 Nucleotide and amino acid sequences of the Magnet domain insertion region of p(+)MV-EGFP-L_{LMH}. MeV L, pMag, nMagHigh1, and HA tag regions are indicated in white, blue, pink, and light green, respectively

6. Trypsin (0.25 w/v%)-EDTA (1 mM) (trypsin/EDTA) solution.
7. 6-well culture plates.
8. Transfection reagents: X-tremeGENE HP (Roche).
9. CO_2 incubator.
10. Light irradiation device: ISLM-150×150-RB (CCS, Kyoto, Japan).
11. Constant current power supply ISC-201-2 (CCS, Kyoto, Japan).

2.5 Rescue of Optically Controllable RABV from Plasmids

1. Full-length RABV genome plasmids: p(+)RABV-EGFP-L_{LMH2} and -L_{DMH2} (Figs. 4 and 5) [13]. Plasmid sequences are available upon request (*see* **Note 3**).
2. Support plasmids for RABV rescue: pH-N, pH-P, pH-G, and pH-L [17].
3. BHK-T7/9 cells.
4. Vero cells: Our preexisting stocks of the cells were originally obtained from American Type Culture Collection (ATCC) and have been maintained for more than 20 years in our laboratory.
5. PBS.
6. Trypsin/EDTA solution.
7. 6-well culture plates.
8. Transfection reagents: X-tremeGENE HP.
9. CO_2 incubator.
10. Light irradiation device: ISLM-150×150-RB (CCS, Kyoto, Japan).
11. Constant current power supply ISC-201-2.

2.6 Rescue of Optically Controllable BPIV3 from Plasmids

1. Full-length BPIV3 genome plasmid: p(+)57BPIV3-EGFP-L-Mag3 (Fig. 6) [18]. Plasmid sequences are available upon request (*see* **Note 3**).
2. Support plasmids for BPIV3 rescue: pCAGGS-BPIV3-N, pCAGGS-BPIV3-P, pCAGGS-BPIV3-L, and a codon-optimized T7 pol plasmid (pCAGGS-T7opt #65974 from Addgene). Plasmid sequences are available upon request.
3. 293T cells: Our preexisting stocks of the cells were originally obtained from ATCC and have been maintained for more than 20 years in our laboratory.
4. Madin-Darby bovine kidney (MDBK) cells: Our preexisting stocks of the cells were originally obtained from ATCC and have been maintained for more than 20 years in our laboratory.
5. PBS.

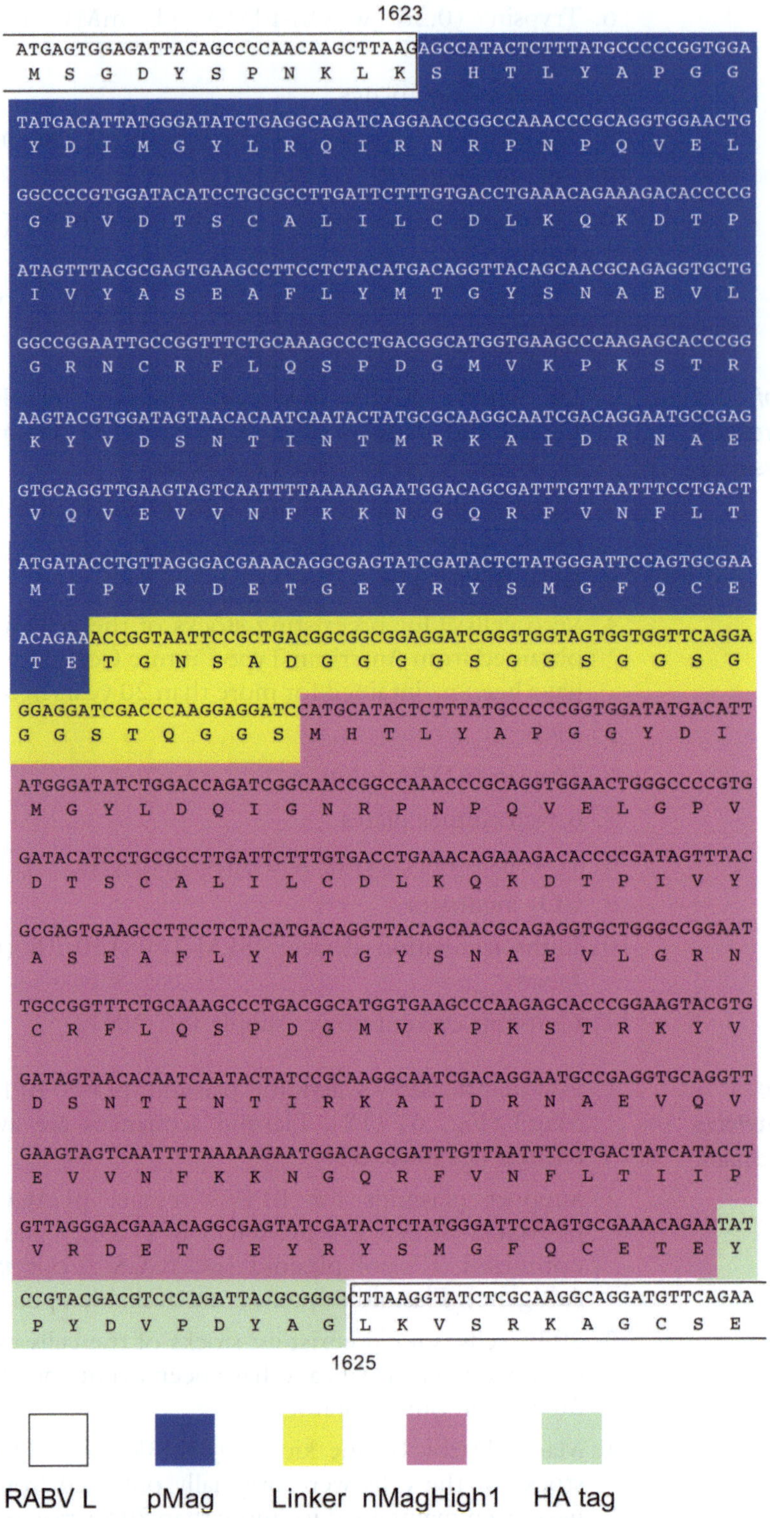

Fig. 4 Nucleotide and amino acid sequences of the Magnet domain insertion region of p(+)RABV-EGFP-L_{LMH2}. Rabies virus L gene (RABV L), pMag, Linker, nMagHigh1, and HA tag regions are indicated in white, blue, yellow, pink, and light green, respectively

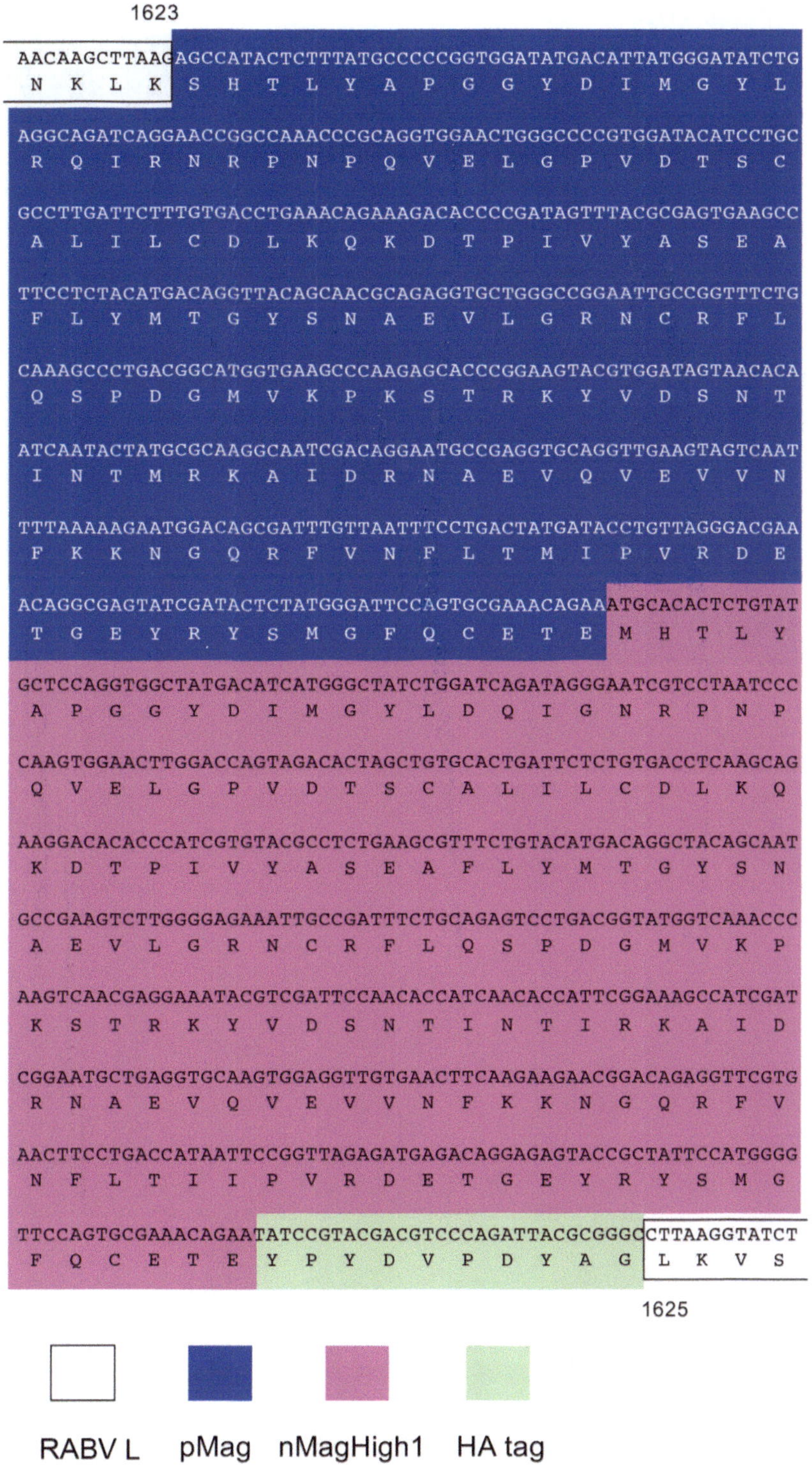

Fig. 5 Nucleotide and amino acid sequences of the Magnet domain insertion region of p(+)RABV-EGFP-L_{DMH2}. RABV L, pMag, nMagHigh1, and HA tag regions are indicated in white, blue, pink, and light green, respectively

```
1720
AATGAAAATCAAAGCCATACTCTTTATGCCCCCGGTGGATATGACATTATGGGATATCTG
 N  E  N  Q  S  H  T  L  Y  A  P  G  G  Y  D  I  M  G  Y  L

AGGCAGATCAGGAACCGGCCAAACCCGCAGGTGGAACTGGGCCCCGTGGATACATCCTGC
 R  Q  I  R  N  R  P  N  P  Q  V  E  L  G  P  V  D  T  S  C

GCCTTGATTCTTTGTGACCTGAAACAGAAAGACACCCCGATAGTTTACGCGAGTGAAGCC
 A  L  I  L  C  D  L  K  Q  K  D  T  P  I  V  Y  A  S  E  A

TTCCTCTACATGACAGGTTACAGCAACGCAGAGGTGCTGGGCCGGAATTGCCGGTTTCTG
 F  L  Y  M  T  G  Y  S  N  A  E  V  L  G  R  N  C  R  F  L

CAAAGCCCTGACGGCATGGTGAAGCCCAAGAGCACCCGGAAGTACGTGGATAGTAACACA
 Q  S  P  D  G  M  V  K  P  K  S  T  R  K  Y  V  D  S  N  T

ATCAATACTATGCGCAAGGCAATCGACAGGAATGCCGAGGTGCAGGTTGAAGTAGTCAAT
 I  N  T  M  R  K  A  I  D  R  N  A  E  V  Q  V  E  V  V  N

TTTAAAAAGAATGGACAGCGATTTGTTAATTTCCTGACTATGATACCTGTTAGGGACGAA
 F  K  K  N  G  Q  R  F  V  N  F  L  T  M  I  P  V  R  D  E

ACAGGCGAGTATCGATACTCTATGGGATTCCAGTGCGAAACAGAAATGCACACTCTGTAT
 T  G  E  Y  R  Y  S  M  G  F  Q  C  E  T  E  M  H  T  L  Y

GCTCCAGGTGGCTATGACATCATGGGCTATCTGGATCAGATAGGGAATCGTCCTAATCCC
 A  P  G  G  Y  D  I  M  G  Y  L  D  Q  I  G  N  R  P  N  P

CAAGTGGAACTTGGACCAGTAGACACTAGCTGTGCACTGATTCTCTGTGACCTCAAGCAG
 Q  V  E  L  G  P  V  D  T  S  C  A  L  I  L  C  D  L  K  Q

AAGGACACACCCATCGTGTACGCCTCTGAAGCGTTTCTGTACATGACAGGCTACAGCAAT
 K  D  T  P  I  V  Y  A  S  E  A  F  L  Y  M  T  G  Y  S  N

GCCGAAGTCTTGGGGAGAAATTGCCGATTTCTGCAGAGTCCTGACGGTATGGTCAAACCC
 A  E  V  L  G  R  N  C  R  F  L  Q  S  P  D  G  M  V  K  P

AAGTCAACGAGGAAATACGTCGATTCCAACACCATCAACACCATTCGGAAAGCCATCGAT
 K  S  T  R  K  Y  V  D  S  N  T  I  N  T  I  R  K  A  I  D

CGGAATGCTGAGGTGCAAGTGGAGGTTGTGAACTTCAAGAAGAACGGACAGAGGTTCGTG
 R  N  A  E  V  Q  V  E  V  V  N  F  K  K  N  G  Q  R  F  V

AACTTCCTGACCATAATTCCGGTTAGAGATGAGACAGGAGAGTACCGCTATTCCATGGGG
 N  F  L  T  I  I  P  V  R  D  E  T  G  E  Y  R  Y  S  M  G

TTCCAGTGCGAAACAGAATATCCGTACGACGTCCCAGATTACGCGGCGAAGAGAAATAAG
 F  Q  C  E  T  E  Y  P  Y  D  V  P  D  Y  A  A  K  R  N  K
                                                      1721
```

BPIV3 L   pMag   nMagHigh1   HA tag

Fig. 6 Nucleotide and amino acid sequences of the Magnet domain insertion region of p(+)57BPIV3-EGFP-L-Mag3. Bovine parainfluenza virus type 3 L gene (BPIV3 L), pMag, nMagHigh1, and HA tag regions are indicated in white, blue, pink, and light green, respectively

6. Trypsin/EDTA solution.
7. 6-well culture plates.
8. Transfection reagents: lipofectamine LTX and PLUS reagent (Invitrogen).
9. CO_2 incubator.
10. Light irradiation device: ISL-150×150-HBB.
11. Constant current power supply ISC-201-2.

2.7 Virus Titration of MeV and RABV

1. 12-well cluster plates.
2. Vero/hSLAM cells for MeV.
3. Vero cells for RABV.
4. CO_2 incubator.
5. PBS.
6. Agarose ME (IWAI CHEMICALS).
7. Eagle's MEM "Nissui" 1.
8. Penicillin/streptomycin solution: 10,000 unit/mL penicillin, 10,000 μg/mL streptomycin (Thermo Fisher Scientific, Gibco).
9. 200 mM L-glutamine (Thermo Fisher Scientific, Gibco).
10. 7.5% $NaHCO_3$ (Thermo Fisher Scientific, Gibco).
11. Light irradiation device: ISLM-150×150-RB.
12. Constant current power supply ISC-201-2.
13. Fluorescence microscope Axio Observer 01 (ZEISS).

2.8 Virus Titration of BPIV3

1. 96-well cluster plates.
2. MDBK cells.
3. CO_2 incubator.
4. PBS.
5. DMEM containing 10% FBS.
6. Light irradiation device: ISL-150×150-HBB.
7. Constant current power supply ISC-201-2.
8. Fluorescence microscope Axio Observer 01 (ZEISS).

2.9 Replication Kinetics Assay of Recombinant MeV, RABV, and BPIV3

1. Vero/hSLAM cells for MeV.
2. Vero cells for RABV.
3. MDBK cells for BPIV3.
4. 6-well cluster plates for MeV and RABV.
5. 12- and 96-well cluster plates for BPIV3.

6. Light irradiation device: ISLM-150×150-RB for MeV and RABV and ISL-150×150-HBB for BPIV3.
7. Constant current power supply ISC-201-2 (CCS, Kyoto, Japan).

2.10 Character Writing by Optically Controllable MeV and BPIV3

1. Vero/hSLAM cells for MeV.
2. MDBK cells for BPIV3.
3. Optically controllable MeV (rMeV^{EGFP}-L$_{DMH}$) [13].
4. Optically controllable BPIV3 (rBPIV3-L-Mag3) [18].
5. Black shading film (Vessel).
6. Black 24-well plate with glass bottom (Greiner) for MeV.
7. Black 6-well plate with glass bottom (IWAKI) for BPIV3.
8. Light irradiation device: ISLM-150×150-RB for MeV and ISL-150×150-HBB for BPIV3.
9. Constant current power supply ISC-201-2.
10. All-in-one fluorescence microscope BZ-X800 (Keyence).
11. Keyence's analysis software.

3 Methods

3.1 Preparation of Support Plasmids for MeV, RABV, and BPIV3 Rescue

1. Mix plasmid with the competent cells and incubate on ice for 1 h.
2. Heat shock at 42 °C for 30 s and cool rapidly on ice.
3. Incubate the cells with a small amount of LB media for 1 h.
4. Plate transformed cells on LB plates containing 100 μg/mL ampicillin.
5. Incubate plates overnight at 37 °C.
6. Pick isolated colonies and inoculate into 50 mL of LB medium containing 100 μg/mL ampicillin.
7. Incubate overnight at 37 °C.
8. Extract plasmids using NucleoBond Xtra Midi kit.
9. (Optional) To confirm the plasmid, perform appropriate restriction digestion or sequence the critical regions of the plasmid.

3.2 Preparation of Full-Length MeV and RABV Genome Plasmids

1. Mix plasmid with the competent cells and incubate on ice for 1 h.
2. Heat shock at 42 °C for 30 s and cool rapidly on ice.
3. Incubate the cells with a small amount of LB media for 1 h.

4. Plate transformed cells on an LB plate containing 100 μg/mL ampicillin.
5. Incubate the plates overnight at 37 °C (*see* **Note 4**).
6. Pick isolated colonies and inoculate into 100 mL of LB medium containing 100 μg/mL ampicillin (*see* **Note 5**).
7. Incubate at 37 °C for 16–24 h (*see* **Note 6**).
8. Extract plasmids using NucleoBond Xtra Midi kit.
9. (Optional) To confirm the plasmid, perform appropriate restriction digestion or sequence the critical regions of the plasmid.

3.3 Preparation of Full-Length BPIV3 Genome Plasmids

1. Mix plasmid with the competent cells and incubate on ice for 1 h.
2. Heat shock at 42 °C for 30 s and cool rapidly on ice.
3. Incubate the cells with a small amount of LB media for 1 h.
4. Plate transformed cells on an LB plate containing 100 μg/mL ampicillin.
5. Incubate the plates overnight at 37 °C (*see* **Note 4**).
6. Pick isolated colonies and inoculate into 100 mL of LB medium containing 100 μg/mL ampicillin (*see* **Note 5**).
7. Incubate at 37 °C for 16–24 h (*see* **Note 6**).
8. Extract plasmids using NucleoBond Xtra Midi kit.
9. (Optional) To confirm the plasmid, perform appropriate restriction digestion or sequence the critical regions of the plasmid.

3.4 Rescue of Optically Controllable MeV from Plasmids

1. Prepare BHK-T7/9 cells in 6-well culture plates (seeded to 70–80% confluency).
2. Transfect these BHK-T7/9 cells with 5.0 μg of the full-genome MeV plasmid (p(+)MV-EGFP-L_{LMH} or p(+)MV-EGFP-L_{DMH}), 0.8 μg of pCITE-IC-N, 0.6 μg of pCITE-IC-PΔC, and 1.6 μg of pCITE-ko-9301B-L using X-tremeGENE HP according to the manufacturer's instructions (*see* **Note 7**).
3. Twenty-four hours after transfection, replace the culture medium with BHK/T7–9 cell maintenance medium, and culture the cells under blue light irradiation (470 ± 20 nm) using the LED Light Unit ISLM-150×150-RB with power supply ISC-201-2 at low output (Fig. 7) (*see* **Note 8**).
4. Forty-eight hours after transfection, remove the cultured cells, wash once with PBS, and detach BHK-T7/9 cells by adding 0.2 mL of trypsin/EDTA solution.
5. Add 2 mL of Vero/hSLAM cell maintenance medium to make a cell suspension of detached BHK-T7/9 cells.

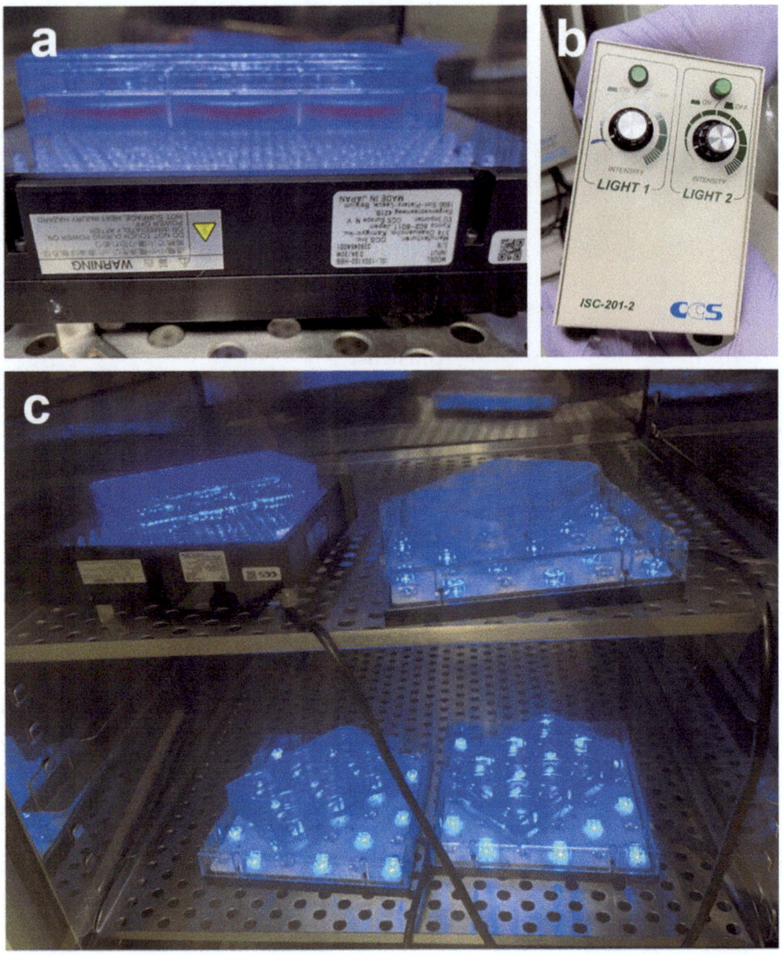

Fig. 7 Blue light irradiation of 6-well plates with cells in culture. (**a**) A 6-well plate in cell culture on ISL-150×150-HBB (side view). (**b**) The power supply ISC-201-2 set at low output. (**c**) 6-well plates in cell culture on ISL-150×150-HBB or ISLM-150×150-RB (diagonal view)

6. Transfer the BHK-T7/9 cell suspension onto Vero/hSLAM cells in a 10 cm culture dish (seeded to 70–80% confluency).
7. Culture these cells under blue light irradiation (470 ± 20 nm) using the LED Light Unit ISLM-150×150-RB with power supply ISC-201-2 at low output (*see* **Note 9**).

3.5 Rescue of Optically Controllable RABV from Plasmids

1. Prepare BHK-T7/9 cells in 6-well culture plates (seeded to 70–80% confluency).
2. Transfect these BHK-T7/9 cells with 5.0 μg of the full-genome RABV plasmid (p(+)RABV-EGFP-L_{LMH2} or p(+)RABV-EGFP-L_{DMH2}), 0.8 μg of pH-N, 0.6 μg of pH-P, 0.5 μg of pH-G, and 1.6 μg of pH-L using X-tremeGENE HP according to the manufacturer's instructions.

3. Twenty-four hours after transfection, replace the culture medium with BHK/T7–9 cell maintenance medium, and culture the cells under blue light irradiation (470 ± 20 nm) using the LED Light Unit ISLM-150×150-RB with power supply ISC-201-2 at low output (*see* **Note 8**).
4. Forty-eight hours after transfection, remove the cultured cells, wash once with PBS, and detach BHK-T7/9 cells by adding 0.2 mL of Trypsin/EDTA solution.
5. Add 2 mL of Vero cell maintenance medium to make a cell suspension of detached BHK-T7/9 cells.
6. Transfer the BHK-T7/9 cell suspension onto Vero cells in a 10 cm culture dish (seeded to 70–80% confluency).
7. Culture these cells under blue light irradiation (470 ± 20 nm) using the LED Light Unit ISLM-150×150-RB with power supply ISC-201-2 at low output.

3.6 Rescue of Optically Controllable BPIV3 from Plasmids

1. Prepare 293T cells in 6-well culture plates (seeded to 70–80% confluency).
2. Transfect these 293T cells with 2.0 μg of the full-genome BPIV3 plasmid (p(+)57BPIV3-EGFP-L-Mag3), 1.9 μg of pCAGGS-BPIV3-N, 2.4 μg of pCAGGS-BPIV3-P, 0.8 μg of pCAGGS-BPIV3-L, and 4.0 μg of a codon-optimized T7 pol plasmid using lipofectamine LTX and PLUS reagent according to the manufacturer's instructions.
3. Twenty-four hours after transfection, replace the culture medium with a 293T cell maintenance medium, and culture the cells under blue light irradiation (470 ± 20 nm) using the LED Light Unit ISL-150×150-HBB with power supply ISC-201-2 at low output (*see* **Note 8**).
4. Forty-eight hours after transfection, transfer the 1 mL of MDBK cell suspension in MEM containing 2% FBS onto transfected 293T cells.
5. Culture these cells under blue light irradiation (470 ± 20 nm) using the LED Light Unit ISL-150×150-HBB with power supply ISC-201-2 at low output.

3.7 Preparation of 2× MEM

1. Dissolve 9.4 g of Eagle's MEM "Nissui" 1 in 500 mL of dH_2O.
2. Sterilize by using an autoclave at 121 °C for 15 min.
3. Cool at room temperature.
4. Add 10 mL of penicillin/streptomycin solution.
5. Add 10 mL of L-glutamine and 20 mL $NaHCO_3$ (the final concentrations are 4 mM and 0.3%, respectively).
6. Add 20 mL FCS (the final concentration is 4%).

3.8 Preparation of Agarose-Containing MEM

1. Melt 1% agarose ME in dH_2O using an autoclave at 121 °C for 15 min.
2. Keep in a water bath set at 45 °C.
3. Warm 2×MEM containing 4% FBS to 45 °C in a water bath.
4. Mix 1% agarose ME solution with the same volume of the 2×MEM containing 4% FBS.
5. Use immediately or keep in a water bath at 45 °C (*see* **Note 10**).

3.9 Virus Titration of MeV and RABV

1. Seed 1 mL of a Vero/hSLAM cell suspension for MeV and a BHK cell suspension for RABV at a concentration of ~300,000 cells/mL into each well of 12-well cluster plates.
2. Incubate the cells in a cell culture incubator at 37 °C with 5% CO_2.
3. After cells form a monolayer (16–36 h post seeding), remove the culture media, and add 0.1 mL of serially diluted virus solutions.
4. Incubate the cells with the virus solution for 1 h in a cell culture incubator at 37 °C with 5% CO_2 with gentle agitation every 15 min.
5. After washing with PBS, add 1 mL of agarose-containing MEM which is kept in a water bath at 45 °C (*see* **Note 10**).
6. Cells were cultured under blue light irradiation (470 ± 20 nm) using the LED Light Unit ISLM-150×150-RB with power supply ISC-201-2 at low output.
7. Incubate the cells for 4 days in a cell culture incubator at 37 °C with 5% CO_2.
8. Count the number of foci expressing EGFP using a fluorescence microscope.

3.10 Virus Titration of BPIV3

1. Seed 0.1 mL of a MDBK cell suspension at a concentration of ~200,000 cells/mL into each well of 96-well cluster plates.
2. Incubate the cells in a cell culture incubator at 37 °C with 5% CO_2.
3. After cells form a monolayer (16–36 h post seeding), add 0.1 mL of serially diluted virus solutions.
4. Cells were illuminated continuously with blue light (470 ± 20 nm).
5. Incubate the cells for 4 days in a cell culture incubator at 37 °C with 5% CO_2.
6. Count the number of positive wells expressing EGFP using a fluorescence microscope, and calculate the viral titer as 50% tissue culture infectious dose ($TCID_{50}$)/mL.

3.11 Replication Kinetics Assay of Recombinant MeV and RABV

1. Prepare monolayers of Vero/hSLAM cells for MeV and BHK cells for RABV in 6-well cluster plates (cell number approximately 400,000 cells per well).
2. Remove culture medium.
3. Inoculate each well with 0.2 mL of virus solution ($rMeV^{EGFP}$-L_{LMH} and $rMeV^{EGFP}$-L_{DMH}) adjusted to MOI = 0.01.
4. Incubate the cells with the virus solution for 1 h in a cell culture incubator at 37 °C with 5% CO_2 with gentle agitation every 15 min.
5. After washing with PBS, add 3 mL of Vero/hSLAM maintenance medium per well.
6. Culture cells under blue light irradiation (470 ± 20 nm) or red light irradiation (655 ± 20 nm) using the LED Light Unit ISLM-150×150-RB with power supply ISC-201-2 at a low output or in the dark.
7. Collect cells daily with the culture medium.
8. Freeze and thaw the collected samples three times.
9. Centrifuge at 3000 rpm for 5 min and collect the supernatant.
10. Determine infection titer using Vero/hSLAM cells.

3.12 Replication Kinetics Assay of Recombinant BPIV3

1. Prepare monolayers of MDBK cells in 12-well cluster plates (cell number approximately 100,000 per well).
2. Remove culture medium.
3. Inoculate each well with 0.25 mL of virus solution (rBPIV3-EGFP-L-Mag3) adjusted to 10,000 $TCID_{50}$/mL (MOI = 0.1).
4. Incubate the cells with the virus solution for 1 h in a cell culture incubator at 37 °C with 5% CO_2 with gentle agitation every 15 min.
5. After washing with PBS, add 1 mL of MDBK maintenance medium per well.
6. Culture cells under blue light irradiation (470 ± 20 nm) using the LED Light Unit ISL-150×150-HBB with power supply ISC-201-2 at a low output or in the dark.
7. Collect cells daily with the culture medium.
8. Freeze and thaw the collected samples three times.
9. Centrifuge at 5000 rpm for 5 min and collect the supernatant.
10. Determine infection titer using MDBK cells.

3.13 Character Writing by Optically Controllable MeV

1. Prepare monolayers of Vero/SLAM cells in a black 24-well plate with a glass bottom.

Fig. 8 Character writing by optically controllable MeV. (**a**) A 24-well plate with a black shading film with character-shaped slits applied to the bottom. (**b**) A 24-well plate illuminated by blue light from the bottom of the plate using ISLM-150×150-RB. The culture plate's top and side were also covered by black shading film. (**c**) EGFP fluorescence images captured using a fluorescence microscope after removing the black shading film with letter-shaped slits

2. Infect Vero/hSLAM cells with rMeV^{EGFP}-L$_{DMH}$ to an MOI of 0.01.
3. Place a black shading film with letter-shaped slits at the bottom of the black 24-well plate with a glass bottom (Fig. 8a).
4. Incubate under blue light (470 ± 20 nm) irradiation for 5 days (Fig. 8b).
5. After incubation, remove the black shading film with letter-shaped slits, and capture EGFP fluorescence images using an all-in-one fluorescence microscope BZ-X800 (magnification, 4×) (Fig. 8c).

3.14 Character Writing by Optically Controllable BPIV3

1. Prepare monolayers of MDBK cells in a black 6-well plate with a glass bottom (*see* **Note 4**).
2. Infect MDBK cells with rBPIV3-EGFP-L-Mag3 to an MOI of 10.
3. Place a black shading film with letter-shaped slits at the bottom of the black 6-well plate with a glass bottom.
4. Incubate under blue light (470 ± 20 nm) irradiation for 2–3 days (Fig. 9).

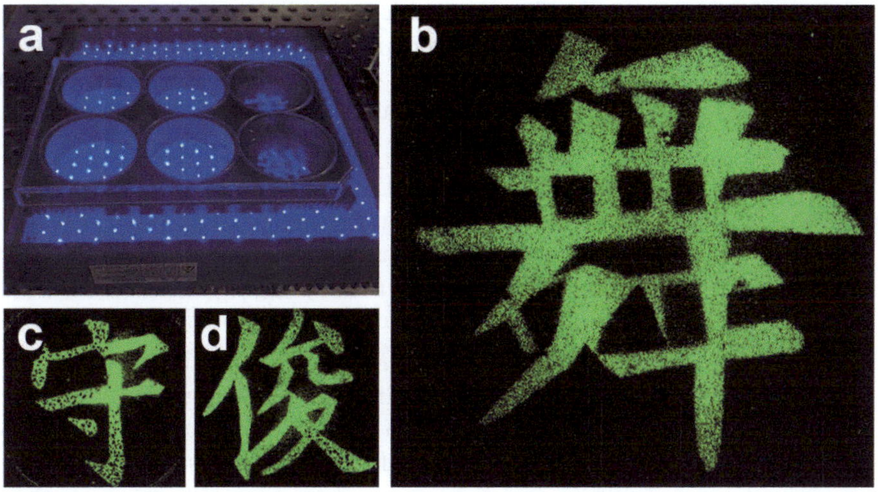

Fig. 9 Character writing by optically controllable BPIV3. (**a**) A 6-well plate with a black shading film with character-shaped slits applied to the bottom. The plate was illuminated by blue light from the bottom of the plate using ISL-150×150-HBB. (**b–d**) EGFP fluorescence images captured using a fluorescence microscope after removing the black shading film with letter (Japanese Kanji)-shaped slits

5. After incubation, remove the black shading film with letter-shaped slits, and capture EGFP fluorescence using an all-in-one fluorescence microscope BZ-X800 (magnification, 4×) (Fig. 9).

4 Notes

1. BHK/T7–9 is a cell line expressing the T7 RNA polymerase cloned using the hygromycin resistance gene as a drug selection marker. Therefore, to maintain stable expression of T7 RNA polymerase, it is better to have hygromycin in the culture medium. However, making as much working stock as possible is more critical to avoid repeated cell passages. Adding hygromycin in the culture medium is necessary only if the number of passages is too long.
2. Vero/hSLAM is a human SLAM-expressing cell line cloned using the neomycin resistance gene as a drug selection marker. Therefore, to maintain stable expression of human SLAM, it is better to have neomycin in the culture medium. However, it has been confirmed that SLAM expression does not decrease even after several dozen or more long-term passages without neomycin in the culture medium.
3. A full-length genome plasmid contains the cDNA for about 16 kilobases of the virus genome. Therefore, they have to be used more carefully than ordinary plasmids. Repeated freeze-

thaw cycles may reduce the efficiency of virus recovery from the full-length genome plasmid, so it is recommended that the plasmid solutions be divided into several aliquots and stored at −80 °C. Even for a day, storage of transformed *E. coli* is not recommended.

4. The viral genome cDNA of about 16 kilobases in the full-length genome plasmid reduces the amplification efficiency of the plasmid. If the transformed *E. coli* is incubated for a long time, the *E. coli* may modify the plasmid. Therefore, the incubation time should be as short as possible.
5. Compared to other *E. coli* colonies, larger colonies are more likely to harbor modified full-length genome plasmids. Therefore, the colonies to be picked should be as small as other colonies. If the transformed *E. coli* is grown in a small 2 mL culture and then stepped up to a 100 mL size, the colony is more likely to have the modified full-length genome plasmid, so the colony should be inoculated directly into the 100 mL medium and grown.
6. Stop incubation as soon as possible after visual confirmation that *E. coli* has grown sufficiently.
7. The ratio of the three support plasmids is critical to the success or failure of virus recovery. Even if the same procedure is followed, the state of the plasmids is not perfectly constant each time, so the quantity ratio may need to be adjusted each time the three support plasmids are obtained from the transformed *E. coli*.
8. Blue light can damage cells and prevent virus recovery; since the sensitivity of the Magnet photoswitch is very high, it is unlikely that the light intensity is too weak to activate virus growth, so light intensity should be as low as possible.
9. The success or failure of virus recovery can be confirmed by the expression of EGFP incorporated into the viral genome. In the case of the measles virus, if multinucleated giant cells (syncytia) expressing EGFP are observed, the virus has been successfully recovered. However, since the optically controllable viruses multiply slowly even under blue light illumination, about a 1-week cell culture may be required to confirm the virus recovery even in case of success. If too many cells are grown during that time, the cells may be passaged to propagate the virus. Removing only BHK/T7–9 cells is possible by adding neomycin to the culture medium (because Vero/hSLAM cells, but not BHK/T7–9, are resistant to neomycin).
10. MEM-containing agarose must be used at a temperature that will not harden and damage the cells.

Acknowledgments

The reverse genetics system for the RABV HEP-Flury strain was provided by Drs. Mutsuyo Takayama-Ito and Masayuki Saijo (Department of Virology 1, National Institute of Infectious Diseases). BHK/T7 cells and Vero/hSLAM cells were provided by Drs. Naoto Ito (Gifu University) and Yusuke Yanagi (Kyushu University), respectively. This study was supported by grants from JSPS under grant number 19K07768 to M. Tahara, 22K08616 to TO, and 21H02744 to M. Takeda, grants from the Naito Foundation and Research Foundation for Opto-Science and Technology to M. Takeda, and grant from the Japan Agency for Medical Research and Development (AMED) under grant number 23wm0325063 to M. Takeda.

References

1. Sharp B, Rallabandi R, Devaux P (2022) Advances in RNA viral vector technology to reprogram somatic cells: the paramyxovirus wave. Mol Diagn Ther 26(4):353–367. https://doi.org/10.1007/s40291-022-00599-x
2. Russell CJ, Hurwitz JL (2021) Sendai virus-vectored vaccines that express envelope glycoproteins of respiratory viruses. Viruses 13(6): 1023. https://doi.org/10.3390/v13061023
3. Tang C, Li L, Mo T, Na J, Qian Z, Fan D, Sun X, Yao M, Pan L, Huang Y, Zhong L (2022) Oncolytic viral vectors in the era of diversified cancer therapy: from preclinical to clinical. Clin Transl Oncol 24(9):1682–1701. https://doi.org/10.1007/s12094-022-02830-x
4. de Graaf JF, de Vor L, Fouchier RAM, van den Hoogen BG (2018) Armed oncolytic viruses: a kick-start for anti-tumor immunity. Cytokine Growth Factor Rev 41:28–39. https://doi.org/10.1016/j.cytogfr.2018.03.006
5. Ogino T, Green TJ (2019) RNA synthesis and capping by non-segmented negative strand RNA viral polymerases: lessons from a prototypic virus. Front Microbiol 10:1490. https://doi.org/10.3389/fmicb.2019.01490
6. Liang B (2020) Structures of the mononegavirales polymerases. J Virol 94(22):e00175-20. https://doi.org/10.1128/JVI.00175-20
7. Te Velthuis AJW, Grimes JM, Fodor E (2021) Structural insights into RNA polymerases of negative-sense RNA viruses. Nat Rev Microbiol 19(5):303–318. https://doi.org/10.1038/s41579-020-00501-8
8. Dochow M, Krumm SA, Crowe JE Jr, Moore ML, Plemper RK (2012) Independent structural domains in paramyxovirus polymerase protein. J Biol Chem 287(9):6878–6891. https://doi.org/10.1074/jbc.M111.325258
9. Kawano F, Suzuki H, Furuya A, Sato M (2015) Engineered pairs of distinct photoswitches for optogenetic control of cellular proteins. Nat Commun 6:6256. https://doi.org/10.1038/ncomms7256
10. Zoltowski BD, Vaccaro B, Crane BR (2009) Mechanism-based tuning of a LOV domain photoreceptor. Nat Chem Biol 5(11): 827–834. https://doi.org/10.1038/nchembio.210
11. Duprex WP, Collins FM, Rima BK (2002) Modulating the function of the measles virus RNA-dependent RNA polymerase by insertion of green fluorescent protein into the open reading frame. J Virol 76(14):7322–7328. https://doi.org/10.1128/jvi.76.14.7322-7328.2002
12. Ruedas JB, Perrault J (2009) Insertion of enhanced green fluorescent protein in a hinge region of vesicular stomatitis virus L polymerase protein creates a temperature-sensitive virus that displays no virion-associated polymerase activity in vitro. J Virol 83(23):12241–12252. https://doi.org/10.1128/JVI.01273-09
13. Tahara M, Takishima Y, Miyamoto S, Nakatsu Y, Someya K, Sato M, Tani K, Takeda M (2019) Photocontrollable mononegaviruses. Proc Natl Acad Sci USA 116(24): 11587–11589. https://doi.org/10.1073/pnas.1906531116
14. Seki F, Yamada K, Nakatsu Y, Okamura K, Yanagi Y, Nakayama T, Komase K, Takeda M (2011) The SI strain of measles virus derived from a patient with subacute sclerosing

panencephalitis possesses typical genome alterations and unique amino acid changes that modulate receptor specificity and reduce membrane fusion activity. J Virol 85(22): 11871–11882. https://doi.org/10.1128/JVI.05067-11

15. Ito N, Takayama-Ito M, Yamada K, Hosokawa J, Sugiyama M, Minamoto N (2003) Improved recovery of rabies virus from cloned cDNA using a vaccinia virus-free reverse genetics system. Microbiol Immunol 47(8):613–617. https://doi.org/10.1111/j.1348-0421.2003.tb03424.x

16. Ono N, Tatsuo H, Hidaka Y, Aoki T, Minagawa H, Yanagi Y (2001) Measles viruses on throat swabs from measles patients use signaling lymphocytic activation molecule (CDw150) but not CD46 as a cellular receptor. J Virol 75:4399–4401. https://doi.org/10.1128/jvi.75.9.4399-4401.2001

17. Inoue K, Shoji Y, Kurane I, Iijima T, Sakai T, Morimoto K (2003) An improved method for recovering rabies virus from cloned cDNA. J Virol Methods 107(2):229–236. https://doi.org/10.1016/S0166-0934(02)00249-5

18. Okura T, Tahara M, Otsuki N, Sato M, Takeuchi K, Takeda M (2023) Generation of a photocontrollable recombinant bovine parainfluenza virus type 3. Microbiol Immunol 67(4):204–209. https://doi.org/10.1111/1348-0421.13052

Chapter 5

Generation of Defective Interfering Particles of Morbilliviruses Using Reverse Genetics

Linda J. Rennick, W. Paul Duprex, and Natasha L. Tilston-Lunel

Abstract

RNA viruses generate defective genomes naturally during virus replication. Defective genomes that interfere with the infection dynamics either through resource competition or by interferon stimulation are known as defective interfering (DI) genomes. DI genomes can be successfully packaged into virus-like-particles referred to as defective interfering particles (DIPs). Such DIPs can sustainably coexist with the full-length virus particles and have been shown to negatively impact virus replication in vitro and in vivo. Here, we describe a method to generate a clonal DI genome population by reverse genetics. This method is applicable to other RNA viruses and will enable assessment of DIPs for their antiviral properties.

Key words Defective interfering particles, Defective genomes, Reverse genetics, Morbillivirus, Canine distemper virus, Single-step attenuation, Vaccines, Antivirals

1 Introduction

Defective interfering particles (DIPs) contain defective (i.e., incomplete) genomes. DIPs were first identified in the late 1940s when Preben Von Magnus discovered the loss of influenza virus infectivity under high titer passage conditions [1, 2]. Since then, DIPs have been identified across almost every RNA virus family. Virus stocks with high DIP content typically demonstrate pronounced reduction in viral titers. At least two plausible mechanisms exist for such DIP-mediated restriction: (1) direct interference affecting viral replication, assembly, and egress and/or (2) indirect interference through stimulation of the antiviral response. Direct interference is likely due to competition for viral and host resources between the defective and full-length genomes [3–9]. Historically, DIPs were considered artifacts of in vitro virus propagation. However, this view is changing as DI sequences have been identified from clinical infections [10, 11]. DIPs have now been suggested for use in vaccines as adjuvants or even as a type of coevolving vaccine

Dzwokai Z. Ma and Christian K. Pfaller (eds.), *Measles and Related Morbilliviruses: Methods and Protocols*, Methods in Molecular Biology, vol. 2808, https://doi.org/10.1007/978-1-0716-3870-5_5,

[12–20]. Considering the shift in DIP importance, we have developed a methodology to generate DIPs by reverse genetics. This will be useful for producing clonal DIPs and assessing their antiviral properties without the confounding effects of other DI genomes or the full-length virus.

Two well-characterized DI genomes are the deletion and copyback DI genomes [21]. Deletion genomes form when the viral RNA-dependent RNA polymerase (RdRp, L protein) skips portions of the genome template during replication. This results in deletions within the viral genome. Copyback genomes, on the other hand, are formed when the RdRp translocates from the replication template strand onto the nascent strand. As a result, the RdRp copies back onto itself forming a highly complementary genome-antigenome hybrid. Deletion and copyback DI genomes with intact replication and packaging signals can successfully persist within viral populations. Sequences on the viral genome where the RdRp dissociates and reinitiates replication are called break points and reinitiation points, respectively. Identifying these sequences in naturally occurring DI genome populations is important to accurately recreate them synthetically. Our in vitro evolution experiments using the morbillivirus canine distemper virus (CDV) demonstrated that certain DI genomes are more successful than others in coexisting with the full-length genome [22]. These DI genomes were ultimately responsible for viral interference, and like the full-length virus, their genome lengths followed the rule-of-six. Morbilliviruses typically require hexamer phasing for efficient genome replication [21, 23–25]. In this chapter, we describe a method to generate clonal morbillivirus DI genomes. The protocol assumes that a DI sequence of interest is available and utilizes an attenuated "producer virus" (i.e., helper virus) to package the DI genomes into DIPs. The producer virus is subsequently inactivated. DI sequences in viral stocks can be identified using conventional reverse transcription-polymerase chain reaction (RT-PCR)/Sanger sequencing and/or next generation sequencing (NGS) techniques [22, 26] or can be designed synthetically.

2 Materials

2.1 General Equipment

2.1.1 Routine Molecular Biology Equipment

1. Thermal cycler.
2. Heat block.
3. Agarose gel electrophoresis equipment.
4. Micro and benchtop centrifuges.
5. Bacterial incubator.
6. NanoDrop spectrophotometer.

Table 1
Examples of exposure values as displayed on the CX-2000 Crosslinker

Dose (mJ/cm^2)	Dose (μJ/cm^2)	Displayed value
240	240,000	2400
120	120,000	1200
60	60,000	600
6	6000	60
0.6	600	6

7. Ultraviolet (UV) crosslinker (e.g., CX-2000 Crosslinker (UVP). The preset exposure on the CX-2000 delivers 120,000 microjoules (mJ)/cm^2 in a single run which is displayed as 1200 on the LCD. Multiply this number by 100 to obtain the exposure value.) (Table 1).
8. Genomics software (e.g., DNASTAR).

2.1.2 Routine Tissue Culture Equipment

1. Class II biological safety cabinet (BSC).
2. CO_2 incubators.
3. Phase-contrast microscope.
4. Water bath.
5. Tissue culture grade plastics.

2.2 Molecular Biology Reagents

1. TRIzol LS reagent.
2. Nuclease-free water.
3. NEBuilder® HiFi DNA assembly.
4. Ethidium bromide.
5. High fidelity DNA polymerase.
6. DNA ladder.
7. One-step RT-qPCR kit.
8. T7 transcription kit.
9. Various restriction and modifying enzymes.
10. Oligonucleotides, gene fragments, plasmids.

2.3 Bacterial Culture

1. LB agar (*see* **Note 1**).
2. LB broth (*see* **Note 1**).
3. Ampicillin. Stock solution at 100 mg/mL.

2.4 Cell Culture

1. Vero cells (*see* **Note 2**).
2. Dulbecco's Modified Eagle Medium (DMEM).
3. Fetal bovine serum (FBS, heat inactivated).
4. GlutaMAX-I.
5. 0.25% trypsin-EDTA.
6. Dulbecco's phosphate-buffered saline (DPBS), no calcium, no magnesium.
7. Lipofectamine 2000.
8. Staining solution: crystal violet stain (20% v/v ethanol, 1% v/v methanol, 0.1% w/v crystal violet, diluted in water).
9. 4% paraformaldehyde.

3 Methods

Prepare and store all reagents according to manufacturer's suggestions. Obtain institutional biosafety committee (IBC) approval for all procedures. Perform all procedures using appropriate personal protection equipment. Methods requiring the use of infectious virus must be carried out in a Class II BSC. Follow appropriate institutional waste disposal regulations when treating and discarding waste. Treat all liquid from tissue culture and all plasticware before disposal through the appropriate waste stream.

3.1 Generate a Producer Morbillivirus

3.1.1 Rationally Attenuate the Producer Virus

1. Measles virus (MV) and CDV have previously been attenuated by inserting a foreign sequence within the hinge region of the L protein. Align the polymerase protein sequences against MV, CDV, and other morbilliviruses to identify the hinge 2 region [27, 28].
2. Insert a foreign sequence of choice within the hinge 2 region (cloning of such a virus is beyond the scope of this protocol; *see* **Note 3**).

3.1.2 Culture a Low Multiplicity of Infection Producer Virus

1. Seed Vero cells at 5×10^6 in a medium flask (75 cm^2) or 1×10^7 in a large flask (150 cm^2).
2. When the cells reach 70–80% confluence, infect with the producer virus at a low multiplicity of infection (MOI) of 0.01–0.005 in DMEM, 2% FBS (*see* **Note 4**).
3. Incubate for 1 h at 37 °C with occasional rocking.
4. Discard the inoculum and rinse cells with DPBS.
5. Add 15 mL (75 cm^2) or 30 mL (150 cm^2) of DMEM, 2% FCS.
6. Incubate at 37 °C until complete cytopathic effect (CPE) [29], and then freeze the flask at −80 °C.

7. Prechill a 15 mL tube on ice and precool centrifuge to 4 °C.
8. Thaw the flask and scrape the cells with a cell scraper.
9. Transfer the contents into the 15 mL tube, and centrifuge for 15 min at 3000 rpm/4 °C, to clarify the virus.
10. Aliquot the virus containing supernatant in 1 mL volumes and store at −80 °C.

3.1.3 Titrate Producer Virus

1. Plate 2×10^4 cells per well of a 96-well tray in 100 μL of growth medium (*see* **Note 5**).
2. Prepare tenfold virus dilutions (10^{-1} to 10^{-8}) in Opti-MEM. Aliquot 900 μL Opti-MEM into required number of wells of a 96-well assay block. Add 100 μL of virus to be diluted into row A, and pipette up and down to mix. Using a multichannel pipette to transfer 100 μL from row A to row B, pipette to mix and discard tips into disinfectant. Use new tips to transfer 100 μL from row B to row C, pipette to mix, and discard tips into disinfectant. Repeat by moving down the rows of the block until dilutions are complete.
3. Using a multichannel pipette, add 200 μL of each virus dilution to 4 wells of the 96-well trays containing cells.
4. Incubate the plates at 37 °C, 5% CO_2, for 5–7 days.
5. Evaluate the wells for CPE. Calculate the Tissue Culture Infections Dose 50 ($TCID_{50}$) based on the methods described by Reed and Muench [30] (*see* **Note 6**).

3.2 Generate a Morbillivirus DIPs

3.2.1 Clone DI Genome Sequences

1. Use bioinformatics software of choice (e.g., DNASTAR Lasergene) to build a virtual "computer clone" representing the plasmid DNA sequence that will be generated in the lab.
   (a) Insert the desired DI genome sequence into a plasmid containing morbillivirus sequences surrounded by the components (T7 RNA polymerase promoter/terminator sequences and a hepatitis delta ribozyme sequence) necessary to allow T7 RNA polymerase-mediated transcription of DI genomes from the plasmid (Fig. 1; *see* **Notes** 7 and **8**).
   (b) Morbillivirus genomes/DI genomes will not tolerate additional nucleotides on their termini, so the minimal T7 RNA polymerase promoter directly abuts the terminus of the DI genome.
   (c) Similarly, the hepatitis delta ribozyme directly abutting the terminus at the other end of the DI genome self-cleaves ensuring no additional nucleotides on this terminus.
   (d) Make sure the nucleotide length of DI genomes is divisible by 6 to allow efficient replication [31].

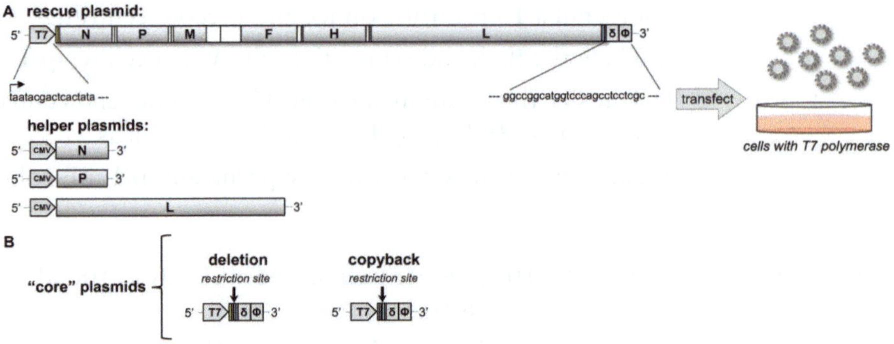

Fig. 1 Schematic of a morbillivirus reverse genetics system. (**a**) The entire morbillivirus genome is encoded from a T7-driven plasmid consisting of a hepatitis delta ribozyme (δ) sequence and T7 terminator (θ) at the 3′ end. Proteins N, P, and L are encoded from CMV-driven expression plasmids. When transfected into cells expressing T7 polymerase (cell line/plasmid/virus), recombinant virus can be generated 5–7 days later. (**b**) "Core" plasmids for deletion and copyback DI genome cloning enable a "plug-and-play" approach to generating several DI genome variants

2. Once the desired sequence has been built virtually, use the bioinformatics software to design a cloning strategy.
   (a) Analyze the plasmid sequence to identify suitable restriction sites in the existing plasmid that will have the insert cloned in. Usually, restriction enzymes with unique recognition sequences in the cloning plasmid are chosen.
   (b) Design the required insert sequence that needs to be inserted between the restriction enzyme sites chosen for the cloning. For cloning by Gibson Assembly, the insert sequence must contain overlaps with the overhangs generated by cleavage of the cloning plasmid with the chosen restriction enzymes (*see* **Note 9**).
   (c) Order the designed insert as synthetic DNA (*see* **Notes 10** and **11**).
3. Prepare vector by digesting plasmid DNA with the enzyme/s determined in the cloning strategy design:
   (a) Combine 5 μg plasmid DNA with 2 μL 10× reaction buffer and 10U each enzyme in a final volume of 20 μL with nuclease-free water in a nuclease-free 1.5 mL tube, and incubate at 37 °C for at least 4 h (*see* **Notes 12–14**).
4. Isolate and purify the vector DNA:
   (a) Prepare a 1% agarose gel containing 50 μg/mL ethidium bromide by standard techniques (dissolve 1 g of agarose per 100 mL ×1 tris-borate-EDTA buffer (Boston Bioproducts), and supplement with ethidium bromide before pouring).
   (b) Mix the digested vector DNA with ×6 loading dye, and load the entire mixture into one well of the agarose gel.

(c) Leave one well as spacer, and then load 1 kb DNA standard for sizing.
(d) Run the gel.
(e) View the gel using a low power transilluminator (*see* **Note 15**).
(f) Excise the band corresponding to the vector DNA.
(g) Purify the vector DNA using a Monarch DNA Gel Extraction Kit according to manufacturer's instructions (*see* **Note 16**).

5. Gibson assembly
   (a) Resuspend gBlock/s in nuclease-free water according to manufacturer's instructions at a concentration of 10 ng/μL.
   (b) Combine 50 ng vector DNA with 3× picomolar equivalent of gBlock insert/s and 5 μL 2× Hifi reaction buffer in a final volume of 10 μL with nuclease-free water in a nuclease-free 1.5 mL tube, and incubate at 50 °C for 1 h. A control reaction containing vector DNA, but no insert DNA, should also be set up (*see* **Notes 17–19**).
6. Transform 1 μL of Gibson assembly reactions into competent cells, plate on agar containing appropriate antibiotic, and incubate according to manufacturer's instructions.
7. Count number of bacterial colonies on each agar plate; there should be at least 100 colonies on the vector + insert/s plate and fewer than 10 colonies on the vector only control plate. At this ratio, it would be expected to recover at least one correct clone by screening six colonies.
8. Perform small-scale plasmid DNA isolation and purification using a QIAprep Spin Miniprep Kit (Qiagen) according to manufacturer's instructions.
9. Set up digests using the same enzymes used for the cloning and run on 1% agarose gel to check for the presence of insert.
10. Confirm the sequence of the insert by performing Sanger sequencing (Azenta Life Sciences) of the entire region; the sequences of the regions upstream and downstream of the restriction sites should also be sequence confirmed (*see* **Note 20**).
11. When a positive clone is identified, perform large-scale plasmid DNA isolation and purification using a HiSpeed Plasmid Maxi Kit (Qiagen) according to manufacturer's instructions.

3.3 Rescue and Culture DIPs

1. Plate Vero cells at 5×10^5 cells per well of a 6-well tray in 2 mL of growth medium. Incubate the plates at 37 °C, 5% CO_2.
2. The next day, infect the cells with MVA-T7 in Opti-MEM (Gibco) at an MOI of 1 for 45 min at 37 °C.

3. Remove medium from the cell monolayers, and transfect cells with 2 μg of defective genome-expressing plasmid along with plasmids expressing the N, P, and L proteins. Incubate the plates at 37 °C, 5% CO_2.
4. The next day, remove the transfection mix from the cell monolayers, and infect with producer virus. Use a low MOI to avoid generation of new DIPs from the producer virus. MOI of 0.001 is recommended.
5. Two days post producer virus infection, detach the cell monolayer using trypsin. Passage the entire cell population to a T75 flask. Monitor for CPE development.
6. Upon maximal CPE, freeze the cells in supernatant at −80 °C.
7. Thaw the cell/supernatant mixture and transfer to a sterile 15 mL tube. Centrifuge at 3000 RPM at 4 °C for 10 min in a benchtop centrifuge. Harvest the supernatant as DIP-containing stock, and aliquot in smaller workable volumes.
8. When required, thaw the DIP stocks, and then UV-inactivate the producer virus at 60 mJ/cm^2 as outlined below.

3.4 Selectively UV-Inactivate the Producer Virus

1. Transfer 1 mL of DIP stock into a well of a 6-well plate, spreading the liquid into a thin layer coating the entire well.
2. Pull the UV exposure chamber out and place the plate on the chamber without the lid on. Close the chamber.
3. Preset exposure of 120,000 mJ/cm^2 is displayed on the screen as 1200. To change this exposure value to 60 mJ/cm^2 key in the number 600 using the touch pad, press *Enter* and then *Start*. The machine counts down and beeps when complete.
4. Take the plate out of the chamber and place the lid back on (*see* **Note 21**).

3.5 Confirm the DIP Inference

1. Plate Vero cells at 5 × 10^4 cells per well in a 24-well plate in 0.5 mL of growth medium. Incubate at 37 °C, 5% CO_2, for 18 h.
2. Once the producer virus is UV-inactivated (Subheading 3.4), split the DIP stock into two.
3. Retain one-half. This half only has the producer virus inactivated (active DIPs).
4. UV-inactivate the other half at 240 mJ/cm^2. This will inactivate the DIPs and serve as a control (inactive DIPs).
5. Prepare 1:10 dilutions, as for producer virus titration (Subheading 3.1.3), of the active and inactive DIP samples.
6. Transfer DIP dilutions onto the Vero cell monolayers, and allow to infect for 1 h at 37 °C.

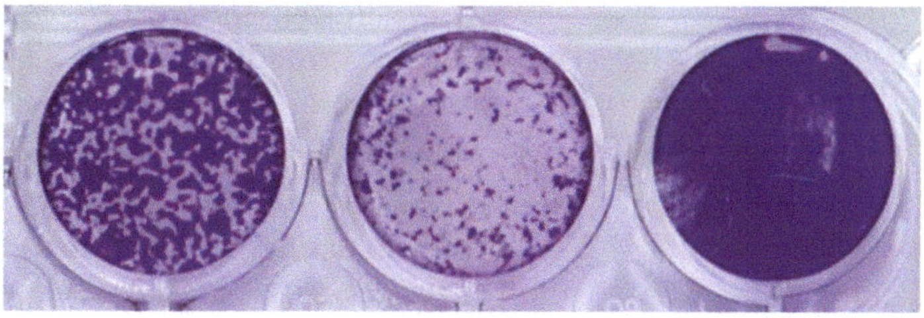

Fig. 2 DIP interference. Virus cytopathic effect in the presence (+) and absence (−) of DIPs is visualized by staining the cell monolayer with crystal violet stain

7. Remove supernatant and wash cells with DPBS three times.
8. Superinfect the cells with full-length virus at a standard MOI (e.g., 0.1) for 1 h at 37 °C.
9. Remove supernatant and wash cells with DPBS three times.
10. Add DMEM, incubate for 48 h, and then freeze at −80 °C for at least 4 h (*see* **Note 22**).
11. Thaw the cell/supernatant mix, and centrifuge for 15 min at 3000 rpm/4 °C to remove cell debris.
12. Quantify infectious virus output from the supernatant of each sample by TCID (method as in Subheading 3.1.2).
13. Alternatively, fix the cell monolayers in 4% PFA and stain with crystal violet. The extent of virus induced CPE can be easily visualized via the crystal violet stain (Fig. 2).
    (a) Remove supernatant and wash cells with DPBS three times.
    (b) Cover cells with excess 4% PFA and incubate for 10 min.
    (c) Remove PFA to chemical waste and wash cells with DPBS three times.
    (d) Stain cells using crystal violet stain for 10 min, and then rinse with water.
    (e) Dry and then visualize.

4 Notes

1. Bacterial cultures on LB agar and in broth are grown overnight at 37 °C.
2. Our CDV DIP studies [22] utilized Vero cells expressing the canine receptor CD150 (Vero-cCD150) [32]. Depending on the virus being used, Vero cells expressing the appropriate CD150 receptor may need to be generated/obtained.

Maintain cell monolayers for routine use in 75 cm^2 tissue culture vented flasks at 37 °C in a 5% CO_2 humidified atmosphere.

3. Alternatively, vaccine or attenuated laboratory-adapted viruses, if available, can be used as producer virus. Rationally attenuated recombinant viruses may offer an advantage for use as producer virus since low passage stocks can be used; if passaged carefully from rescue, these should contain minimal endogenous DI genomes. Producer virus stocks should be screened for the presence of such endogenous DI genomes which may outcompete or otherwise interfere with generation of the desired DI genome/DIPs.
4. Vero-cCD150 cells usually reach 70–80% confluence within 12 h; other cell lines may have varied incubation times.
5. Allow the trays to sit at room temperature in the BSC while the viral dilutions are being prepared. This promotes even cellular coverage in the wells.
6. Various online calculators are available to determine titers by this method although we recommend independently validating the calculator equations before use.
7. Internal deletion DI genomes are generally relatively straightforward to clone by standard molecular biology methods. Copyback DI genomes are trickier due to their complementary sequences. In either case, DI genomes can be cloned using RT-PCR products (which may be generated as part of the identification phase) or using synthetic DNA (essential when only DI sequences are available, e.g., when identified by NGS). When using synthetic DNA to clone copyback DI genomes, it is best to split the genome into two fragments for efficient in vitro synthesis. Depending on the number of DI genomes to be cloned, it may be beneficial to design "core shuttle vector/s." Copyback and deletion core vectors would contain sequences that are common among the various DI genomes. Core vectors can be linearized to allow a plug-and-play approach for insertion of unique DI genome insert cassettes.
8. This protocol assumes that the necessary components of a morbillivirus reverse genetics system are in place. These systems have been published for many morbilliviruses and are adaptable to other members. At minimum, a plasmid containing T7 RNA polymerase promoter and terminator sequences and a hepatitis delta ribozyme sequence (Fig. 1a, b) and expression plasmids encoding morbillivirus N, P, and L proteins are required. A source of T7 RNA polymerase (virus [22], plasmid [33], or stable cell-derived [34]) is also required. This protocol describes reverse genetics mediated recovery ("rescue") of DI RNPs using a host range restricted modified vaccinia Ankara

(MVA) virus encoding T7 RNA polymerase [35]. Reverse genetics systems based on other technologies, e.g., cellular RNA polymerase II [36], are available.

9. Overlaps of 20 bases work well.
10. IDT gBlocks work well.
11. Sequence repeats and complementarity, as seen in copyback DI sequences, can be problematic for in vitro DNA synthesis. In these cases, the synthetic DNA can be split into two overlapping gBlocks to separate these features. Again, overlaps of 20 bases work well.
12. Final volume of reaction can be increased dependent on plasmid DNA concentration; adjust volume of reaction buffer accordingly.
13. Check restriction enzyme documentation for any additional considerations, e.g., sensitivity to DNA methylation, star activity, nonstandard incubation temperature, reaction buffer incompatibility when using more than one restriction enzyme, etc. Common issues, e.g., different restriction enzyme incubation temperatures and reaction buffer incompatibility, can be addressed by performing sequential digests; for reaction buffer incompatibility, the lower salt concentration reaction should be performed first to eliminate an intermediate purification step.
14. Incubation time can be varied up to overnight without deleterious effects (unless contraindicated in restriction enzyme documentation).
15. Eyes and skin must not be exposed to UV light—appropriate personal protection equipment must be used.
16. This kit allows elution in small volumes to maximize vector concentration.
17. Online calculators, e.g., Promega biomath calculators, are convenient for calculating the amount of insert required.
18. gBlock resuspension concentrations can vary depending on how much is required in the Gibson assembly reaction.
19. Final volume of the reaction can be increased if vector concentration is low; the volume of 2× HiFi master mix should be increased accordingly.
20. We recommend sequencing at least 100 bases upstream and downstream of the restriction sites when cloning by Gibson assembly.
21. 60 mJ/cm^2 was determined as the optimal dose for inactivation of rCDV producer virus with minimal effects on DI genomes; optimization may be required for other viruses.
22. Once frozen, samples can remain at −80 °C for at least 3 months.

Acknowledgments

This work was funded by the Defense Advanced Research Projects Agency (DARPA) INTERfering and Co-Evolving Prevention and Therapy (INTERCEPT) program (HR0011940493), Boston University and the University of Pittsburgh.

References

1. Von Magnus P, Gard S (1947) Studies on interference in experimental influenza. II. Purification and centrifugation experiments, Arkiv for Kemi, Mineralogi och Geologi, vol 24. Almqvist & Wiksell, Stockholm
2. Von Magnus P (1951) Propagation of the PR8 strain of influenza A virus in chick embryos. II. The formation of incomplete virus following inoculation of large doses of seed virus. Acta Pathol Microbiol Scand 28(3):278–293. https://doi.org/10.1111/j.1699-0463.1951.tb03693.x
3. Ziegler JE, Lavin GI, Frank Horsfall AL (1944) Interference between the influenza viruses: II. The effect of virus rendered noninfective by ultraviolet radiation upon the multiplication of influenza viruses in the chick embryo. J Exp Med 79(4):379–400. https://doi.org/10.1084/jem.79.4.379
4. Huang AS, Baltimore D (1970) Defective viral particles and viral disease processes. Nature 226(5243):325–327. https://doi.org/10.1038/226325a0
5. Johnston MD (1981) The characteristics required for a Sendai virus preparation to induce high levels of interferon in human lymphoblastoid cells. J Gen Virol 56:175–184. https://doi.org/10.1099/0022-1317-56-1-175
6. Patel AH, Elliott RM (1992) Characterization of Bunyamwera virus defective interfering particles. J Gen Virol 73:389–396. https://doi.org/10.1099/0022-1317-73-2-389
7. Manzoni TB, López CB (2018) Defective (interfering) viral genomes re-explored: impact on antiviral immunity and virus persistence. Future Virol 13(7):493–503. https://doi.org/10.2217/fvl-2018-0021
8. Shingai M, Ebihara T, Begum NA, Kato A, Honma T, Matsumoto K, Saito H, Ogura H, Matsumoto M, Seya T (2007) Differential type I IFN-inducing abilities of wild-type versus vaccine strains of measles virus. J Immunol 179:6123–6133. https://doi.org/10.4049/jimmunol.179.9.6123
9. Yount JS, Kraus TA, Horvath CM, Moran TM, López CB (2006) A novel role for viral-defective interfering particles in enhancing dendritic cell maturation. J Immunol 177: 4503–4513. https://doi.org/10.4049/jimmunol.177.7.4503
10. Vignuzzi M, López CB (2019) Defective viral genomes are key drivers of the virus–host interaction. Nat Microbiol 4(7):1075–1087. https://doi.org/10.1038/s41564-019-0465-y
11. Felt SA, Sun Y, Jozwik A, Paras A, Habibi MS, Nickle D, Anderson L, Achouri E, Feemster KA, Cárdenas AM, Turi KN, Chang M, Hartert TV, Sengupta S, Chiu C, López CB (2021) Detection of respiratory syncytial virus defective genomes in nasal secretions is associated with distinct clinical outcomes. Nat Microbiol 6(5):672–681. https://doi.org/10.1038/s41564-021-00882-3
12. Saira K, Lin X, DePasse JV, Halpin R, Twaddle A, Stockwell T, Angus B, Cozzi-Lepri A, Delfino M, Dugan V, Dwyer DE, Freiberg M, Horban A, Losso M, Lynfield R, Wentworth DEN, Holmes EC, Davey R, Wentworth DEN, Ghedin E (2013) Sequence analysis of in vivo defective interfering-like RNA of influenza A H1N1 pandemic virus. J Virol 87(14):8064–8074. https://doi.org/10.1128/JVI.00240-13
13. Vasilijevic J, Zamarreño N, Carlos Oliveros J, Rodriguez-Frandsen A, Gómez G, Rodriguez G, Pérez-Ruiz M, Rey S, Barba I, Pozo F, Casas I, Nieto A, Falcó A (2017) Reduced accumulation of defective viral genomes contributes to severe outcome in influenza virus infected patients. PLoS Pathog 13(10):e1006650. https://doi.org/10.1371/journal.ppat.1006650
14. Aaskov J, Buzacott K, Thu HM, Lowry K, Holmes EC (2006) Long-term transmission of defective RNA viruses in humans and Aedes mosquitoes. Science 311:236. https://doi.org/10.1126/science.1115030
15. Noppornpanth S, Smits SL, Lien TX, Poovorawan Y, Osterhaus ADME, Haagmans BL (2007) Characterization of hepatitis C virus

deletion mutants circulating in chronically infected patients. J Virol 81(22): 12496–12503. https://doi.org/10.1128/JVI.01059-07

16. Li Q, Tong Y, Xu Y, Niu J, Zhong J (2018) Genetic analysis of serum-derived defective hepatitis C virus genomes revealed novel viral cis elements for virus replication and assembly. J Virol 92(7):e02182-17. https://doi.org/10.1128/JVI.02182-17
17. Sun Y, Jain D, Koziol-White CJ, Genoyer E, Gilbert M, Tapia K, Panettieri RA, Hodinka RL, López CB (2015) Immunostimulatory defective viral genomes from respiratory syncytial virus promote a strong innate antiviral response during infection in mice and humans. PLoS Pathog 11(9):e1005122. https://doi.org/10.1371/journal.ppat.1005122
18. Pesko KN, Fitzpatrick KA, Ryan EM, Shi P-Y, Zhang B, Lennon NJ, Newman RM, Henn MR, Ebel GD (2012) Internally deleted WNV genomes isolated from exotic birds in New Mexico: function in cells, mosquitoes, and mice. Virology 427(1):10–17. https://doi.org/10.1016/j.virol.2012.01.028
19. Li D, Lott WB, Lowry K, Jones A, Thu HM (2011) Defective interfering viral particles in acute dengue infections. PLoS One 6(4): 19447. https://doi.org/10.1371/journal.pone.0019447
20. Dimmock NJ, Easton AJ (2014) Defective interfering influenza virus RNAs: time to reevaluate their clinical potential as broad-spectrum antivirals? J Virol 88(10): 5217–5227. https://doi.org/10.1128/JVI.03193-13
21. Murphy SK, Parks GD (1997) Genome nucleotide lengths that are divisible by six are not essential but enhance replication of defective interfering RNAs of the paramyxovirus simian virus 5. Virology 232(1):145–157. https://doi.org/10.1006/viro.1997.8530
22. Tilston-Lunel NL, Welch SR, Nambulli S, de Vries RD, Ho GW, Wentworth D, Shabman R, Nichol ST, Spiropoulou CF, de Swart RL, Rennick LJ, Duprex WP (2021) Sustained replication of synthetic canine distemper virus defective genomes in vitro and in vivo. bioRxiv:2021.2006.2011.448162-442021.448106.448111.448162. https://doi.org/10.1101/2021.06.11.448162
23. Wignall-Fleming EB, Vasou A, Young D, Short JAL, Hughes DJ, Goodbourn S, Randall RE (2020) Innate intracellular antiviral responses restrict the amplification of defective virus genomes of parainfluenza virus 5. J Virol 94(13): 246–266. https://doi.org/10.1128/jvi.00246-20
24. Komarova AV, Combredet C, Sismeiro O, Dillies MA, Jagla B, David RYS, Vabret N, Coppeé JY, Vidalain PO, Tangy F (2013) Identification of RNA partners of viral proteins in infected cells. RNA Biol 10(6):943–956. https://doi.org/10.4161/rna.24453
25. Sourimant J, Plemper RK (2016) Organization, function, and therapeutic targeting of the morbillivirus RNA-dependent RNA polymerase complex. Viruses 8:251. https://doi.org/10.3390/v8090251
26. Welch SR, Tilston NL, Lo MK, Whitmer SLM, Harmon JR, Scholte FEM, Spengler JR, Duprex WP, Nichol ST, Spiropoulou CF (2020) Inhibition of Nipah virus by defective interfering particles. J Infect Dis 221(Suppl 4): S460–S470. https://doi.org/10.1093/infdis/jiz564
27. Duprex WP, Collins FM, Rima BK (2002) Modulating the function of the measles virus RNA-dependent RNA polymerase by insertion of green fluorescent protein into the open reading frame. J Virol 76(14):7322–7328. https://doi.org/10.1128/JVI.76.14.7322-7328.2002
28. Silin D, Lyubomska O, Ludlow M, Duprex WP, Rima BK (2007) Development of a challenge-protective vaccine concept by modification of the viral RNA-dependent RNA polymerase of canine distemper virus. J Virol 81(24):13649–13658. https://doi.org/10.1128/jvi.01385-07
29. Mazure G, Grundy JE, Nygard G, Hudson M, Khan K, Srai K, Dhillon AP, Pounder RE, Wakefield AJ (1994) Measles virus induction of human endothelial cell tissue factor procoagulant activity in vitro. J Gen Virol 75(Pt 11): 2863–2871. https://doi.org/10.1099/0022-1317-75-11-2863
30. Armitage P, Allen I (1950) Methods of estimating the LD 50 in quantal response data. J Hyg 48(3):298–322. https://doi.org/10.1017/S0022172400015084
31. Calain P, Roux L (1993) The rule of six, a basic feature for efficient replication of Sendai virus defective interfering RNA. J Virol 67(8): 4822–4830. https://doi.org/10.1128/jvi.67.8.4822-4830.1993
32. Seki F, Ono N, Yamaguchi R, Yanagi Y (2003) Efficient isolation of wild strains of canine distemper virus in Vero cells expressing canine SLAM (CD150) and their adaptability to marmoset B95a cells. J Virol 77(18):9943–9950. https://doi.org/10.1128/jvi.77.18.9943-9950.2003
33. Beaty SM, Park A, Won ST, Hong P, Lyons M, Vigant F, Freiberg AN, tenOever BR, Duprex WP, Lee B (2017) Efficient and robust

Paramyxoviridae reverse genetics systems. mSphere 2(2):e00376-00316. https://doi.org/10.1128/msphere.00376-16

34. Buchholz UJ, Finke S, Conzelmann KK (1999) Generation of bovine respiratory syncytial virus (BRSV) from cDNA: BRSV NS2 is not essential for virus replication in tissue culture, and the human RSV leader region acts as a functional BRSV genome promoter. J Virol 73(1):251–259. https://doi.org/10.1128/jvi.73.1.251-259.1999

35. Sutter G, Ohlmann M, Erfle V (1995) Non-replicating vaccinia vector efficiently expresses bacteriophage T7 RNA polymerase. FEBS Lett 371(1):9–12. https://doi.org/10.1016/0014-5793(95)00843-x

36. Martin A, Staeheli P, Schneider U (2006) RNA polymerase II-controlled expression of antigenomic RNA enhances the rescue efficacies of two different members of the Mononegavirales independently of the site of viral genome replication. J Virol 80(12):5708–5715. https://doi.org/10.1128/jvi.02389-05

Chapter 6

Molecular Analysis of Copy-Back Defective Interfering RNAs of Morbilliviruses

Felix G. M. Andres and Christian K. Pfaller

Abstract

Copy-back defective interfering RNAs are major contaminants of viral stock preparations of morbilliviruses and other negative strand RNA viruses. They are hybrid molecules of positive sense antigenome and negative sense genome. They possess perfectly complementary ends allowing the formation of extremely stable double-stranded RNA panhandle structures. The presence of the 3′-terminal promoter allows replication of these molecules by the viral polymerase. They thereby negatively interfere with replication of standard genomes. In addition, the double-stranded RNA stem structures are highly immunostimulatory and activate antiviral cell-intrinsic innate immune responses. Thus, copy-back defective interfering RNAs severely affect the virulence and pathogenesis of morbillivirus stocks. We describe two biochemical methods to analyze copy-back defective interfering RNAs in virus-infected samples, or purified viral RNA. First, we present our Northern blotting protocol that allows accurate size determination of defective interfering RNA molecules and estimation of the relative contamination level of virus preparations. Second, we describe a PCR approach to amplify defective interfering RNAs specifically, which allows detailed sequence analysis.

Key words *Paramyxoviridae*, Polymerase, Replication, Innate immune response, Attenuation, Defective viral genome, DVG, DI RNA, C protein, Virulence, Pathogenesis

1 Introduction

Defective viral genomes (DVGs) are major contaminants in virus stock preparations and tissue culture preparations of most negative strand RNA viruses [1–3]. These molecules arise from aberrant replication of the viral genome and antigenome molecules by the RNA-dependent RNA polymerase [4]. Figure 1 includes a schematic of the mechanism of viral RNA synthesis by morbilliviruses, which in principle is common to all non-segmented negative strand RNA viruses [5]. Following cell entry, the viral polymerase complex consisting of the multi-catalytic subunit Large protein (L) and its tetrameric cofactor phosphoprotein (P) initiates the transcription of individual mRNAs from the six genes along the linear (−)strand

Dzwokai Z. Ma and Christian K. Pfaller (eds.), *Measles and Related Morbilliviruses: Methods and Protocols*,
Methods in Molecular Biology, vol. 2808, https://doi.org/10.1007/978-1-0716-3870-5_6,

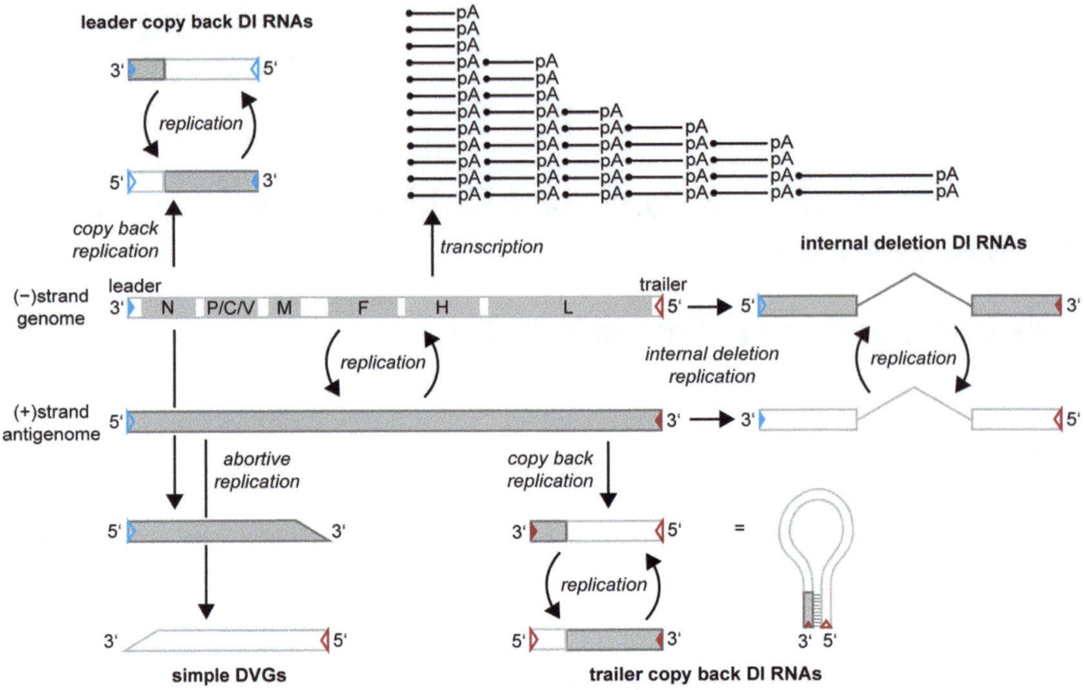

Fig. 1 Mechanism of morbillivirus RNA synthesis and generation of different types of defective RNAs. The (−) strand RNA genome [white with light gray boxes representing transcription units for nucleoprotein (N), phosphoprotein (P), C protein (C), V protein (V), matrix protein (M), fusion protein (F), hemagglutinin (H), and large protein (L)] serves as template for transcription of individual capped and polyadenylated (pA) mRNAs (top right). It also serves as template for replication of (+)strand antigenome (gray). Leader promoter (light blue triangle) and trailer promoter (red triangle) are located at the 3′ termini of the genome and antigenome, respectively. Their complements (light blue and red open triangles) are located at the 5′ ends of the respective complementary strands. Simple defective viral genomes (DVGs) arise through abortive replication (bottom left). They lack the 3′ promoter and thus are not replication competent. Defective interfering (DI) RNAs possess the 3′ promoters and thus are replication competent. They can contain internal deletions (center right) or have copy-back characteristics (bottom right and top left). Copy-back DI RNAs are hybrids of (−)strand and (+) strand RNA with perfectly complementary 5′ and 3′ ends that can form panhandle structures (bottom right)

genome, which is encapsidated by the nucleoprotein (N), forming a tight helical nucleocapsid [6]. These mRNAs are capped at their 5′ ends by L. While transcription initiation only occurs at the 3′-end of the genome (called the "leader"), individual polyadenylation and transcription reinitiation sites between the genes allow subsequent transcription of downstream genes [6]. However, the transcription reinitiation frequency is below 100%, and frequently the polymerase leaves the template before reinitiating the next transcript. This leads to the generation of a mRNA gradient, meaning that transcript abundance of the individual genes is reduced with the distance from the leader [6]. With free N from the translation of its mRNA becoming available, the polymerase switches from the transcription mode to a replication mode. Now, newly synthetized

RNA will be immediately encapsidated with N, the latter of which is delivered to the RNA exit site in L by P [6]. This signal causes the polymerase to skip transcription reinitiation sites and to generate a complementary (+)strand full length RNA called the antigenome. Vice versa, antigenome contains an entry site at its 3′-end (called the "trailer"), allowing replication (but not transcription) of the antigenome, giving rise to new (−)strand genomes, which are also immediately encapsidated by N, closing the viral replication cycle. Since promoter initiation from the trailer is more efficient than from the leader [6], and also because leader is utilized both for transcription and replication, whereas trailer is only utilized for replication [6], more (−)strand genomes than (+)strand antigenomes are synthetized.

The viral replication process is error prone. Various types of DVGs have been reported [1, 2], which can be classified in several groups with specific features (Fig. 1). The simplest form of DVGs arises from abortive replication and comprises only a partially replicated genome (Fig. 1, bottom left). These DVGs reduce the virulence of progeny virus, as they lack essential genes required to finish the viral life cycle. However, since they lack either the leader (in case of the genome) or the trailer (in case of the antigenome), they will not be further replicated, and thus their abundance in a virus-infected cell is self-limiting. In contrast, various types of defective interfering (DI) RNAs exist, which have very strong inhibitory effects on viral replication. These include internal deletion genomes (Fig. 1, center right), in which central parts of the genome sequence are deleted but which still contain both leader and trailer sequences making them replication competent. Hence, they possess the name defective interfering RNA. They form by a mechanism in which the polymerase prematurely aborts genome synthesis (breakpoint) but binds back at a downstream internal position and reinitiates synthesis from there (reinitiation site). The most abundant DI RNA forms are so-called copy-back DI RNAs (Fig. 1, top left and bottom right), which were first described for vesicular stomatitis virus and for Sendai virus [7, 8]. Here, after abortion of genome synthesis at the breakpoint (Fig. 2a, position X), the polymerase binds back to the newly emerging RNA strand reaching out of the exit tunnel and reinitiates synthesis using this strand as a new template (Fig. 2a, reinitiation site Y) [9, 10]. This generates hybrid RNA molecules partially consisting of (−)strand and (+) strand viral RNA (Fig. 2b). The ends are perfectly complementary, and thus the naked copy-back DI RNA molecules are able to form panhandle double-stranded RNA structures (Fig. 1, bottom right), which can activate antiviral innate immune responses [4, 11–14]. Whereas copy-back DI RNAs from the 5′-end of the genome containing parts of the L sequence and the trailer are frequently observed (Fig. 1, bottom right), analogous 3′-copy-back DI RNAs containing parts of N and the leader (Fig. 1, top left) may also exist

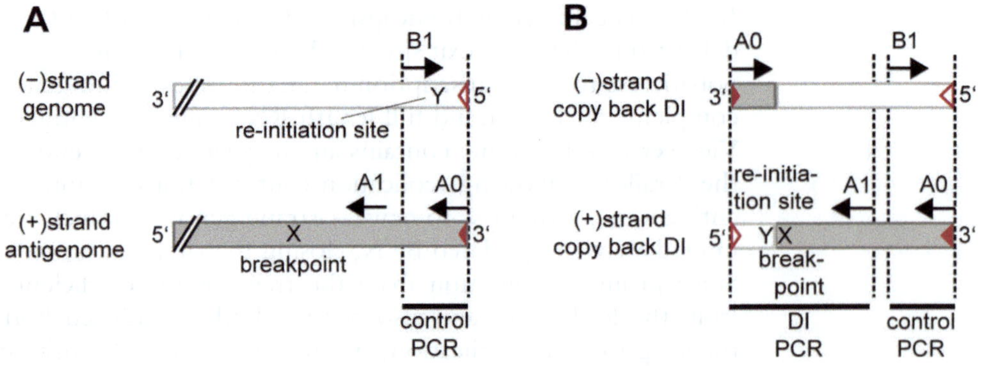

Fig. 2 Mechanism of copy-back DI RNA generation and principle of specific amplification of DI RNAs by PCR. (**a**) During (−)strand genome synthesis, the polymerase (not shown) binds the (+)strand antigenome 3′ promoter (red triangle on grey bar) and synthesizes the complementary RNA strand (white bar). Synthesis is interrupted randomly at breakpoint X due to polymerase detaching from the template. Reinitiation occurs on the attached partial genome strand at position Y, leading to synthesis of the complementary (+)strand RNA. Two primers A0 and A1 both binding to the (+)strand antigenome do not yield a PCR product when only standard genomes are present. Standard genomes can be amplified only with primers A0 and B1, binding to opposite strands. (**b**) Schematic presentation of the product copy-back DI molecule from the process described in (**a**). Top shows the immediate product, which possesses a 3′ promoter (red triangle) and thus serves as a template for generation of the complementary molecule shown at the bottom. This also can be further replicated due to the presence of a 3′ promoter. Amplification of copy-back DI RNA by primers A0 and A1 is possible due to the (+)strand/(−)strand hybrid character of DI RNA. Control PCR with primers A0/B1 will work, as long as the B1 binding site is still encompassed within the copy-back DI RNA molecule. X and Y indicate breakpoint and reinitiation site of the DI RNA. They can be identified by sequence analysis of the A0/A1 amplicon

(*see* **Note 1**). Copy-back DI RNAs are replication competent, since they contain a viral promoter for replication at their 3′-end. Due to the perfect complementarity of their 5′- and 3′-ends, the progeny antisense DI RNA will again harbor a 3′-promoter and can serve as a new template for replication. Compared to 3′-copy-back DI RNAs, 5′-copy-back DI RNAs more efficiently interfere with viral replication and outcompete standard genomes for two reasons. First, the strongly reduced size correlates with faster synthesis, and second, the duplication of the strong trailer promoter allows equally strong replication of sense and antisense DI RNA. Hence, copy-back DI RNAs interfere with viral replication by two independent mechanisms: first, by activating antiviral innate immunity [4, 11–14] and, second, by directly competing with standard genome replication [2, 8, 15, 16]. For measles virus, copy-back DI RNAs were first described to amplify easily in tissue culture infections at high multiplicity [17–19]. They were later found in batches of the live-attenuated vaccine [3, 19]. In addition, deletion of the viral C protein leads to frequent generation of DI RNAs [14, 20], which was confirmed with canine distemper virus (CDV) [21], suggesting that the morbillivirus C protein is a quality assurance factor of the viral polymerase [4, 22].

Since copy-back DI RNAs are associated with viral attenuation and innate immunity activation and frequently arise in tissue culture during serial passaging of virus stocks [3, 19, 20], it may be required to assess their presence in virus stocks used for pathogenesis or virulence studies. Here, we will describe two methods to analyze DI RNAs in CDV infections. First, we will present a nonradioactive protocol for Northern blotting optimized for the detection of CDV DI RNAs and the assessment of their abundance relative to standard RNAs. Second, we will describe a method to generate cDNA libraries of DI RNAs, which allows subcloning and analysis of individual DI RNA sequences using a DI RNA-specific PCR reaction (Fig. 2). Combined, these two methods allow comprehensive evaluation of 5′-copy-back DI RNAs of morbilliviruses.

2 Materials

2.1 *In Vitro Runoff Transcription*

1. Thermocycler.
2. Nuclease-free reaction vials (PCR tubes; *see* **Note 2**).
3. Ultrapure, nuclease-free H_2O (e.g., DEPC-treated; *see* **Note 2**).
4. 5× labeling mix: 5 mM ATP, 5 mM CTP, 5 mM GTP, 3.25 mM UTP, 1.75 mM DIG-11-UTP, pH 7.5 (*see* **Note 3**).
5. 5× transcription buffer: 200 mM Tris-HCl, pH 8.0, 30 mM $MgCl_2$, 50 mM DTT, 10 mM spermidine (*see* **Note 3**).
6. 20 U/μL T7 RNA polymerase (*see* **Note 3**).
7. 20 U/μL SP6 RNA polymerase (*see* **Note 3**).
8. DNase I (RNase-free) (*see* **Note 3**).
9. Pure, linearized plasmid DNA (Fig. 3) containing T7 and SP6 RNA polymerase promoter sequences and unique restriction sites (*Hin*dIII/*Xho*I) flanking the 250 nt L/trailer-terminal sequence (nt 15,440–15,689) of the CDV genome (*see* **Note 4**).
10. 0.2 M EDTA-NaOH, pH 8.0.

2.2 *Denaturing RNA Gel Electrophoresis*

1. Horizontal nucleic acid gel electrophoresis chamber (*see* **Note 2**) and power supply.
2. Microwave.
3. Thermocycler.
4. Ice bath.
5. 10× MOPS: 200 mM MOPS-NaOH, pH 7.0, 80 mM sodium acetate, 10 mM Titriplex III.
6. Denaturing agarose gel: 1% agarose, 1× MOPS, 2% formaldehyde (*see* **Note 5**).

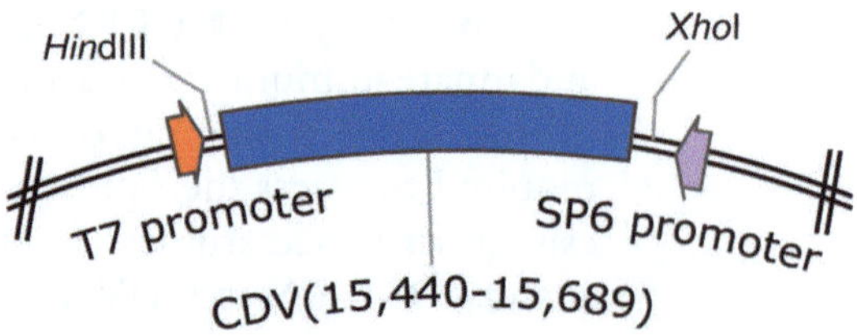

Fig. 3 Schematic representation of the DNA cassette for in vitro transcription of single-stranded RNA probes. A 250 nt fragment of the CDV genome (nt positions 15,440–15,689, blue box) is cloned between a T7 promoter (orange arrow) and a SP6 promoter (purple arrow). For generation of (−)strand probe, the plasmid is linearized with *Xho*I and in vitro transcribed using T7 RNA polymerase. For generation of (+)strand probe, the plasmid is linearized with *Hin*dIII and in vitro transcribed with SP6 RNA polymerase

7. RNA loading buffer: 50% formamide, 6% formaldehyde, 1× MOPS, 10% glycerol, 0.05% bromophenol blue (*see* **Note 6**).
8. RNA samples from CDV-infected cells (total RNA), concentrated virus particles, or CsCl-density gradient-purified CDV ribonucleocapsids (*see* **Note** 7).
9. Optional: RNA molecular weight marker, DIG-labeled (*see* **Note 3**).

2.3 Capillary Transfer

1. Plastic or glass tray (*see* **Note 2**).
2. Bowl and support (*see* **Note 2**).
3. Support weight.
4. Roller device, for example, a sterile 10 mL pipet.
5. Nylon membrane, positively charged. Size according to the used agarose gel.
6. Whatman paper 3 MM.
7. Parafilm.
8. 20× SSC: 0.3 M sodium citrate-HCl, pH 7.0, 3 M NaCl.
9. Absorbent paper (e.g., paper towels).

2.4 Crosslinking and Methylene Blue Staining

1. Plastic or glass tray (*see* **Note 2**).
2. UV illuminator (e.g., UV StratalinkerTM 1800).
3. Orbital shaker.
4. 2× SSC (diluted from 20× SSC).
5. 0.2% methylene blue solution in H_2O (*see* **Note 8**).
6. Ultrapure, nuclease-free H_2O (*see* **Note 2**).
7. RNA membranes.

2.5 Probe Hybridization

1. Hybridization oven.
2. Thermocycler.
3. Ice bath.
4. Hybridization tubes or temperature resistant, sealable plastic/ glass tubes (*see* **Note 2**).
5. DIG Easy-Hyb hybridization buffer (*see* **Notes 3** and **9**).
6. DIG-labeled RNA probes.
7. Crosslinked RNA membranes.

2.6 Probe Detection

1. Hybridization oven.
2. Tube roller.
3. Hybridization tubes or temperature resistant, sealable plastic/ glass tubes (*see* **Note 2**).
4. Sealable hybridization bag.
5. Chemiluminescence imager.
6. Hybridization-Wash buffer 1: 2× SSC, 0.1% SDS.
7. Hybridization-Wash buffer 2: 0.1× SSC, 0.1% SDS.
8. 10× blocking buffer (*see* **Notes 3**, **9**, and **10**).
9. Maleic acid buffer: 0.1 M maleic acid-NaOH, pH 7.5, 0.15 M NaCl (*see* **Note 3**).
10. Washing buffer: 0.1 M maleic acid-NaOH, pH 7.5, 0.15 M NaCl, 0.3% Tween-20 (*see* **Notes 3** and **11**).
11. Detection buffer: 0.1 M Tris-HCl, pH 9.5, 0.1 M NaCl (*see* **Note 3**).
12. Anti-DIG Fab fragment, alkaline phosphatase-conjugated (*see* **Note 3**).
13. CDP-Star detection reagent: 0.25 mM disodium 2-chloro-5-(4-methoxyspiro{1,2-dioxetane-3,2′-(5′-chloro)tricyclo [3.3.1.13,7]decan}-4-yl)-1-phenyl phosphate (*see* **Note 3**).

2.7 DI RNA-Specific RT-PCR

1. Thermocycler.
2. Ice bath.
3. Nuclease-free reaction vials (PCR tubes; *see* **Note 2**).
4. Ultrapure, nuclease-free H_2O (*see* **Note 2**).
5. Superscript III reverse transcriptase.
6. 5× First-strand buffer.
7. 100 mM DTT.
8. RNase OUT, or a different RNase inhibitor.
9. dNTP mix, 10 mM each dATP, dCTP, dGTP, dTTP.

10. 10 μM stock solutions of DI RNA-specific primers A0 (ATG CGC ATG CGG CCG CAG ACA AAG CTG GGT ATG ATA ACT TAT TG) and A1 (ATG CGC ATG TTT AAA CCT TTA TCA GGT AAT GAC TCG ATA TCC G) or 50 μM stock solution of random hexamer primers.
11. RNA samples from CDV-infected cells (total RNA), concentrated virus particles, or CsCl-density gradient-purified CDV ribonucleocapsids (*see* **Note 7**).

2.8 PCR Amplification and Gel Electrophoresis

1. Thermocycler.
2. Horizontal nucleic acid gel electrophoresis chamber and power supply.
3. UV transilluminator.
4. Nuclease-free reaction vials (PCR tubes; *see* **Note 2**).
5. Ultrapure, nuclease-free H_2O (*see* **Note 2**).
6. DNA polymerase (e.g., Pfu Ultra II, or Taq).
7. 10× PCR reaction buffer.
8. dNTP mix, 10 mM each dATP, dCTP, dGTP, dTTP.
9. 10 μM stock solutions of DI RNA-specific primers A0 (ATG CGC ATG CGG CCG CAG ACA AAG CTG GGT ATG ATA ACT TAT TG) and A1 (ATG CGC ATG TTT AAA CCT TTA TCA GGT AAT GAC TCG ATA TCC G) (*see* **Note 12**).
10. 1× TAE buffer: 40 mM Tris-acetate, 1 mM EDTA.
11. 2% agarose gel in TAE.
12. GelRed DNA dye (*see* **Note 13**).
13. DNA low range molecular weight marker.
14. 6× DNA loading dye: 10 mM Tris-HCl, pH 7.6, 0.03% bromophenol blue, 0.03% xylene cyanol FF, 60% glycerol, 60 mM EDTA.
15. 1 μL of cDNA from RT-PCR step.

3 Methods

3.1 In Vitro Transcription of DIG-Labeled (−)Strand and (+)Strand Probes for Northern Blotting

1. In a PCR tube, bring 1 μg linearized template DNA to a final volume of 10 μL with H_2O (*see* **Notes 4** and **14**).
2. On ice, add 4 μL 5× labeling mix, 4 μL 5× transcription buffer, and 2 μL of the RNA polymerase of your choice (*see* **Note 4**).
3. Vortex and pulse spin briefly.
4. Incubate at 42 °C for 1 h in a thermocycler.
5. Add 2 μL DNase I (RNase-free), and incubate for at 37 °C for 15 min in a thermocycler.

6. The reaction can be stopped by adding 2 μL 0.2 M EDTA solution. Alternatively, RNA can be purified from the reaction by either phenol-chloroform precipitation or column purification, followed by reconstitution/elution in H_2O. Store labeled RNA in aliquots at −80 °C.

3.2 Denaturing RNA Gel Electrophoresis

1. Mix 1.5 g agarose with 15 mL 10× MOPS and 126 mL H_2O (*see* **Note 5**). Weigh mixture and heat in a microwave until the agarose is completely dissolved. Add H_2O to adjust for the loss by evaporation.
2. Wait until the gel has cooled down to about hand warm, and add 8.1 mL of 37% formaldehyde (*see* **Note 5**). Cast gel using an appropriate comb (1–1.5 mm thickness is recommended).
3. Prepare fresh RNA loading buffer (*see* **Note 6**).
4. Dilute 3–5 μg total RNA, 0.1–1 μg viral RNA (*see* **Note 7**), or 50 ng DIG-labeled RNA marker in H_2O to a total volume of 5 μL, and add 15 μL fresh RNA loading buffer.
5. Heat-denature samples at 65 °C in a thermocycler for 10 min.
6. Directly place samples into an ice bath for 1 min (*see* **Note 15**).
7. Load samples into the gel pockets. Fill electrophoresis chamber with 1× MOPS, and perform electrophoresis at a low voltage of 3–6 V/cm for 3–4 h or until the bromophenol blue band has migrated to around three-quarters of the gel (*see* **Note 16**).

3.3 Capillary Transfer

1. Place agarose gel in plastic tray, and wash it twice in 20× SSC for 15 min each.
2. Cut nylon membrane and two pieces of Whatman paper to the size of your agarose gel (15 × 15 cm). Cut one additional Whatman paper as "bridge" to a size of 15 × 45 cm.
3. Prepare four parafilm strips with a length of 15–20 cm each.
4. Soak the membrane and all Whatman papers in 20× SSC. Mark the membrane by indention at a corner or other method of choice to identify correct orientation of the membrane afterward.
5. Assemble the transfer setup as shown in Fig. 4 (*see* **Note 17**). Make sure to place agarose gel upside down on top of the bridge to ensure correct orientation of the samples on the membrane afterward. During the assembly, make sure to squeeze out any residual air bubbles between each layer with an appropriate roller device, for example, a sterile 10 mL pipette.
6. Let capillary transfer occur overnight (>12 h).

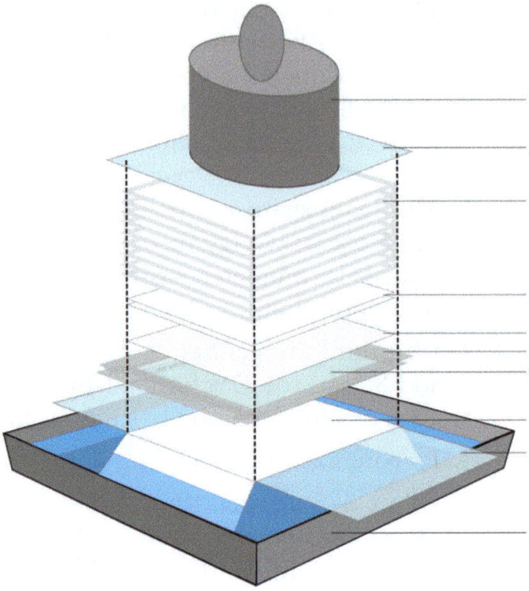

Fig. 4 Assembly of the capillary transfer system. The transfer system is assembled from bottom to top. Filter paper bridge, nylon membrane, and two additional filter papers are soaked in 20× SSC prior to assembly. Remove air bubbles between each layer using a roller device, for example, a sterile 10 mL pipette. Place the agarose gel upside down onto the filter paper bridge, such that the back part of the gel is closest to the nylon membrane. Parafilm strips around the outer edges of the gel act as insulators to avoid buffer bypassing the gel. Use enough dry absorbent paper, such as paper towels, to allow buffer to be soaked up from the reservoir. About 5–10-cm-thick layer is recommended. Use a second glass or plexiglass plate to distribute the top weight evenly. About 0.5–1 kg of weight is recommended

3.4 Crosslinking and Methylene Blue Staining

1. Take membrane out of the transfer setup, and place on a piece of Whatman paper soaked with 2× SSC, RNA side up.
2. Place in a UV crosslinker and irradiate the membrane with 254 nm UV light at 0.125 J/cm^2. Alternatively, the membrane can be baked at 80 °C for 2 h, or at 120 °C for 30 min, to fix RNAs to the membrane.
3. To stain total RNA with methylene blue, place the membrane in a plastic tray, and add an appropriate amount 0.2% methylene blue solution to cover the membrane.
4. Incubate on shaker incubator for 1–5 min or until blue bands appear.
5. Remove methylene blue solution (*see* **Note 8**), and wash the membrane several times with H_2O until the ribosomal RNA bands are clearly visible (Fig. 5a).

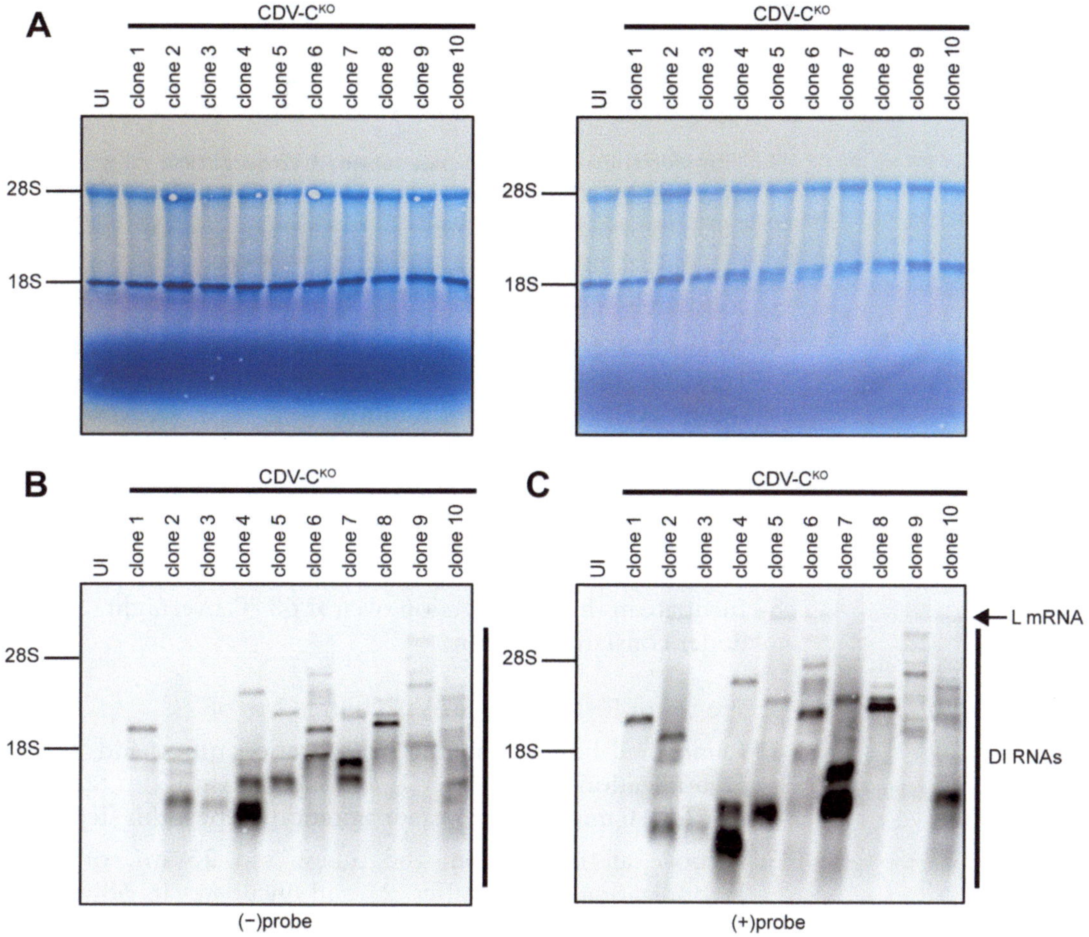

Fig. 5 Example Northern blots with total RNA from CDV-C^{KO}-infected cells. Northern blots were performed with 4 μg total RNA per sample. Each lane shows a different CDV-C^{KO} clone. UI = uninfected control sample. (**a**) Methylene blue staining of two membranes subsequently used for Northern blot probing. 28S and 18S rRNA appear as strong blue bands. Bromophenol blue from the RNA loading buffer appears as darker blue cloud in the bottom part. (**b**) Northern blot using the (−)probe detecting (−)strand portions of DI RNA molecules with lower molecular weight. Note that the different band patterns for each clone indicate individual DI RNA species with different molecule sizes ranging from >500 to <4000 nt (range indicated by black bar). (**c**) Northern blot using the (+)probe detecting (+)strand viral RNA, including L mRNA and (+)strand portions of DI RNA molecules (range indicated by black bar) with lower molecular weight. Note that copy-back DI RNA molecules are detected with both (+)probe and (−)probe, indicating their hybrid RNA character, while L mRNA is only detected by (+)probe

6. For size reference, the membrane is marked on the side at the height of the 28S and 18S rRNA bands. These bands have a size of approximately 5 kb and 1.9 kb, respectively.
7. Cut small pieces from the edge of the membrane at position of the rRNA bands.
8. Take a picture of the membrane at this point, since the methylene blue staining will disappear in the following steps.

3.5 Probe Hybridization

1. Cut the membrane into smaller pieces, if you would like to incubate with different probes.
2. Pre-warm hybridization oven and hybridization buffer to 68 °C.
3. Transfer the membrane pieces into hybridization tubes (make sure the RNA side is facing inward), and add about 10 mL of warm hybridization buffer.
4. Incubate in the hybridization oven at 68 °C for 30 min under constant agitation.
5. Dilute 2.5–5 μL DIG-labeled RNA probes (equivalent to 1.25 μg labeled RNA) in 10 μL H_2O, and denature in a thermocycler at 95 °C for 5 min.
6. Immediately put the probes on ice for 5 min (*see* **Note 15**).
7. Dilute the probes in 3.5 mL pre-warmed hybridization buffer, and replace buffer on the membrane with the probe-containing buffer.
8. Incubate in the hybridization oven at 68 °C overnight (>12 h) under constant agitation.

3.6 Probe Detection

1. Pre-warm Hybridization-Wash buffer 2 to 68 °C.
2. Remove all liquid from the hybridization tubes, add 25 mL Hybridization-Wash buffer 1, and incubate at room temperature for 10 min under constant agitation. Repeat this step once.
3. Remove all liquid from the tubes, add 25 mL of warm Hybridization-Wash buffer 2, and incubate at 68 °C for 10 min under constant agitation. Repeat this step once.
4. Remove all liquid, add 20 mL washing buffer per tube, and incubate at room temperature for 2 min under constant agitation.
5. Prepare 1× blocking solution by diluting 10× blocking buffer in maleic acid buffer (*see* **Note 10**).
6. Remove all liquid, add 10 mL of 1× blocking solution, and incubate at room temperature for 30 min under constant agitation.
7. Prepare antibody solution by diluting AP-conjugated anti-DIG Fab fragment 1:10,000 in 1× blocking solution.
8. Remove all liquid, add 10 mL antibody solution, and incubate at room temperature for 30 min under constant agitation.
9. Remove all liquid, add 20 mL washing buffer, and incubate at room temperature for 15 min under constant agitation. Repeat this step once.
10. Remove all liquid, add 10 mL detection buffer, and equilibrate at room temperature for 15 min under constant agitation.

11. Place the membrane in a hybridization bag with the RNA side up, and add an appropriate amount of CDP-Star substrate to evenly soak the membrane (about 1 mL for a 100 cm^2 membrane is sufficient). Immediately cover the membrane with a second sheet, and make sure to avoid air bubbles. Incubate at room temperature for 5 min.
12. Squeeze out excess liquid and seal the edges of the hybridization bag. Expose the membrane for 5–20 min to an imaging device or X-ray film. Adjust exposure time if necessary. Example blots are shown in Fig. 5b, c.

3.7 DI RNA-Specific RT-PCR

1. In a PCR tube, mix 1 μg of total RNA (*see* **Note 7**) with 1 μL dNTP mix and 1 μL each of A0 and A1 primers or 1 μL of random hexamer primers. Adjust the volume to 13 μL with H_2O.
2. Incubate mixture in a thermocycler at 98 °C for 10 min (*see* **Note 18**).
3. Prepare enzyme mix. Per sample, combine 4 μL of 5× first-strand buffer, 1 μL of 100 mM DTT, 1 μL of RNase OUT, and 1 μL of SuperScript III reverse transcriptase. Store on ice.
4. Immediately place samples into an ice bath and incubate for 10 min (*see* **Note 15**).
5. Add 7 μL of enzyme mix to each RNA sample.
6. In a thermocycler, start the following program:
   (a) 25 °C for 5 min
   (b) 50 °C for 1 h
   (c) 70 °C for 15 min
   (d) 4 °C at hold
7. cDNA can be stored at −20 °C for short term (a few weeks). For longer storage, we recommend −80 °C.

3.8 PCR Amplification and Gel Electrophoresis

1. Prepare reaction mix (*see* **Note 19**). Per sample, combine 39.75 μL of H_2O, 5 μL of 10× PCR reaction buffer, 1.25 μL of dNTP mix, 1 μL of each primer (A0 + A1, or A0 + B1; *see* **Note 12**), and 1 μL of DNA polymerase.
2. Distribute 49 μL of reaction mix into PCR tubes.
3. Add 1 μL of respective cDNA template.
4. Start PCR using the following program:
   (a) Initial denaturation: 98 °C for 2 min.
   (b) Denaturation: 98 °C for 30 s.
   (c) Annealing: 48 °C for 30 s.
   (d) Elongation: 72 °C for 1–2 min.

Repeat **steps (b)–(d)** for 30 cycles.

(e) 72 °C for 5 min.

(f) 4 °C at hold.

5. Prepare 2% agarose gel with 1× TAE, and add gel dye at recommended concentration.
6. Add 10 μL of 6× DNA loading dye to each PCR reaction and mix well.
7. Load DNA marker and samples into the gel pockets.
8. Perform electrophoresis in 1×TAE at 100–130 V for 30–60 min, or until the dye band has migrated about three-quarters into the gel.
9. Visualize bands on an UV transilluminator. Example gel images are shown in Fig. 6a, b (*see* **Note 20**).

4 Notes

1. The protocols described here focus on detection and analysis of 5′-copy-back DI RNAs. However, designing adequate probes for Northern blotting and primers for RT-PCR will also allow study of 3′-copy-back DI RNAs.
2. All solutions are prepared using ultrapure, nuclease-free water (e.g., DEPC-treated), unless indicated otherwise. Only use analytical grade reagents and verified nuclease-free plastic ware. Equipment should be cleaned with RNase Away and rinsed with ultrapure H_2O prior to use.
3. These reagents can be purchased ready to use as part of the DIG-Northern product series (Roche).
4. With a template as shown in Fig. 3, in vitro transcription with T7 RNA polymerase will yield the (−)strand probe for detection of genomic RNA and requires plasmid linearization with *Xho*I. Likewise, in vitro transcription with SP6 RNA polymerase will generate the (+)strand probe for detection of antigenomic RNA and requires linearization with *Hin*dIII. A suitable plasmid for this purpose is, for example, pCR3. If choosing other backbones and restriction enzymes, it is advisable to use enzymes creating 5′ overhangs or blunt ends to prevent undesirable sequences.
5. For a large, 15 × 15 cm agarose gel, we recommend preparing 150 mL gel solution. If you choose to operate with a smaller size gel, adjust accordingly. The gel still has to be liquid but also safe to touch when adding formaldehyde to prevent the formation of hazardous gases and degradation of formaldehyde.

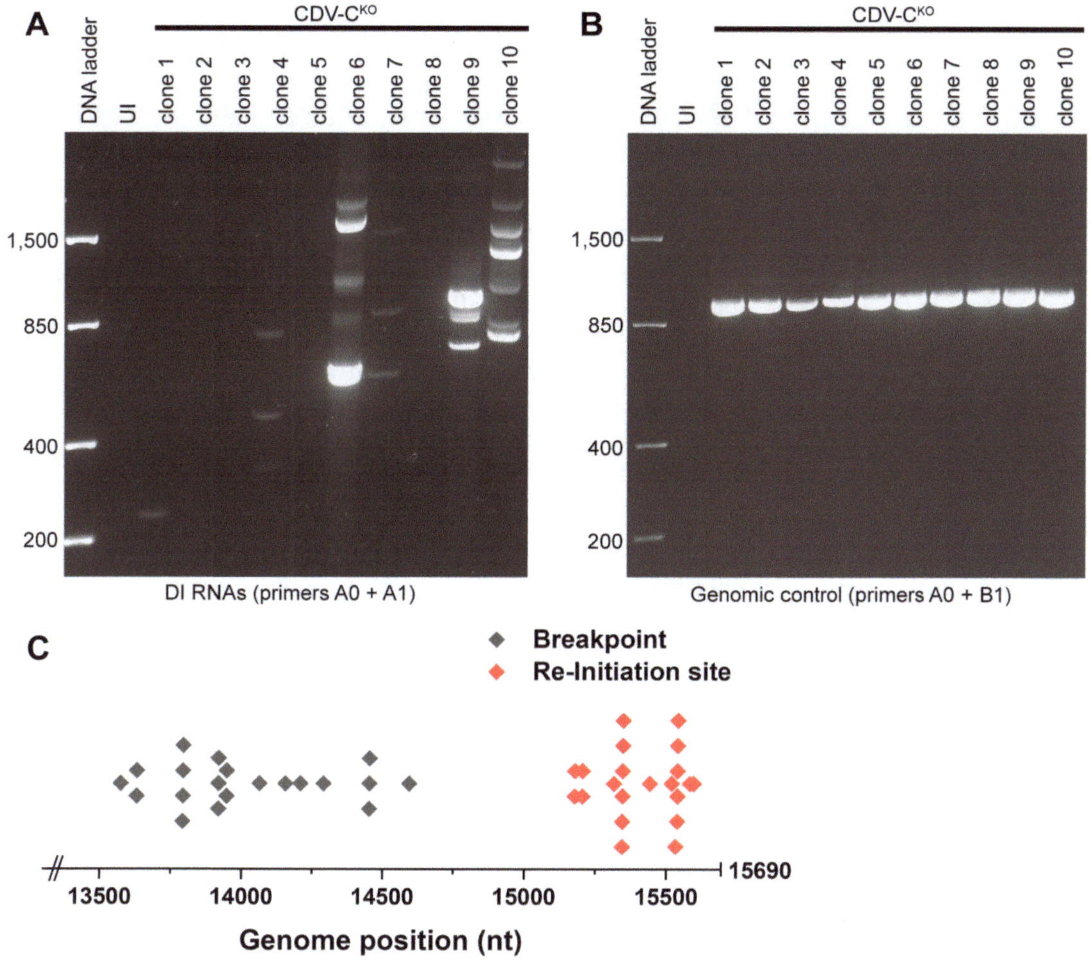

Fig. 6 Example agarose gel of PCR detection of CDV-C^{KO}-derived copy-back DI RNA. 1 μg total RNA from infected cells was used for RT-PCR, and 1 μL of cDNA was used as template for further PCR amplification. (**a**) Gel image of PCR products amplified with DI RNA-specific primers A0 and A1. Each lane represents a different virus clone. UI = uninfected control sample. Note that the observed band patterns indicate different DI RNA species in each clone. (**b**) Control PCR using primers A0 and B1 yields a single PCR product of 976 bp length, indicating positive CDV infection. (**c**) Analysis of breakpoint (gray diamonds) and reinitiation site (red diamonds) positions of DI RNA products after subcloning into pCR3 vector, Sanger sequencing, and alignment against the CDV 5804P reference genome (GenBank AY386316)

6. The RNA loading buffer has always to be prepared freshly, since formamide and formaldehyde will degrade in aqueous solution over time. For more accurate pipetting of the glycerol, it can help to premix 10× MOPS and 100% glycerol and add the respective volume of this mixture.
7. For Northern blotting, we recommend 3–5 μg of total RNA from CDV-infected cells [20], 1 μg of RNA from sucrose-purified CDV stocks [20], or 100–500 ng of CsCl-density gradient-purified CDV ribonucleocapsids [23]. For RT-PCR,

we recommend 0.1–1 μg of total RNA from CDV-infected cells [20], 10–100 ng of RNA from sucrose-purified CDV stocks [20], or 10–100 ng of CsCl-density gradient-purified CDV ribonucleocapsids [23]. For preparation of these RNAs, we recommend Trizol purification. For details, see references.

8. Methylene blue solution can be reused multiple times.
9. Composition proprietary to manufacturer.
10. 10× Blocking solution should be thawed once; 5 mL aliquots should be prepared in 50 mL conical tubes and stored at −20 ° C. For preparation of 1× blocking solution, simply add 45 mL maleic acid buffer to the frozen aliquot, and thaw at room temperature. Mix well by inverting the tube multiple times. 50 mL of 1× blocking solution is sufficient for blocking of a 15 × 15 cm membrane.
11. Tween-20 in the washing buffer may phase-separate over time. Before use, mix well by gently inverting the storage bottle multiple times.
12. As a control, we recommend to perform an additional PCR substituting primer A1 with primer B1 (ATG CGC ATG TTT AAA CGA TAT TAG GGA AGG TAG GGT CAG TG). This will yield a product of 976 bp length indicating presence of standard CDV genomic RNA in the RNA samples (Figs. 2 and 6). Primer A1 can also be substituted with alternative primers binding more upstream in the CDV genome. This will allow analysis of larger DI RNA molecules with breakpoints further upstream of the primer binding site. When designing these alternative primers, make sure to match the annealing temperature of primer A0.
13. Any DNA-intercalating dye will work, for example, ethidium bromide or SYBR Safe.
14. The procedure described here is designed for 1 μg of linear template DNA, which will yield ~20 μg labeled RNA in a 1-h reaction. Higher yields can be achieved by scaling up the reaction mix accordingly.
15. Directly placing the samples from the hot heat block into an ice bath is better than cooling down the heat block, as the cold shock will prevent the internal dsRNA structures in the sample RNAs from reannealing. This is critical for probe and primer annealing in the later steps.
16. Low voltages and longer runtimes are preferred to ensure even migration of the RNAs. For a 15 × 15 cm gel, 80 V gives optimal results.
17. The parafilm strips in the setup act as insulators at the outer border of the gel and membrane to prevent buffer from bypassing the gel. This is crucial for efficient transfer of the RNAs from the gel to the membrane.

18. Heat denaturation at 98 °C is important to dissolve efficiently all rigid secondary structures in DI RNAs, since they form very stable intramolecular panhandle structures (Fig. 1).
19. Reaction volumes can be scaled down to 10–15 μL per sample, if only a qualitative analysis of the band patterns by gel electrophoresis is planned.
20. For a more in-depth analysis of DI RNAs, the primers used in the DI RNA detection PCR have restriction sites *Not*I (primer A0, underlined sequence) and *Pme*I (primers A1, B1, underlined sequence) at their 5′ ends for subsequent cloning and sequencing. Alternatively, the PCR products can be gel-purified and sequenced directly using these primers. This approach will allow determination of breakpoints and reinitiation sites of individual DI RNAs (Fig. 6c).

References

1. Lazzarini RA, Keene JD, Schubert M (1981) The origins of defective interfering particles of the negative-strand RNA viruses. Cell 26:145–154. https://doi.org/10.1016/0092-8674(81)90298-1
2. Ziegler CM, Botten JW (2020) Defective interfering particles of negative-strand RNA viruses. Trends Microbiol 28:554–565. https://doi.org/10.1016/j.tim.2020.02.006
3. Bellocq C, Mottet G, Roux L (1990) Wide occurrence of measles virus subgenomic RNAs in attenuated live-virus vaccines. Biologicals 18:337–343. https://doi.org/10.1016/1045-1056(90)90039-3
4. Siering O, Cattaneo R, Pfaller CK (2022) C proteins: controllers of orderly paramyxovirus replication and of the innate immune response. Viruses 14:137. https://doi.org/10.3390/v14010137
5. Lamb RA (2013) Mononegavirales. In: Knipe DM, Howley PM (eds) Fields virology, vol 1, 6th edn. Wolters Kluwer; Lippincott Williams & Wilkins, Philadelphia, pp 880–884
6. Lamb RA, Parks GD (2013) Paramyxoviridae. In: Knipe DM, Howley PM (eds) Fields virology, vol 1, 6th edn. Wolters Kluwer; Lippincott Williams & Wilkins, Philadelphia, pp 957–995
7. Perrault J, Leavitt RW (1978) Characterization of snap-back RNAs in vesicular stomatitis defective interfering virus particles. J Gen Virol 38:21–34. https://doi.org/10.1099/0022-1317-38-1-21
8. Kolakofsky D (1976) Isolation and characterization of Sendai virus DI-RNAs. Cell 8:547–555. https://doi.org/10.1016/0092-8674(76)90223-3
9. Calain P, Curran J, Kolakofsky D et al (1992) Molecular cloning of natural paramyxovirus copy-back defective interfering RNAs and their expression from DNA. Virology 191:62–71. https://doi.org/10.1016/0042-6822(92)90166-M
10. Leppert M (1977) Further characterization of Sendai virus DI-RNAs: a model for their generation. Cell 12:539–552. https://doi.org/10.1016/0092-8674(77)90130-1
11. Linder A, Bothe V, Linder N et al (2021) Defective interfering genomes and the full-length viral genome trigger RIG-I after infection with vesicular stomatitis virus in a replication dependent manner. Front Immunol 12:595390. https://doi.org/10.3389/fimmu.2021.595390
12. Genoyer E, López CB (2019) The impact of defective viruses on infection and immunity. Annu Rev Virol 6:547–566. https://doi.org/10.1146/annurev-virology-092818-015652
13. Sánchez-Aparicio MT, Garcin D, Rice CM et al (2017) Loss of Sendai virus C protein leads to accumulation of RIG-I immunostimulatory defective interfering RNA. J Gen Virol 98:1282–1293. https://doi.org/10.1099/jgv.0.000815
14. Pfaller CK, Radeke MJ, Cattaneo R et al (2014) Measles virus C protein impairs production of defective copyback double-stranded viral RNA and activation of protein kinase R. J Virol 88:456–468. https://doi.org/10.1128/JVI.02572-13

15. Rao DD, Huang AS (1982) Interference among defective interfering particles of vesicular stomatitis virus. J Virol 41:210–221. https://doi.org/10.1128/JVI.41.1.210-221.1982
16. Bay PH, Reichmann ME (1982) In vitro and in vivo inhibition of primary transcription of vesicular stomatitis virus by a defective interfering particle. J Virol 41:172–182. https://doi.org/10.1128/JVI.41.1.172-182.1982
17. Hall WW, Martin SJ, Gould E (1974) Defective interfering particles produced during the replication of measles virus. Med Microbiol Immunol 160:155–164. https://doi.org/10.1007/BF02121722
18. Whistler T, Bellini WJ, Rota PA (1996) Generation of defective interfering particles by two vaccine strains of measles virus. Virology 220: 480–484. https://doi.org/10.1006/viro.1996.0335
19. Calain P, Roux L (1988) Generation of measles virus defective interfering particles and their presence in a preparation of attenuated live-virus vaccine. J Virol 62:2859–2866. https://doi.org/10.1128/JVI.62.8.2859-2866.1988
20. Pfaller CK, Mastorakos GM, Matchett WE et al (2015) Measles virus defective interfering RNAs are generated frequently and early in the absence of C protein and can be destabilized by adenosine deaminase acting on RNA-1-like hypermutations. J Virol 89:7735–7747. https://doi.org/10.1128/JVI.01017-15
21. Siering O, Sawatsky B, Pfaller CK (2020) C protein is essential for canine distemper virus virulence and pathogenicity in ferrets. J Virol 95:e01840-20. https://doi.org/10.1128/JVI.01840-20
22. Pfaller CK, Bloyet L-M, Donohue RC et al (2020) The C protein is recruited to measles virus ribonucleocapsids by the phosphoprotein. J Virol 94:e01733-19. https://doi.org/10.1128/JVI.01733-19
23. Pfaller CK, Donohue RC, Nersisyan S et al (2018) Extensive editing of cellular and viral double-stranded RNA structures accounts for innate immunity suppression and the proviral activity of ADAR1p150. PLoS Biol 16: e2006577. https://doi.org/10.1371/journal.pbio.2006577

Chapter 7

Applying Reverse Genetics to Study Measles Virus Interactions with the Host

Heidy Vera-Peralta, Valerie Najburg, Chantal Combredet, Thibaut Douché, Quentin Giai Gianetto, Mariette Matondo, Frédéric Tangy, Marie Mura, and Anastassia V. Komarova

Abstract

The study of virus-host interactions is essential to achieve a comprehensive understanding of the viral replication process. The commonly used methods are yeast two-hybrid approach and transient expression of a single tagged viral protein in host cells followed by affinity purification of interacting cellular proteins and mass spectrometry analysis (AP-MS). However, by these approaches, virus-host protein-protein interactions are detected in the absence of a real infection, not always correctly compartmentalized, and for the yeast two-hybrid approach performed in a heterologous system. Thus, some of the detected protein-protein interactions may be artificial. Here we describe a new strategy based on recombinant viruses expressing tagged viral proteins to capture both direct and indirect protein partners during the infection (AP-MS in viral context). This way, virus-host protein-protein interacting co-complexes can be purified directly from infected cells for further characterization.

Key words Reverse genetics, Recombinant virus, Virus-host interactions, Measles virus, Tagged proteins, Affinity purification

1 Introduction

Measles virus (MV) is a negative-sense single-stranded RNA virus from the *Paramyxoviridae* family that encodes eight proteins, two of which (V and C) are virulence factors that are transcribed from the phosphoprotein (P) gene. Characterization of host protein partners of these factors is of major importance as they are susceptible to exert multiple regulatory functions during the viral replication cycle.

Reverse genetics makes it possible to express tagged viral proteins directly from a viral genome. Numerous reverse genetics protocols allow the generation of infectious viruses from plasmids

Dzwokai Z. Ma and Christian K. Pfaller (eds.), *Measles and Related Morbilliviruses: Methods and Protocols*, Methods in Molecular Biology, vol. 2808, https://doi.org/10.1007/978-1-0716-3870-5_7,

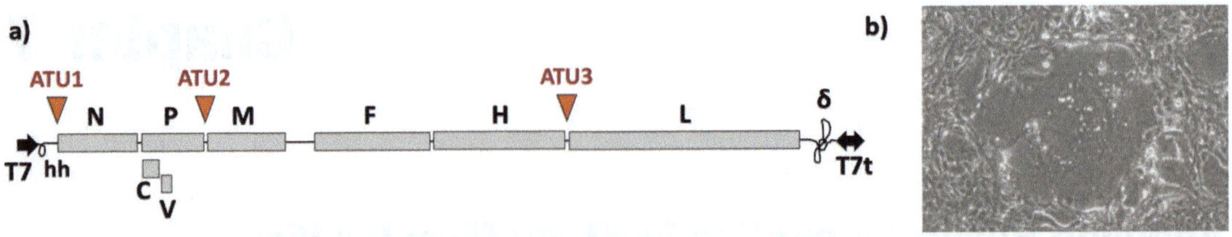

Fig. 1 (**a**) Schematic representation of cloned measles virus vector genome (pTM-MVSchw) containing additional transcription units (ATUs). Structural protein genes (N, nucleoprotein; P, phosphoprotein; M, matrix protein; F, fusion protein; H, hemagglutinin; L, large polymerase) and accessory protein genes (C and V proteins) are shown. T7 is a T7 RNA polymerase promoter sequence, hh is a hammerhead ribozyme, T7t is a T7 RNA polymerase termination signal, δ is a hepatitis delta ribozyme. (**b**) Syncytia formation on Vero cells infected with a recombinant measles virus

coding for genomic cDNA [1–7]. For MV, a noteworthy and pioneer work performed in the Billeter laboratory allowed to clone an infectious cDNA corresponding to the antigenome of MV and established a reverse genetics procedure to rescue the virus [8]. We adapted this protocol by engineering an infectious cDNA from the Schwarz/Moraten strain (pTM-MVSchw), a widely used MV vaccine [9]. pTM-MVSchw plasmid was further modified to introduce additional transcription units (ATUs) at different positions of the genome for the expression of foreign genes [9]. ATUs enable the cloning of supplementary sequences into the MV genome. Three different positions of ATUs on the MV genome can be considered, ATU1 (before the N gene), ATU2 (between P and M genes), and ATU3 (between H and L genes) (Fig. 1a) (*see* **Note 1**) [4].

We used a reverse genetic protocol to generate recombinant MV strains expressing tagged viral proteins. This provided MV-V and MV-C purification directly from infected cells using a modified one-step affinity purification approach. In this protocol, viral proteins of interest (baits) are labeled with a recombinant fusion tag, which is then used to purify baits and their respective interacting partners (preys). Examples of tags frequently used in virus-host protein-protein interactions studies are FLAG-tag, 6-His-Tag, Myc-Tag, HA-tag, and STrEP-Tag [10–13]. We set a protocol to study virus-host protein-protein interactions in infected cells using recombinant MVs expressing One-STrEP tagged proteins (Fig. 2). We chose STrEP-Tag because being a short peptide, none or minimal interference with viral replication is expected [14–18]. If little knowledge of a viral protein of interest is available, N-terminal (N-ter) and C-terminal (C-ter) recombinant fusion tags should be considered. We successfully cloned and produced recombinant MVs expressing N-terminally tagged V protein and nucleoprotein (N) from ATU2 [19] and N- and C-terminally tagged MV-C protein from ATU3 [20] (*see* **Note 1**). Upon infection with MV

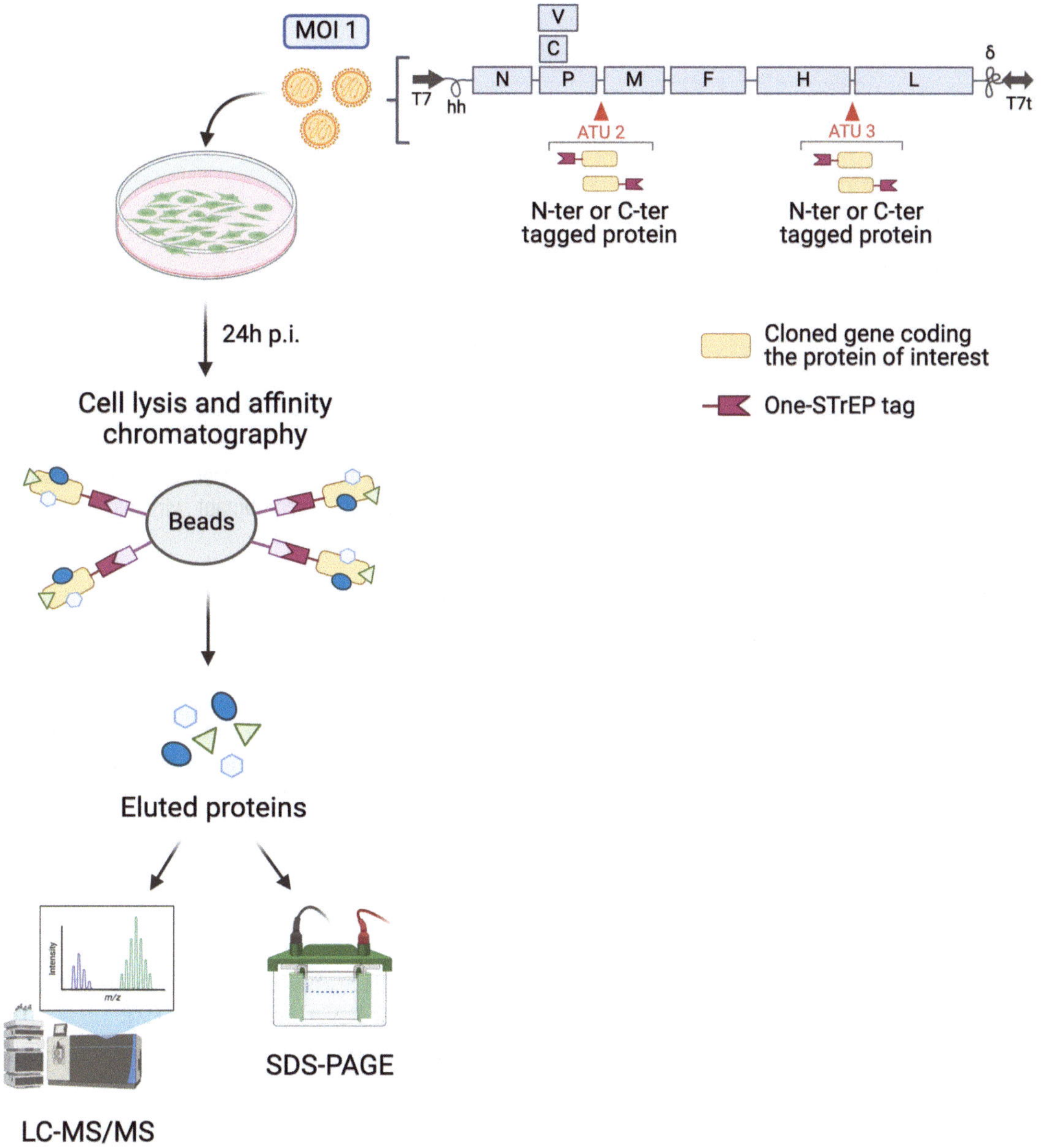

Fig. 2 Summary of the protocol used to detect protein-protein interactions. Suitable cell lines are infected at MOI 1 with recombinant measles virus expressing One-STrEP tagged proteins. Twenty-four hours postinfection, protein co-complexes are purified by affinity chromatography using StrepTactin Sepharose beads. Protein complexes are released from the beads using biotin buffer. Eluted proteins are analyzed by LC-MS/MS or resolved by SDS-PAGE

recombinant tagged viruses, cells are lysed and viral protein-specific co-complexes are purified by affinity chromatography for further study by bottom-up mass spectrometry-based proteomics (LC-MS/MS) and protein-protein interaction validation by Western blotting (WB).

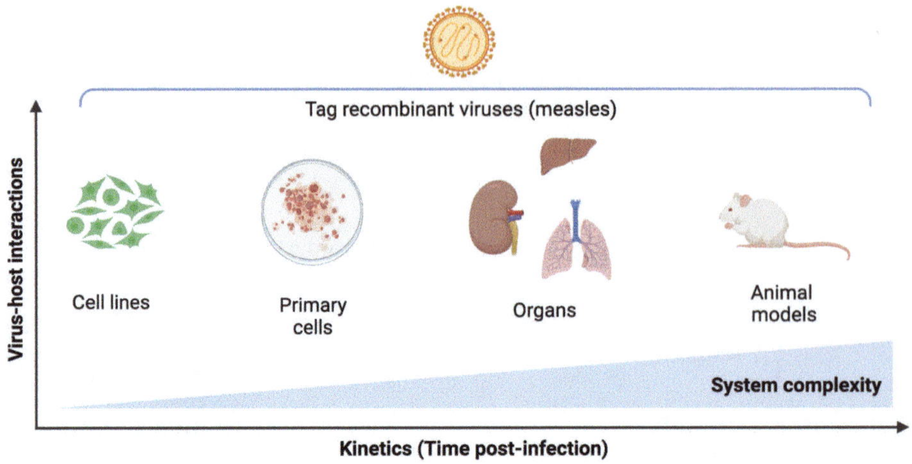

Fig. 3 Global diagram illustrating the variety of biological samples that can be used to study virus-host interactions with AP-MS described method following infection with recombinant viruses expressing One-STrEP tagged proteins of interest

This powerful approach can be applied to different cell lines, primary cells, organ cultures, or even animal model organisms to study protein-protein interactions in a viral infectious context (Fig. 3). Moreover, this protocol can be adjusted to study specific RNA-protein interactions [21]. Finally, by taking samples at specific time points, the kinetics of interactions can be analyzed giving important insights into host-viral interplay during a viral replication cycle.

2 Materials

2.1 Cell Culture

2.1.1 Cell Lines

1. Helper HEK293-T7-NP cells stably expressing N and P proteins from MV and phage T7 RNA polymerase [9].
2. Vero-NK cells (ATCC #CCL-81).
3. HEK293T (ATCC #CRL-3216) (*see* **Note 2**).

2.1.2 Media Culture

1. Dulbecco's Modified Eagle's Medium (DMEM) high glucose.
2. 100X penicillin-streptomycin (#15140122, Gibco).
3. Fetal bovine serum (FBS) (#MT35015CV, Corning).
4. Opti-MEM Reduced Serum Medium, GlutaMAX™ Supplement (#51985026, Gibco).
5. 1X DPBS.

2.2 Production of Recombinant MV

2.2.1 Plasmids

1. pTM2-STrEP-V: plasmid encoding for MV-V protein with an amino-terminal One-STrEP tag inserted into the pTM-MVSchw plasmid containing MV genome [19] (*see* **Note 3**).

2. pTM2-STrEP-CH: plasmid encoding for mCherry protein with an amino-terminal One-STrEP tag inserted into the pTM-MVSchw plasmid containing MV genome [19] (*see* **Notes 3** and **4**).
3. pEMC-La expressing the MV polymerase L gene [9].

2.2.2 Transfection Reagents

1. HEPES 2X (#51558, Sigma) passed by a 0.2 μm filter.
2. $CaCl_2$ 1 M.
3. DNase-/RNase-free distilled water.

2.3 Extraction of Protein Co-complexes

1. Recombinant tagged MV strains: rMV STrEP-V and rMV STrEP-CH.
2. Lysis buffer: 20 mM MOPS-KOH pH 7.4, 120 mM of KCl, 0.5% Igepal, 2 mM β-mercaptoethanol, supplemented with Complete Protease Inhibitor Cocktail (#11873580001, Roche) (*see* **Note 5**).
3. StrepTactin High Performance Sepharose beads (#28935599, GE Healthcare).
4. Washing buffer: 20 mM MOPS-KOH pH 7.4, 120 mM KCl, 2 mM β-mercaptoethanol supplemented with Complete Protease Inhibitor Cocktail (#11873580001, Roche) (*see* **Note 5**).
5. 10X StrepTactin elution buffer with Biotin (#2-1000-025, IBA): prepare 1X solution by diluting 1 volume of 10X StrepTactin elution buffer with nine volumes of DNase-/RNase-free distillated water.
6. 12% trichloroacetic acid (TCA).
7. Acetone.
8. 8 M urea solution.
9. 4% SDS solution.
10. DNase-/RNase-free distilled water.

2.4 Mass Spectrometry Analysis and Database Search

1. Denaturation buffer: 4 M urea (#U4883-6X25ML, Sigma-Aldrich), 100 mM NH_4HCO_3 (#09830, Fluka).
2. Sonicator: Ultrasound Technology UP200St equipped with the Vialtweeter Sonotrode, Hielscher.
3. Reduction of disulfide bond: 500-mM tris (2-carboxyethyl) phosphine, TCEP (#646547, Sigma-Aldrich).
4. Alkylation of Cysteine: 500-mM iodoacetamide (#I114, Sigma-Aldrich).
5. Digestion 1: ratio 50:1 (protein/enzyme) rLys-C (#V1671, Promega).
6. Digestion 2: ratio 50:1 (protein/enzyme) Sequencing Grade Modified Trypsin (#V5111, Promega).

7. Thermostatic enclosure: ThermoMixer® C equipped with a ThermoTop®, Eppendorf.
8. Stop solution: 4% formic acid (FA) (#94318, Fluka).
9. Elution buffer: 80% acetonitrile (ACN) (#412042, Carlo Erba Reagents), 20% water (#412092, Carlo Erba Reagents), 0.1% formic acid (FA) (#94318, Fluka).
10. Loading buffer: 2% acetonitrile (ACN) (#412042, Carlo Erba Reagents)—0.1% formic acid (FA) (#94318, Fluka)—98% water (#412092, Carlo Erba Reagents).
11. Speedvac: Concentrator plus, Eppendorf.
12. C18 Spin Columns Pierce™ (#89870, Thermo Fisher Scientific).
13. Q Exactive™ Plus Mass Spectrometer (Thermo Fisher Scientific).
14. Proxeon EASY-nLC 1000 (Thermo Fisher Scientific).
15. Column oven for Nano-ESI (#PRSO-V1, Sonation).
16. Protein LoBind® 1.5 mL tubes (#0030108116, Eppendorf).
17. C18 column homemade (#ReproSil-Pur Basic C18, 1.9 μm, Dr. Maisch GmbH).
18. Maxquant software using Andromeda as a search engine (freely available on https://www.maxquant.org/).
19. R packages: cp4p, imp4p, DAPAR, and limma, freely available on https://cran.rstudio.com/web/packages/cp4p/index.html, https://cran.rstudio.com/web/packages/imp4p/index.html, https://www.bioconductor.org/packages/release/bioc/html/DAPAR.html, and https://www.bioconductor.org/packages/release/bioc/html/limma.html.

2.5 Protein Extraction Validation

1. SDS-PAGE on 4–12% NuPAGE Bis-Tris gels (#WG1401BOX, Thermo Fisher Scientific).
2. Protein Ladder 10–250 kDa (#26619, Thermo Fisher Scientific).
3. MOPS Migration buffer (#NP0001, Thermo Fisher Scientific).
4. NuPAGE LDS Sample Buffer 4X (#NP0007, Thermo Fisher Scientific).
5. Antioxidant NuPAGE (#NP0005, Thermo Fisher Scientific).
6. NuPAGE Sample Reducing agent 10X (#NP0004, Thermo Fisher Scientific).
7. Cellulose membranes (#10600018, GE Healthcare).
8. Whatman paper Grade 3MM Chr (#3030-917, GE Healthcare).

9. TBS solution: 20 mM Tris pH 7.4, 140 mM NaCl, 0.1% Tween20.
10. Blocking buffer: TBS solution with 5% of skimmed milk.
11. Transfer buffer: 25 mM Tris pH 8.3, 192 mM glycine, 20% ethanol. Prechill transfer buffer to 4 °C.
12. High-current power supply.
13. Vertical protein electrophoresis chamber.
14. Ice block.
15. Magnetic stir bar.
16. Magnetic stirrer.
17. Wet transfer chamber with cassettes for the gel/membrane setup.
18. Fiber pads.
19. Monoclonal Anti-β-Actin antibody (#A5441, Sigma).
20. Strep-tag antibody (#34850, Qiagen).
21. Goat Anti-Mouse Immunoglobulins/HRP (#P0447, Dako).
22. SuperSignal West Pico PLUS Chemiluminescent Substrate (#34580, Thermo Fisher Scientific).

3 Methods

3.1 Rescue of Recombinant MVs by $CaCl_2$ Transfection

1. The day before transfection, plate HEK293T7-NP helper cells into 6-well plates. Cells need to be around 60% confluence the day of transfection (*see* **Note 6**).
2. Place all the reagents at room temperature before starting transfection.
3. Prepare the transfection reaction in a 1.5 mL Eppendorf tube by mixing the following:
   (a) 20 ng of pEMC-La plasmid and 5 μg of pTM2-STrEP-V or pTM2-STrEP-CH plasmid in a final volume of 131 μL with 44 μL of $CaCl_2$ 1 M (125 mM final concentration).
   (b) Add 175 μL of HEPES 2X.
4. Incubate for ≈ 1 min at room temperature.
5. Add the transfection mixture dropwise to the cells and incubate overnight (15 h) at 37 °C in a 5% CO_2 atmosphere.
6. The day after transfection, replace media culture with fresh DMEM supplemented with 10% FBS and 1% PS.
7. Perform a heat shock by incubating the cells at 42 °C for 3 h, and immediately return them to a 37 °C incubator.

8. Two days post transfection, prepare two 100 mm dishes with Vero cells to obtain 70–80% confluency the day after seeding.
9. Three days post transfection, proceed to a coculture step, and transfer HEK293T7-NP cells to 100 mm dishes with monolayers of Vero cells, and incubate for 2–7 days before collecting virus stock as soon as viral syncytia are observed in Vero cells (Fig. 1b).

3.2 Extraction of Virus-Host Protein Co-complexes

1. Seed two T150 flasks per condition 1 day before the experiment.
2. Infect HEK293T cells (3×10^7 cells per flask) with recombinant MV at multiplicity of infection (MOI) of 1 for 24 h (*see* **Note 7**). Include two T150 flasks with mock-infected cells (treated with media alone). Incubate at 37 °C in a 5% CO_2 atmosphere.
3. Twenty-four hours postinfection, place T150 flasks on ice, eliminate growth media, and wash cells twice with 10 mL of cold 1X DPBS.
4. Collect the infected cells by scraping them into 4 mL of lysis buffer per 1× T150 flask, and combine lysates for each sample in one 15 mL Falcon tube. Incubate tubes on ice for 20 min with gentle mixing every 5 min.
5. Clarify cell lysates by centrifugation at 16,000*g* for 15 min at 4 °C, and transfer clarified lysates to a new 15 mL Falcon tube. Take two 100 μL aliquots for SDS-PAGE analysis and mass spectrometry analysis of the input protein samples (*see* **Note 8**). Store samples at −80 °C.
6. Incubate clarified cell lysates for 2 h on a spinning wheel at 4 °C with 200 μL of StrepTactin Sepharose High Performance beads for 6×10^7 lysed cells (2× T150 flasks).
7. Collect Sepharose beads by centrifugation (1600*g* for 5 min at 4 °C), and wash twice for 5 min on a spinning wheel with 5 mL of washing buffer.
8. After the last wash, discard supernatant leaving 2 mL of it in the tube with the beads. Resuspend beads and transfer to 2 mL Eppendorf tubes, and then centrifuge again and proceed with elution step.
9. Perform a first elution with 600 μL of elution buffer per one 2 mL Eppendorf. Incubate for 15 min at 4 °C on a spinning wheel.
10. Centrifuge at 1600*g* for 5 min at 4 °C to collect the eluate.
11. Perform a second elution with 600 μL of elution buffer per one 2 mL Eppendorf. Repeat the previous steps and then mix the two eluates (total volume = 1200 μL).

12. Aliquot protein samples into three 1.5 mL Eppendorf tubes (400 μL in each tube).
13. Precipitate eluted proteins overnight at 4 °C with 55 μL of trichloroacetic acid (12% final concentration).
14. Wash protein pellets twice with ice-cold acetone and store on −80 °C (output protein samples for SDS-PAGE and mass spectrometry).

3.3 Assessment of Protein Interactors by LC-MS/MS

3.3.1 Digestion

1. Add 90 μL of denaturing buffer to the pellets to denature proteins and sonicate them.
2. Reduce with 10 μL of the reduction buffer and alkylate with 11.1 μL of the alkylation buffer for 1 h each at room temperature in the dark.
3. Digest proteins with digestion buffer 1 for 3 h/1000 rpm/37 °C, and then digest with digestion buffer 2 overnight/1000 rpm/37 °C.
4. Stop digestion with 22.2 μL of the stop buffer.
5. Desalt resulting peptides using C18 Spin Columns Pierce™, and centrifuge for 3 min at 1500*g*.
6. Elute peptides twice with 50 μL of elution buffer in 1.5 mL Protein LoBind® Eppendorf tubes, and centrifuge for 3 min at 1500*g*.
7. Dry peptides using the SpeedVac, and then resuspend in 15 μL of loading buffer.

3.3.2 MS Analysis

1. Perform an LC-MS/MS analysis of protein co-complexes on a Q ExactiveTM Plus Mass Spectrometer coupled with a Proxeon EASY-nLC 1000. Set the column temperature to 60 °C on the column oven for Nano-ESI.
2. Inject peptides onto a homemade 50 cm C18 column, and elute with a multistep gradient from 2% to 27% ACN in 105 min, 27% to 50% ACN in 40 min, and 50% to 60% ACN in 10 min, at a flow rate of 250 nL/min over 185 min.
3. Acquire MS data using Xcalibur software with a Top10 data-dependent method.
4. Set MS scans at a resolution of 70,000 with an AGC of 3E+06 and a maximum injection time of 20 ms.
5. Set MS/MS scans at a resolution of 17,500 with an AGC of 1E+06 and a maximum injection time of 60 ms. The isolation window is set to 1.6 m/z and normalized collision energy of 28 for HCD fragmentation.
6. Set an underfill ratio of 1.0% for an intensity threshold of 1.7E+05.

7. Reject unassigned precursor ion charge states as well as 1, 7, 8, and >8 charged states.
8. Disable peptide match.
9. Set a dynamic exclusion of 45 s.

3.3.3 Database Search

1. Analyze RAW data using MaxQuant software including the appropriate database of the virus and the host. Proteomes fasta files are available on Uniprot: UP000180902 for MV Schwarz vaccine strain (https://www.uniprot.org/proteomes/UP000180902) and UP000005640 for human reference proteome (https://www.uniprot.org/proteomes/UP000005640).
2. Set variable (methionine oxidation and N-terminal acetylation) and fix (cysteine carbamidomethylation) modifications.
3. Allow a maximum of two missed cleavages for theoretical tryptic peptides with a minimum peptide length of seven amino acids.
4. Adjust the false discovery rate (FDR) for peptide and protein identification to 0.01. A unique peptide per protein group is required for the identification of protein.
5. Set the main search peptide tolerance to 4.5 ppm and to 20 ppm for the MS/MS match tolerance.
6. Enable the second peptides to identify co-fragmentation events.
7. Select the match between runs (MBR) option with a match time window of 0.7 min for an alignment time window of 20 min (*see* **Note 9**).
8. Perform the quantification using the XIC-based LFQ algorithm with the Fast LFQ mode. Accept unique and razor peptides, included modified peptides, with at least two ratio counts for quantification (*see* **Note 10**).

3.3.4 Statistical Analysis

1. Remove the proteins categorized as "reverse" and "potential contaminants" in MaxQuant outputs.
2. Only consider proteins with at least three quantified values in an experimental condition for further analysis (*see* **Note 11**).
3. Process intensities of proteins using a log2 transformation, and normalize them by median centering within conditions or any other normalization method available in the R package DAPAR.
4. Consider proteins that are quantified in only one condition out of the two compared conditions as "differentially abundant proteins."

5. Fill missing values of the remaining proteins using the R package imp4p.
6. Perform the statistical analysis using a paired t-test with the limma R package.
7. Perform an adaptive Benjamini-Hochberg correction of the *p* values using the R package cp4p.
8. Consider proteins that are associated to a fold change above 2.0 and to an adjusted *p*-value inferior to FDR of 1% as "significantly differentially abundant proteins."
9. Generate the lists of "differentially abundant proteins": the proteins that appear to be only quantified in a condition (**Step 4**) and the proteins that meet the fold change and FDR thresholds (**Step 8**).

3.4 Protein-Protein Interaction Validation by WB Analysis

SDS-PAGE analysis of protein extracts before (input: sample acquired in Subheading 3.2 after **step 5**) and after (output: sample acquired in Subheading 3.2 after **step 14**) affinity chromatography purification.

1. Resuspend output sample in 30 μL 8 M urea, 4% SDS final solution.
2. Prepare protein samples for loading by mixing 5 μL of sample buffer with 2 μL of reducing agent and 13 μL of the output or input samples (20 μL final volume).
3. Heat for 5 min at 98 °C.
4. Load 20 μL of the sample onto 4–12% NuPAGE Bis-Tris gel and 10 μL of a Protein Ladder.
5. Add 1 mL of antioxidant for each 400 mL of MOPS 1× running buffer, and run the gel at 100–120 V for 1–1.5 h.
6. Transfer proteins to cellulose membrane at 100 V for 1 h. Pre-wet in chilled transfer buffer each component before setting up the gel/membrane. Set up the gel/membrane sandwich by placing into the cassette in this order (from cathode to anode): a fiber pad, a Whatman paper, the gel, the cellulose membrane, a Whatman paper, and a fiber pad (Fig. 4) (*see* **Note 12**). Place the transfer chamber with a magnetic stir bar and an ice block inside, on the top of a magnetic stirrer, and procced to transfer.
7. Block membranes in TBS buffer with 5% skimmed milk for 30 min at room temperature.
8. Incubate for 1 h with α-Strep-Tag antibody (1:10,000 dilution in TBS 5% skimmed milk) at room temperature and gentle agitation (*see* **Note 13**).
9. Wash membranes twice with TBS buffer for 5 min at room temperature.

Fig. 4 Diagram of Western blot transfer stack

10. Incubate with HRP-conjugated secondary antibody (1:10,000 dilution in TBS 5% skimmed milk) for 1 h at room temperature with gentle agitation.
11. Wash membranes twice with TBS buffer for 5 min at room temperature.
12. Prepare SuperSignal West Pico PLUS Chemiluminescent Substrate by mixing equal parts of both components.
13. Incubate the membrane with the substrate working solution for 5 min at room temperature.
14. Image the membrane using an imaging system or an X-ray film.

4 Notes

1. ATU positions differ in their expression level due to the paramyxovirus (and other mononegavirales [22, 23]) mechanism of transcription. Viral genes are transcribed in succession from the 5′ to 3′ extremities. L polymerase stops and restarts transcription at intergenic regions. However, as the efficiency of re-initialization is not 100%, mRNAs are produced in decreasing amounts from the 5′ to 3′ end of the genome. Consequently, a gradient of viral mRNA transcription is observed from N to L genes [24, 25]. For this reason, cloning in the ATU1 position can lead to gene overexpression and interfere with the replication of the recombinant MV (lower viral fitness compared to the parental Schwarz strain). ATU2 position is higher expressed than ATU3 but less than ATU1, and these aspects need to be considered when deciding position of a gene insertion into the MV vector.
2. The protocol can be adapted to other cell lines as HeLa cells (ATCC #CCL-2) or A549 cells (ATCC #CCL-185).
3. Measure DNA concentration by spectrophotometry, and adjust to about 1 μg/μL. It is important to obtain high-purity transfection-grade plasmids, since residual ethanol, solvents, or other compounds can inhibit the successful rescue of recombinant viruses.

4. Negative control virus expressing a nonrelevant protein (e.g., mCherry) should be used in parallel in each experiment to assess nonspecific protein-protein interactions.
5. MOPS Lysis and washing buffers should be kept at 4 °C protected from the light. β-Mercaptoethanol and Complete Protease Inhibitor Cocktail should be added just before the experiment.
6. The confluence of HEK293T7-NP cells during transfection for virus rescue should be between 50% and 70%. This ensures that they can still grow exponentially.
7. Additional setting experiments to determine the percentage of infected cells with different MOIs should be performed to have from 50% to 90% infected cells 24 h postinfection (*see* **Note 2**).
8. The input sample for mass spectrometry should be treated following **steps 13** and **14** of Subheading 3.2.
9. Set a distinct fraction number between the two conditions to avoid misleading identification with MBR option.
10. Because the bait and mock conditions are not similar, it is better to use skip normalization.
11. Each experimental condition (peptides derived from rMV STrEP-V-infected cells, peptides derived from rMV STrEP-CH-infected cells, and peptides derived from noninfected cells) should be tested in triplicate, and proteins with at least three quantified intensity values should be retained for analysis.
12. When putting the cellulose membrane on top of the gel, be sure to eliminate any air bubbles that may be trapped underneath.
13. Use Anti-β-Actin at 1:10,000 as a total protein loading control.

Acknowledgments The authors would like to thank all past and present members of the Innovation Lab: Vaccines and the RNA and immunity laboratory. This work was supported by Institut Pasteur grants and BECAL Paraguay PhD grant to HV. Figures 2 and 3 were created with BioRender.com.

Conflict of Interest The authors declare no conflicts of interest.

References

1. Fodor E, Devenish L, Engelhardt OG, Palese P, Brownlee GG, Garcia-Sastre A (1999) Rescue of influenza A virus from recombinant DNA. J Virol 73(11): 9679–9682. https://doi.org/10.1128/jvi.73.11.9679-9682.1999
2. Nakaya T, Cros J, Park MS, Nakaya Y, Zheng HY, Sagrera A, Villar E, Garcia-Sastre A, Palese P (2001) Recombinant Newcastle disease virus as a vaccine vector. J Virol 75(23): 11868–11873. https://doi.org/10.1128/jvi.75.23.11868-11873.2001

3. Yun SI, Kim SY, Rice CM, Lee YM (2003) Development and application of a reverse genetics system for Japanese encephalitis virus. J Virol 77(11):6450–6465. https://doi.org/10.1128/jvi.77.11.6450-6465.2003
4. Billeter MA, Naim HY, Udem SA (2009) Reverse genetics of measles virus and resulting multivalent recombinant vaccines: applications of recombinant measles viruses. Curr Top Microbiol Immunol 329:129–162. https://doi.org/10.1007/978-3-540-70523-9_7
5. Isel C, Munier S, Naffakh N (2016) Experimental approaches to study genome packaging of influenza A viruses. Viruses 8(8):218. https://doi.org/10.3390/v8080218
6. Neumann G, Kawaoka Y (2001) Reverse genetics of influenza virus. Virology 287(2): 243–250. https://doi.org/10.1006/viro.2001.1008
7. Stobart CC, Moore ML (2014) RNA virus reverse genetics and vaccine design. Viruses 6(7):2531–2550. https://doi.org/10.3390/v6072531
8. Radecke F, Spielhofer P, Schneider H, Kaelin K, Huber M, Dotsch C, Christiansen G, Billeter MA (1995) Rescue of measles viruses from cloned DNA. EMBO J 14(23):5773–5784. https://doi.org/10.1002/j.1460-2075.1995.tb00266.x
9. Combredet C, Labrousse V, Mollet L, Lorin C, Delebecque F, Hurtrel B, McClure H, Feinberg MB, Brahic M, Tangy F (2003) A molecularly cloned Schwarz strain of measles virus vaccine induces strong immune responses in macaques and transgenic mice. J Virol 77(21): 11546–11554. https://doi.org/10.1128/jvi.77.21.11546-11554.2003
10. Mahmoudi Gomari M, Saraygord-Afshari N, Farsimadan M, Rostami N, Aghamiri S, Farajollahi MM (2020) Opportunities and challenges of the tag-assisted protein purification techniques: applications in the pharmaceutical industry. Biotechnol Adv 45:107653. https://doi.org/10.1016/j.biotechadv.2020.107653
11. Ohashi M, Holthaus AM, Calderwood MA, Lai C-Y, Krastins B, Sarracino D, Johannsen E (2015) The EBNA3 family of Epstein-Barr virus nuclear proteins associates with the USP46/USP12 deubiquitination complexes to regulate lymphoblastoid cell line growth. PLoS Pathog 11(4):e1004822. https://doi.org/10.1371/journal.ppat.1004822
12. Fan J, Xiao P, Kong D, Liu X, Meng L, An T, Cai X, Wang H, Yu L (2022) Engineering His-tagged Senecavirus A for one-step purification of viral antigens. Vaccines (Basel) 10(2): 170. https://doi.org/10.3390/vaccines10020170
13. Ma X, Li C, Xia Q, Zhang Y, Yang Y, Wahaab A, Liu K, Li Z, Li B, Qiu Y, Wei J, Ma Z (2022) Construction of a recombinant Japanese encephalitis virus with a hemagglutinin-tagged NS2A: a model for an analysis of biological characteristics and functions of NS2A during viral infection. Viruses 14(4):706. https://doi.org/10.3390/v14040706
14. Skerra A, Schmidt TGM (1999) Applications of a peptide ligand for streptavidin: the Strep-tag. Biomol Eng 16(1–4):79–86. https://doi.org/10.1016/s1050-3862(99)00033-9
15. Mayer D, Molawi K, Martinez-Sobrido L, Ghanem A, Thomas S, Baginsky S, Grossmann J, Garcia-Sastre A, Schwemmle M (2007) Identification of cellular interaction partners of the influenza virus ribonucleoprotein complex and polymerase complex using proteomic-based approaches. J Proteome Res 6(2):672–682. https://doi.org/10.1021/pr060432u
16. De Maio FA, Risso G, Iglesias NG, Shah P, Pozzi B, Gebhard LG, Mammi P, Mancini E, Yanovsky MJ, Andino R, Krogan N, Srebrow A, Gamarnik AV (2016) The dengue virus NS5 protein intrudes in the cellular spliceosome and modulates splicing. PLoS Pathog 12(8):e1005841. https://doi.org/10.1371/journal.ppat.1005841
17. Fournier G, Chiang C, Munier S, Tomoiu A, Demeret C, Vidalain PO, Jacob Y, Naffakh N (2014) Recruitment of RED-SMU1 complex by Influenza A virus RNA polymerase to control viral mRNA splicing. PLoS Pathog 10(6): e1004164. https://doi.org/10.1371/journal.ppat.1004164
18. Li Y, Frederick KM, Haverland NA, Ciborowski P, Belshan M (2016) Investigation of the HIV-1 matrix interactome during virus replication. Proteomics Clin Appl 10(2): 156–163. https://doi.org/10.1002/prca.201400189
19. Komarova AV, Combredet C, Meyniel-Schicklin L, Chapelle M, Caignard G, Camadro JM, Lotteau V, Vidalain PO, Tangy F (2011) Proteomic analysis of virus-host interactions in an infectious context using recombinant viruses. Mol Cell Proteomics 10(12): M110.007443. https://doi.org/10.1074/mcp.M110.007443
20. Meignie A, Combredet C, Santolini M, Kovacs IA, Douche T, Gianetto QG, Eun H, Matondo M, Jacob Y, Grailhe R, Tangy F, Komarova AV (2021) Proteomic analysis uncovers measles virus protein C interaction with p65-iASPP protein complex. Mol Cell

Proteomics 20:100049. https://doi.org/10.1016/j.mcpro.2021.100049

21. Komarova AV, Combredet C, Sismeiro O, Dillies MA, Jagla B, Sanchez David RY, Vabret N, Coppee JY, Vidalain PO, Tangy F (2013) Identification of RNA partners of viral proteins in infected cells. RNA Biol 10(6): 944–956. https://doi.org/10.4161/rna.24453
22. Iverson LE, Rose JK (1981) Localized attenuation and discontinuous synthesis during vesicular stomatitis virus transcription. Cell 23(2): 477–484. https://doi.org/10.1016/0092-8674(81)90143-4
23. Muhlberger E (2007) Filovirus replication and transcription. Future Virol 2(2):205–215. https://doi.org/10.2217/17460794.2.2.205
24. Cattaneo R, Rebmann G, Schmid A, Baczko K, ter Meulen V, Billeter MA (1987) Altered transcription of a defective measles virus genome derived from a diseased human brain. EMBO J 6(3):681–688. https://doi.org/10.1002/j.1460-2075.1987.tb04808.x
25. Plumet S, Duprex WP, Gerlier D (2005) Dynamics of viral RNA synthesis during measles virus infection. J Virol 79(11):6900–6908. https://doi.org/10.1128/JVI.79.11.6900-6908.2005

Chapter 8

Identification of Host Factors of Paramyxoviruses by siRNA Genome-Wide Screens

Biaoguo Yan, Kristmundur Sigmundsson, Yaw Shin Ooi, and Danielle E. Anderson

Abstract

Measles is a highly infectious disease that continues to spread mainly in developing countries, often resulting in child mortality. Despite the existence of effective vaccines, no specific antivirals are available as targeted therapy to combat measles virus (MeV). The implementation of genome-wide siRNA screens can provide a powerful platform to discover host factors that mediate MeV infection and replication, which could be essential to develop novel therapeutic strategies against this disease. Here, we describe a human genome-wide siRNA screen for MeV.

Key words *Paramyxoviridae*, siRNA, Genome-wide screen, Morbillivirus

1 Introduction

Measles virus (MeV) is a negative-sense, single-stranded RNA virus classified in the family *Paramyxoviridae*, in the genus *Morbillivirus* [1]. Measles is an extremely contagious disease transmitted by respiratory droplets, typically leading to systemic spread in multiple organs such as the hepatic, renal, integumentary, and central nervous system [2, 3]. Despite effective vaccines, measles outbreaks remain a health concern in developing nations. In 2019, the World Health Organization estimated 207,500 measles-related deaths, primarily afflicting children under the age of 5 [4].

As obligate intracellular parasites, viruses depend heavily on host cell machinery for replication and transmission, and genome-wide siRNA screens have provided new insights into the functional interactions between viruses and host cells [5–8]. Comparative genome-wide siRNA screens identified genes coding for factors in vesicular transport, RNA processing, and translation pathways as critical for measles, mumps, and respiratory syncytial virus replication [9]. In particular, the ATP-binding cassette transporter

Dzwokai Z. Ma and Christian K. Pfaller (eds.), *Measles and Related Morbilliviruses: Methods and Protocols*,
Methods in Molecular Biology, vol. 2808, https://doi.org/10.1007/978-1-0716-3870-5_8,

ABCE1 of the translation pathway was found to be a crucial host factor required by these paramyxoviruses. Other siRNA screens have identified essential host factors required for negative-stranded RNA viruses such as vesicular stomatitis virus (VSV), human parainfluenza virus 3 (HPIV3), and Ebola virus (EBOV) [10, 11]. Hence, functional genomic screens using siRNA can narrow knowledge gaps of virus-host interactions and provide an avenue for the discovery of novel cellular targets that can be exploited for developing host-directed therapeutics.

MeV is the prototype morbillivirus with high clinical relevance [1, 12], and this chapter will detail a genome-wide siRNA screen for MeV.

2 Materials

2.1 Equipment

1. Multidrop Combi Dispenser (Thermo Scientific, 5840300).
2. Dispensing cassette—small tube (Thermo Scientific, 24073290).
3. Dispensing cassette—standard tube (Thermo Scientific, 24072670).
4. TC20 automated cell counter.
5. Operetta High Content Microscopy and Imaging System (Perkin Elmer).
6. XL3000 Xeon Fiber-Optic Light Source (Excelitas Technologies).

2.2 Consumables

1. 70% ethanol.
2. 50 mL tubes.
3. A549 (ATCC CCL-185) cells engineered to express human SLAM (*see* **Note 1**).
4. Measles virus (PMID: 31088929) (*see* **Note 2**).
5. Growth media: Dulbecco's Modified Eagle Medium (DMEM) supplemented with 5% FBS and 1× pen/strep.
6. Trypsin 0.25% EDTA.
7. 1X phosphate buffered saline (PBS), without Ca^{2+} and Mg^{2+}.
8. Opti-MEM media.
9. Deionized water (dH_2O).
10. 75 cm^2 tissue culture flasks.

2.3 Transfection

1. Human Genome siRNA library (Invitrogen, 4397926).
2. Invitrogen Silencer Select Negative Control #1 (Invitrogen, 4390843).

3. Black, optically clear bottom 384-well plates (PerkinElmer, 6007460).
4. Lipofectamine RNAiMAX (Invitrogen, 13778150).
5. MeV-N siRNA (antisense sequence: 5′-GGUUCGGAUGG UCGGGAA-3′).
6. MeV-L siRNA (antisense sequence: 5′-GGACAUCAAGGAG AAAAUUAU-3′).
7. Aluminum foil plate seals.

2.4 Fixation and Staining

1. PBS/0.1% Tween-20 (PBS-T).
2. PBS/0.5% Triton-X.
3. 36% formaldehyde.
4. 1% BSA in PBS-T.
5. Hoechst 33342 dye (1:1000 dilution of 1 mg/mL stock in PBS-T).

3 Methods

3.1 siRNA Library

Human genome siRNA libraries are commercially available, and the plating format of the library can be customized. The Silencer Select Human Genome Library is described herein. The Silencer Select Human Genome Library covers 21,584 human genes, with 3 unique siRNAs per gene. The library can be purchased in 384-well plates or 96-well plates. There are 186 384-well plates or 738 96-well plates in the respective library formats. Each 384-well plate has the last two columns empty, while the 96-well plate has the last column empty. This method will describe procedures for 384-well plate library format, using a siRNA volume of 5 μL per well. If the library is purchased in master plates, dilute the siRNA library to the desired concentration to a working volume of 5 μL in deionized water. Do not use DEPC-treated water. Arraying the library into 384-well plates can be performed with robotics, or manually.

3.2 Reverse Transfection

3.2.1 Preparation of Transfection Mixture

1. Remove eight assay plates from the freezer, and defrost at room temperature for 30 min (*see* **Note 3**).
2. Record plate barcodes.

Assay plate barcode tracking:

Plate barcode	1)
Plate barcode	2)
Plate barcode	3)

(continued)

Plate barcode	4)
Plate barcode	5)
Plate barcode	6)
Plate barcode	7)
Plate barcode	8)

3. Warm growth media, trypsin (0.25%), 1X PBS, and OptiMEM at 37 °C.
4. Sterilize biosafety cabinet with 70% ethanol.
5. Aliquot 200 mL 70% ethanol into a bottle. Retrieve a dispensing cassette (small tube). Run 70% ethanol through the dispensing cassette.
6. Centrifuge assay plates at room temperature at 500 × *g* for 1 min.
7. Sterilize the outside of the assay plates with 70% ethanol, and transfer to the biosafety cabinet.
8. Dilute the MeV-N and MeV-L positive controls and Invitrogen Silencer Select Negative Control #1 to the same concentration as the diluted siRNA library with deionized water.
9. Remove the seal assay plate carefully.
10. Add controls to each assay plate in quadruplicate (Fig. 1).
    (a) Add 5 μL NSC siRNA to wells C23, D23, E23, F23.
    (b) Add 5 μL siMeV-N to wells G23, H23, I23, J23.
    (c) Add 5 μL siMeV-L to wells K23, L23, M23, N23.

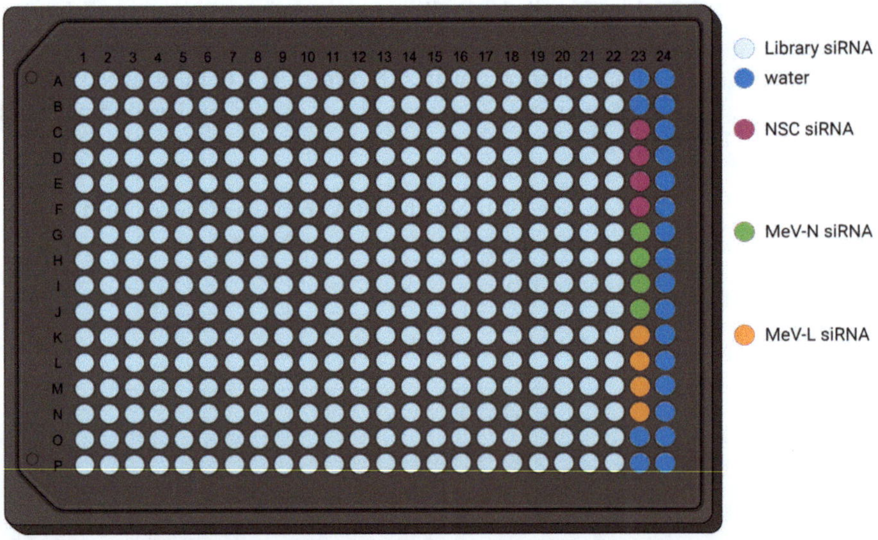

Fig. 1 Plate format. Layout of each 384-well plate indicating which wells contain library, non-silencing control (NSC) and measles virus (MeV) siRNA, or water

11. Centrifuge assay plates 500 × *g* for 1 min.
12. Prepare the transfection mix for eight assay plates:

8 assay plates × 384 wells per plate × 0.01 mL per well = 30.72 mL

Note: Prepare transfection mix in excess to account for dispensing cassette priming:

Calculation for assay plates:

Number of plates	OptiMEM (mL)	RNAi max (μL)	Total volume (mL)
1	3.82	19.2	3.84
9	33.38	172.8	33.55

13. Purge the 70% ethanol from the dispensing cassette.
14. Submerge the tubing from the dispensing cassette 1X PBS, and run 15 mL of 1X PBS through the line to rinse out the remaining ethanol.
15. Before dispensing the transfection mix, ensure there are no air bubbles or obstructions in the tubing.
16. Submerge the tubing from the dispensing cassette into the transfection mix. Press “Prime” until the tubing is completely filled. Minimally prime the line and do not waste transfection mix.
17. Set the Multidrop Combi Dispenser to dispense 10 μL per well for all columns.
18. Load the first assay plate. Ensure that “A1” is in the upper left. Remove the lid and press start.
19. Replace the lid of the first plate and set aside. Repeat steps 18 and 19 for the remaining assay plates.
20. Record the time the transfection mix is dispensed on Plate 1: ________.
21. Clean Multidrop Combi Dispenser cassette.
    (a) Run 15 mL 1X PBS through the dispensing cassette.
    (b) Run 15 mL 70% ethanol through the dispensing cassette, and leave the ethanol in the tubing for 10 min.
    (c) Remove the ethanol from the tubing.
22. Centrifuge all assay plates at 800 × *g* for 20 s.
23. Incubate assay plates for 15 min at room temperature for transfection complex formation.
24. During incubation, proceed to the next step for cell suspension preparation.

3.2.2 Preparation of A549-hSLAM Cell Suspension

1. Retrieve a dispensing cassette (standard tube). Run 70% ethanol through the dispensing cassette and then 15 mL 1X PBS through the lines. Perform all steps in the BSC.
2. Retrieve 75 cm^2 tissue culture flasks containing A549-hSLAM cells.
3. Remove media from cells, and rinse each flask with 4 mL 1X PBS.
4. Add 2 mL of trypsin into each flask and incubate at 37 °C. When cells start to round up and before detachment, gently remove the trypsin.
5. Gently tap the flasks to dislodge the cells.
6. Add 5 mL of DMEM (5% FBS, 1% PS) to each flask. Pipette vigorously several times to disassociate cells.
7. Remove 0.5 mL of cell slurry and count on a hemocytometer or automatic cell counter.

Cell count:

Hemocytometer				
Quadrant I	**Quadrant II**	**Quadrant III**	**Quadrant IV**	**Average**
				______ $\times 10^4$ cells/mL

8. Prepare the cell suspension for eight assay plates:

8 assay plates × 384 wells per plate × 0.05 mL per well = 153.6 mL

9. In a sterile 500 mL bottle, prepare cell suspension in excess to account for dispensing cassette priming:

Cell number calculation:

Number of plates	Cells required (2000 cell/well)	Total volume of cells/mL
1	7.68×10^5	19.2
9	6.91×10^6	172.8

10. Submerge the tubing from the dispensing cassette into the cell suspension. Press "Prime" until the tubing is completely filled. Minimally prime the line.
11. Set the Multidrop Combi Dispenser to dispense 50 μL per well for all columns.
12. Gently mix the cell suspension.

13. Load the first assay plate. Ensure that "A1" is in the upper left. Remove the lid and press start.
14. Repeat **steps 13** and **14** for all remaining assay plates.
15. Record time completed (of the last plate seeded): ________________.
16. Place all assay plates in an incubator at 37 °C, 5% CO_2. To ensure equal heat distribution across the plates, do not stack plates, but incubate in a single layer.
17. Incubate for 48 h at 37 °C, 5% CO_2.
18. Clean Multidrop Combi Dispenser cassette.
    (a) Run 15 mL 1X PBS through the dispensing cassette.
    (b) Run 15 mL 70% ethanol through the dispensing cassette and leave the ethanol in the tubing for 10 min.
    (c) Remove the ethanol from the tubing.

3.3 MeV Infection

1. Retrieve a dispensing cassette (small tube). Run 70% ethanol through the dispensing cassette and then 15 mL 1X PBS through the lines. Perform all steps in the BSC.
2. Set the Multidrop Combi Dispenser to dispense 20 μL per well for all columns.
3. Inspect that there are no air bubbles or obstructions in the cassette tubing.
4. Thaw a vial of MeV on ice (with known titer).
5. Prepare the virus suspension mixture for eight assay plates:

$$8 \text{ assay plates} \times 384 \text{ wells per plate} \times 0.02 \text{ mL per well} = 30.72 \text{ mL}$$

6. In a sterile bottle, prepare diluted virus suspension of the desired MOI in excess to account for dispensing cassette priming (*see* **Note 4**).

$$\text{Volume of virus stock (in mL)} = \frac{\text{Total number wells} \times \text{cells per well after 48 h} \times \text{MOI}}{\text{Virus titre}}.$$

$$\text{Volume of virus stock (in mL)} = \frac{384 \times 4000 \times 0.1}{1,000,000} (\textit{see } \textbf{Note 5}).$$

Virus dilution:

Number of plates	Virus (mL)	Media (mL)	Total volume (mL)
1	0.153	7.52	7.68
9	1.38	67.74	69.12

7. Submerge the tubing from the dispensing cassette into the cell suspension. Press "Prime" until the tubing is completely filled. Minimally prime the line.
8. Set the Multidrop Combi Dispenser to dispense 20 μL per well for all columns.
9. Gently mix the virus suspension.
10. Load the first assay plate. Ensure that "A1" is in the upper left. Remove the lid and press start.
11. Repeat **steps 10** and **11** for all remaining assay plates.
12. Time completed (of the last plate infected): ________________.
13. Place all assay plates in an incubator at 37 °C, 5% CO_2. To ensure equal heat distribution across the plates, do not stack plates, but incubate in a single layer.
14. Incubate for 48 h at 37 °C, 5% CO_2.
15. Clean Multidrop Combi Dispenser cassette.
    (a) Run 15 mL 1X PBS through the dispensing cassette.
    (b) Run 15 mL 70% ethanol through the dispensing cassette, and leave the ethanol in the tubing for 10 min.
    (c) Remove the ethanol from the tubing.

3.4 Fixation

1. Retrieve a dispensing cassette (standard tube). Run 70% ethanol through the dispensing cassette and then 15 mL 1X PBS through the lines. Perform all steps in the BSC until virus is inactivated by PFA.
2. Inspect that there are no air bubbles or obstructions in the cassette tubing.
3. Submerge the tubing from the dispensing cassette into 36% PFA solution. Press "Prime" until the tubing is completely filled. Minimally prime the line.
4. Set the Multidrop Combi Dispenser to dispense 11 μL per well for all columns (*see* **Note 6**).
5. Load the first assay plate. Ensure that "A1" is in the upper left. Remove the lid and press start.
6. Record time completed (of the first plate fixed): ________________.

Note: Start timer after 36% formaldehyde solution is dispensed in the first plate.

7. Repeat **steps 5** and **6** for all remaining assay plates.
8. Fix the plates in the BSC at room temperature for 15 min.
9. Clean Multidrop Combi Dispenser cassette.

(a) Run 15 mL 1X PBS through the dispensing cassette.
(b) Run 15 mL 70% ethanol through the dispensing cassette, and leave the ethanol in the tubing for 10 min.
(c) Remove the ethanol from the tubing.

10. After incubation, flick the 384-well plate to remove liquid from wells (*see* **Note 7**).

3.5 Staining

1. Retrieve a dispensing cassette (standard tube). Run 70% ethanol through the dispensing cassette and then 15 mL 1X PBS through the lines.
2. Inspect that there are no air bubbles or obstructions in the cassette tubing.
3. Submerge the tubing from the dispensing cassette into PBS/0.1% Tween-20 (PBS-T). Press "Prime" until the tubing is completely filled. Minimally prime the line.
4. Set the Multidrop Combi Dispenser to dispense 50 μL per well for all columns.
5. Flick the 384-well plate to remove liquid from wells.
6. Repeat the wash step three times per plate.
7. Run 15 mL 1X PBS through the Multidrop Combi Dispenser cassette.
8. Submerge the tubing from the dispensing cassette into PBS/0.5% Triton-X. Press "Prime" until the tubing is completely filled. Minimally prime the line.
9. Set the Multidrop Combi Dispenser to dispense 50 μL per well for all columns.
10. Permeabilize the plates at room temperature for 30 min.
11. Time completed (of the first plate fixed): ______________.

Note: Start timer after PBS/0.5% Triton-X is dispensed in the first plate.

12. Flick the 384-well plate to remove liquid from wells.
13. Run 15 mL 1X PBS through the Multidrop Combi Dispenser cassette.
14. Submerge the tubing from the dispensing cassette into PBS/0.1% Tween-20 (PBS-T). Press "Prime" until the tubing is completely filled. Minimally prime the line.
15. Set the Multidrop Combi Dispenser to dispense 50 μL per well for all columns.
16. Flick the 384-well plate to remove liquid from wells.
17. Repeat the wash step three times per plate.

18. Run 15 mL 1X PBS through the Multidrop Combi Dispenser cassette.
19. Submerge the tubing from the dispensing cassette into Hoechst (1:1000 dilution of 1 mg/mL stock in PBS-T). Press "Prime" until the tubing is completely filled. Minimally prime the line (*see* **Note 8**).
20. Set the Multidrop Combi Dispenser to dispense 30 μL per well for all columns.
21. Stain the plates at room temperature for 1 h in the dark.
22. Time completed (of the first plate fixed): ________________.

    Note: Start timer after Hoechst is dispensed in the first plate.

23. Flick the 384-well plate to remove liquid from wells.
24. Dispense 50 μL per well of 1% BSA in PBS-T for all columns.
25. Incubate at room temperature for 10 min.
26. Time completed (of the first plate): ________________.

Note: Start timer after 1% BSA in PBST is dispensed in the first plate.

27. Run 15 mL 1X PBS through the Multidrop Combi Dispenser cassette.
28. Submerge the tubing from the dispensing cassette into PBS/0.1% Tween-20 (PBS-T). Press "Prime" until the tubing is completely filled. Minimally prime the line.
29. Set the Multidrop Combi Dispenser to dispense 50 μL per well for all columns.
30. Flick the 384-well plate to remove liquid from wells.
31. Repeat the wash step three times per plate.
32. Run 15 mL 1X PBS through the Multidrop Combi Dispenser cassette.
33. Submerge the tubing from the dispensing cassette into 1X PBS. Press "Prime" until the tubing is completely filled. Minimally prime the line.
34. Set the Multidrop Combi Dispenser to dispense 50 μL per well for all columns.
35. Cover with foil and store at 4 °C until imaging.
36. Clean Multidrop Combi Dispenser cassette.
    (a) Run 15 mL 1X PBS through the dispensing cassette.
    (b) Run 15 mL 70% ethanol through the dispensing cassette, and leave the ethanol in the tubing for 10 min.
    (c) Remove the ethanol from the tubing.

3.6 Imaging

1. Set the high content microscope to image wells (*see* **Note 9**).
2. Set the microscope to a regular mode and a lamp excitation power of 50%. Two fluorescence channels were applied for each image frame, with the following setting shown to provide image data of the optimal signal to noise ratio:

		Bandpass filters (nm)		
Target	**Fluorophore**	**Excitation**	**Emission**	**Exposure (ms)**
Nuclei	Hoechst	360–400	410–480	50
MeV infected cells	GFP	460–490	500–550	100

3. Use a 10X objective to capture nine image fields per well (Fig. 2). This arrangement enables the imaging of the entire area of each well (*see* **Note 10**).

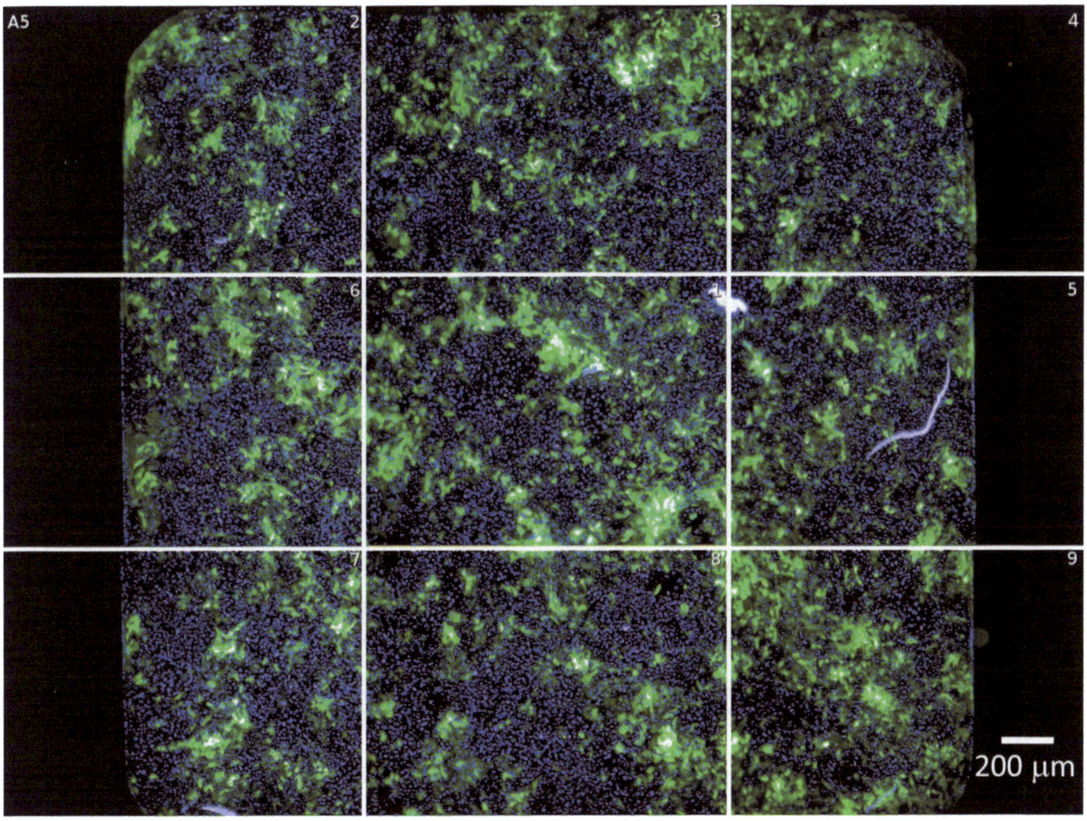

Fig. 2 Overview of 1 well of a 384-well plate. Hoechst stained nuclei are shown in blue, and MeV infected cells (expressing GFP) are shown in green. The combined nine images demonstrate that cell density and infected cells are not equally distributed within the well

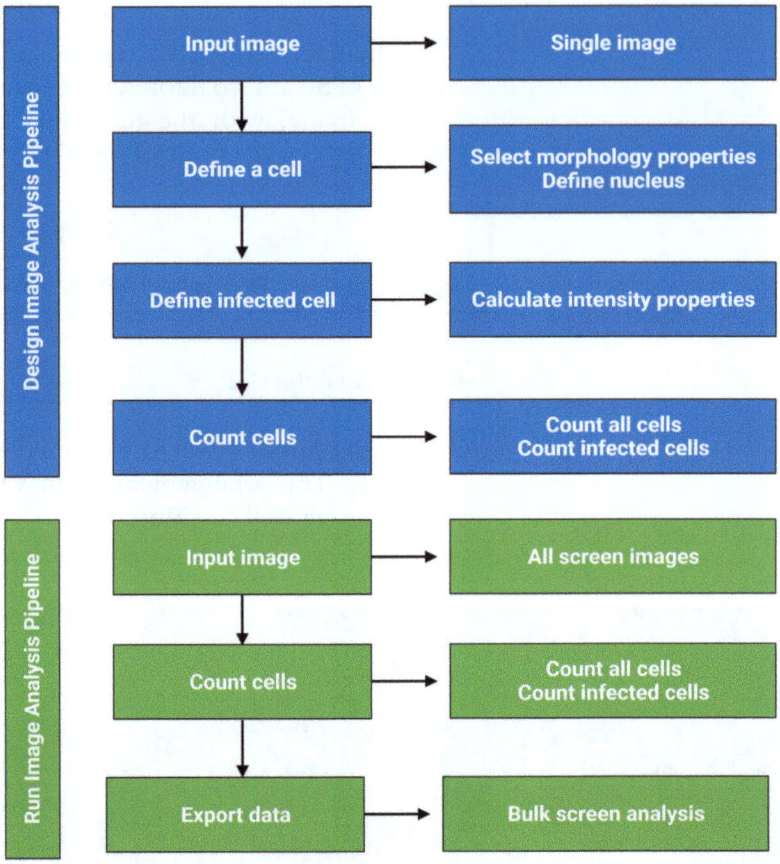

Fig. 3 Overview of an image analysis pipeline

3.7 Image Analysis

1. Generate an image analysis algorithm. An overview of an imaging analysis pipeline is shown in Fig. 3, and specific examples are shown in Table 1 (*see* **Note 11**).
2. Calculate the total number of cells per well (Figs. 4a and 5d).
3. Calculate the total number of infected cells (Figs. 4b and 5e).
4. Apply a module to filter out nonspecific debris (Fig. 5).
5. Calculate the percent infection using $\Sigma(\text{cell}_{\text{GFP+}}/w)/\Sigma(\text{cell}_{\text{Hoechst+}}/w)$, from nine images per well.

4 Notes

1. Cells should have a desirable growth rate and transfection efficiency. Fast-growing cells may outpace virus infection, whereas those that grow too slowly may die before screening. For measles virus infection, permissive cells are required to express the human signaling lymphocyte activation molecule (hSLAM) [13].

Table 1
Image analysis. (A) Classical image analysis modules and their hierarchy. (B) Algorithm for the Columbus program package to analyze infection ratio of MeV infected cells in a siRNA screen

A. Hierarchy of modules	B. Example of a modular image analysis pipe from Columbus
1. *Find objects* Nuclei Cytoplasm Cells Image region 2. *Select region in objects* Cell region	1. Input image (a) Flatfield correction: basic 2. Find nuclei (a) Channel: Hoechst (b) Method: B (i) Common threshold = 10.0 (ii) Area > 30 µm^2 (iii) Splitting coefficient = 6.5 (iv) Individual threshold = 10.0 (v) Contrast > 0.10 (c) Output population: nuclei
3. *Calculate object properties* Intensity Position Morphology Texture	3. Calculate intensity properties (a) Channel: Hoechst (b) Population: nuclei (c) Region: nucleus (d) Method: standard (i) Mean: [X] µm^2 (e) Property prefix IntensityHoecstNuclei 4. Calculate morphology properties: (a) Population: nuclei (b) Region: nucleus (c) Method: standard (i) Area: [X] (ii) Roundness: [X] (d) Property prefix: nuclei morphology
4. *Select a population of objects* Define population(s)	5. Select population (1): (a) Population: nuclei (b) Method: filter by property (i) Filter F1: IntensityHoecstNuclei < 300 (ii) Filter F2: nuclei morphology area [µm^2] < 1200 (c) Output population: nuclei2 6. Select population (2): (a) Population: nuclei2 (b) Method: common filter (i) Remove border objects: [X] (ii) Region: nucleus (c) Output population: nuclei3
5. *Process images* Filter image Calculate image	7. Calculate intensity properties (2): (a) Channel Alexa 488 (b) Population: nuclei3 (c) Region: nucleus (d) Method: standard (i) Mean: [X] (e) Property prefix: IntensityA488Nuclei3 > 100 8. Select population (3): (a) Method: filter by property (i) FilterF1: IntensityA488Nuclei3 > 100 (b) Output population: GFP_Nuclei3

(continued)

Table 1
(continued)

A. Hierarchy of modules	B. Example of a modular image analysis pipe from Columbus
6. *Calculate collated readout values* Define results	9. Define results: (a) Method: list of outputs (i) Population: nuclei3 1. Number of objects: [X] 2. Apply to all: individual selection (ii) Population: GFP_Nuclei3 1. Number of objects: [X] 2. Apply to all: individual selection (b) Method: formula output (i) Formula: *a*/*b* (ii) Variable *a*: GFP+nuclei3 number of objects (iii) Variable *b*: nuclei3 number of objects (iv) Output name: ratio GFP/total

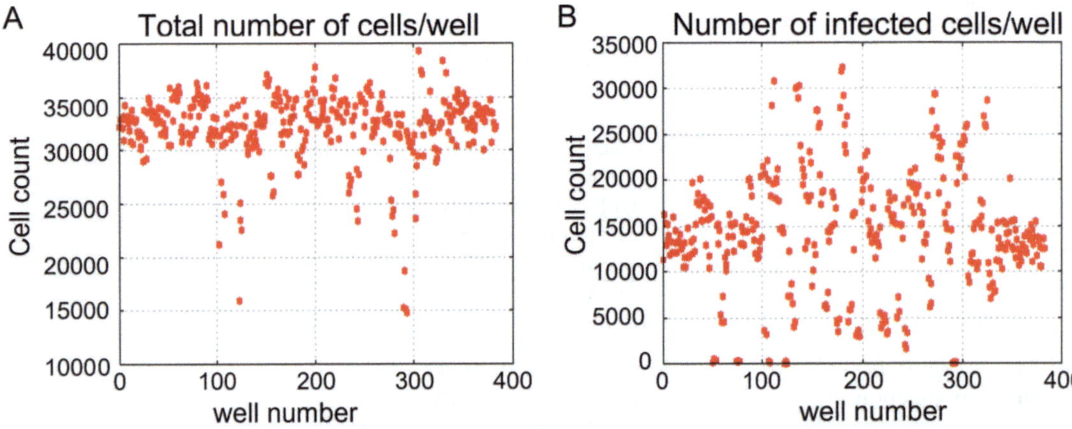

Fig. 4 Cell counts. (**a**) Total number of cells in each well. (**b**) Total number of infected cells in each well

2. The MeV strain used expresses GFP. For viruses lacking a reported gene, antibody staining is required to visualize infected cells. Antibody staining should be optimized prior to screening.
3. The number of assay plates processed per batch can be varied depending on your handling/automation capacity. Generally, there is no limit to the number of plates during reverse transfection/cell seeding/virus infection. However, incubation times affect the number of plates processed per batch during fixing/staining.
4. At the time point of plate reading, infected cells in each well should ideally exhibit 50% cytopathic effect (CPE), which we achieved at 48 h postinfection (hpi) at multiplicity of infection (MOI) of 0.1. An infection optimization to determine the

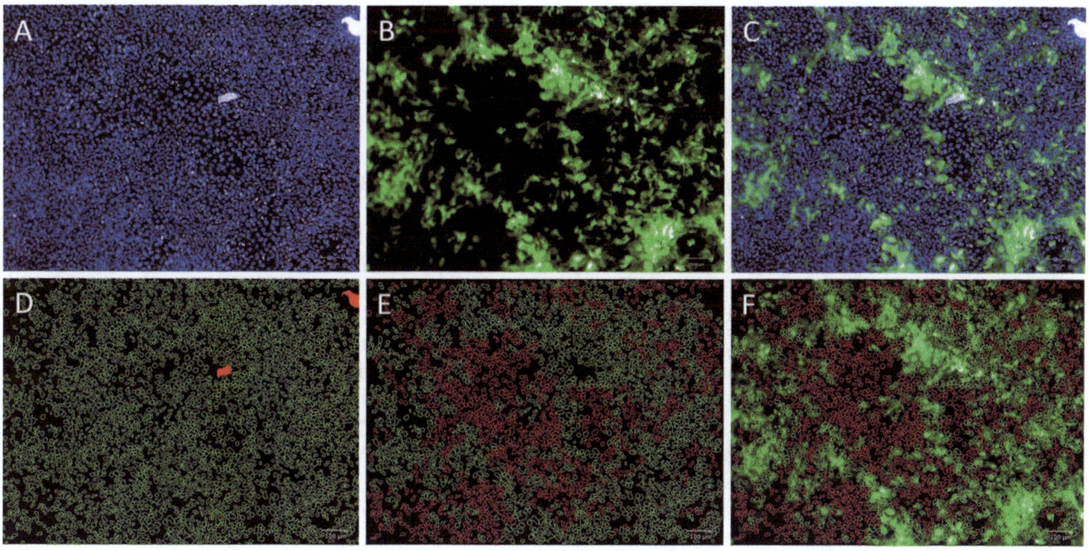

Fig. 5 Image analysis. (**a**) Hoechst staining of nuclei; (**b**) GFP signal from MeV infected cells; (**c**) overlay of (**a**) and (**b**); and (**d**) image filtration. Counted nuclei are shown with green borders, while fluorescent blebs from Hoechst staining are shown in solid red and excluded from analysis. (**e**) Selection of population of nuclei positive for GFP signal, shown with green borders, while nuclei negative for GFP signal are shown with red borders: (**f**) overlay of (**e**) and (**b**)

exact infection time and MOI appropriate for selected cells and virus strain should be performed prior to beginning the screen.

5. Example based on 4000 cells per well and a virus titer of 1×10^6 $TCID_{50}$/mL.
6. The total volume of each well before addition of formaldehyde is 85 μL (includes siRNA, transfection mix, cells, and virus). The final concentration of formaldehyde required in each well is 4%; hence, 10.6 μL 36% PFA is required in each well.
7. Steps using PFA should be performed in a fume cabinet, or an automated plate washer can be used.
8. If antibody staining is required, perform at this step.
9. The Operetta High Content Microscopy and Imaging System, equipped with a XL3000 Xeon Fiber-Optic Light Source, controlled by the Harmony 4.8 software package is described here, but alternate imaging systems can be used.
10. If cells and infection are uniform, a representative field may be used instead of capturing the entire well.
11. Acapella, CellProfiler, Columbus, and Harmony are software for high-throughput cell image analysis. These packages all offer flexible modular approaches to build efficient analysis sequences, tailor designed for specific tasks. The analysis sequence is constructed from building blocks (modules), combined in a hierarchical order.

References

1. Rota PA, Moss WJ, Takeda M, de Swart RL, Thompson KM, Goodson JL (2016) Measles. Nat Rev Dis Primers 2:16049. https://doi.org/10.1038/nrdp.2016.49
2. Laksono BM, de Vries RD, McQuaid S, Duprex WP, de Swart RL (2016) Measles virus host invasion and pathogenesis. Viruses 8(8):210. https://doi.org/10.3390/v8080210
3. Laksono BM, de Vries RD, Verburgh RJ, Visser EG, de Jong A, Fraaij PLA, Ruijs WLM, Nieuwenhuijse DF, van den Ham HJ, Koopmans MPG, van Zelm MC, Osterhaus A, de Swart RL (2018) Studies into the mechanism of measles-associated immune suppression during a measles outbreak in the Netherlands. Nat Commun 9(1):4944. https://doi.org/10.1038/s41467-018-07515-0
4. Chopra M, Bhutta Z, Chang Blanc D, Checchi F, Gupta A, Lemango ET, Levine OS, Lyimo D, Nandy R, O'Brien KL, Okwo-Bele JM, Rees H, Soepardi J, Tolhurst R, Victora CG (2020) Addressing the persistent inequities in immunization coverage. Bull World Health Organ 98(2):146–148. https://doi.org/10.2471/BLT.19.241620
5. Cherry S (2009) What have RNAi screens taught us about viral-host interactions? Curr Opin Microbiol 12(4):446–452. https://doi.org/10.1016/j.mib.2009.06.002
6. Cherry S, Doukas T, Armknecht S, Whelan S, Wang H, Sarnow P, Perrimon N (2005) Genome-wide RNAi screen reveals a specific sensitivity of IRES-containing RNA viruses to host translation inhibition. Genes Dev 19(4):445–452. https://doi.org/10.1101/gad.1267905
7. Panda D, Cherry S (2012) Cell-based genomic screening: elucidating virus-host interactions. Curr Opin Virol 2(6):784–792. https://doi.org/10.1016/j.coviro.2012.10.007
8. Mohr SE, Smith JA, Shamu CE, Neumuller RA, Perrimon N (2014) RNAi screening comes of age: improved techniques and complementary approaches. Nat Rev Mol Cell Biol 15(9):591–600. https://doi.org/10.1038/nrm3860
9. Anderson DE, Pfeffermann K, Kim SY, Sawatsky B, Pearson J, Kovtun M, Corcoran DL, Krebs Y, Sigmundsson K, Jamison SF, Yeo ZZJ, Rennick LJ, Wang LF, Talbot PJ, Duprex WP, Garcia-Blanco MA, von Messling V (2019) Comparative loss-of-function screens reveal ABCE1 as an essential cellular host factor for efficient translation of Paramyxoviridae and Pneumoviridae. mBio 10(3):e00826-19. https://doi.org/10.1128/mBio.00826-19
10. Martin S, Chiramel AI, Schmidt ML, Chen YC, Whitt N, Watt A, Dunham EC, Shifflett K, Traeger S, Leske A, Buehler E, Martellaro C, Brandt J, Wendt L, Muller A, Peitsch S, Best SM, Stech J, Finke S, Romer-Oberdorfer A, Groseth A, Feldmann H, Hoenen T (2018) A genome-wide siRNA screen identifies a druggable host pathway essential for the Ebola virus life cycle. Genome Med 10(1):58. https://doi.org/10.1186/s13073-018-0570-1
11. Panda D, Das A, Dinh PX, Subramaniam S, Nayak D, Barrows NJ, Pearson JL, Thompson J, Kelly DL, Ladunga I, Pattnaik AK (2011) RNAi screening reveals requirement for host cell secretory pathway in infection by diverse families of negative-strand RNA viruses. Proc Natl Acad Sci USA 108(47):19036–19041. https://doi.org/10.1073/pnas.1113643108
12. de Vries RD, Duprex WP, de Swart RL (2015) Morbillivirus infections: an introduction. Viruses 7(2):699–706. https://doi.org/10.3390/v7020699
13. Hashimoto K, Ono N, Tatsuo H, Minagawa H, Takeda M, Takeuchi K, Yanagi Y (2002) SLAM (CD150)-independent measles virus entry as revealed by recombinant virus expressing green fluorescent protein. J Virol 76(3):6743–6749. https://doi.org/10.1128/jvi.76.13.6743-6749.2002

Chapter 9

Dual RNA-seq Analysis of Patients' Cells and Viral Genome After Measles Infection

Vasiliki Pogka, Andreas Mentis, and Timokratis Karamitros

Abstract

During the infection of a host cell by an infectious agent, a series of gene expression changes occurs as a consequence of host-pathogen interactions. Unraveling this complex interplay is the key for understanding of microbial virulence and host response pathways, thus providing the basis for new molecular insights into the mechanisms of pathogenesis and the corresponding immune response. Dual RNA sequencing (dual RNA-seq) has been developed to simultaneously determine pathogen and host transcriptomes enabling both differential and coexpression analyses between the two partners as well as genome characterization in the case of RNA viruses. Here, we provide a detailed laboratory protocol and bioinformatics analysis guidelines for dual RNA-seq experiments focusing on – but not restricted to – measles virus (MeV) as a pathogen of interest. The application of dual RNA-seq technologies in MeV-infected patients can potentially provide valuable information on the structure of the viral RNA genome and on cellular innate immune responses and drive the discovery of new targets for antiviral therapy.

Key words Measles virus, Dual RNA-seq, Host response, RNA viral genome, Ribodepletion, Poly (A) enrichment

1 Introduction

The detection of viruses from human biological samples and the determination of their influence on the host cell transcriptome are crucial for understanding complex human diseases that are initiated or exacerbated by several virus infections. These infections result in large-scale changes in gene expression patterns in both interacting partners. In that respect, a comprehensive understanding of host-pathogen interactions requires a knowledge of the associated gene expression changes in both the pathogen and the host [1].

Traditional, probe-dependent approaches using microarrays have allowed the analysis of host's response to infections. However, the technical difficulties associated with simultaneously determining two often very different transcriptomes, including issues such as probe selection, cross-reactivity, and the required design and cost of

Dzwokai Z. Ma and Christian K. Pfaller (eds.), *Measles and Related Morbilliviruses: Methods and Protocols*, Methods in Molecular Biology, vol. 2808, https://doi.org/10.1007/978-1-0716-3870-5_9,

custom chips, make microarray-based studies challenging and expensive when they are applied to determining both the host and pathogen transcriptomes.

On the other hand, the development of the untargeted RNA sequencing (RNA-seq) approach has begun to revolutionize transcriptomics leading to elucidation of the interactions between the host and pathogen genomes and transcriptomes at the same time (dual RNA-seq) with high accuracy and depth [2]. In this approach, the discrimination between the two partners would take place only at the bioinformatics stage. An unbiased understanding of the host transcriptome and of the pathogen genome can provide new insights into this process by identifying new virulence factors in the pathogen or new pathways in the host cell that respond to the exposure to specific pathogens.

Dual RNA-seq has been previously used to simultaneously map bacterial-host transcriptomes enabling both differential and coexpression analyses between the two partners [3]. Here, we describe a detailed dual RNA-seq protocol for the reconstruction of the measles virus (MeV) genome and the parallel characterization of the host transcriptome using retrospectively analyzed total nucleic acids extracted from pharyngeal swabs of MeV-infected patients, as confirmed via routine real-time PCR testing for viral infections. This approach can serve as proof-of-concept that next-generation mRNA sequencing is a method to evaluate the interplay between the host and the virus which is likely to provide the basis for new molecular insights into the mechanisms of pathogenesis and the corresponding immune response [4].

2 Materials

1. RNase Away.
2. Low-binding PCR tubes.
3. Pipettes.
4. Magnetic stand.
5. 80% ethanol (freshly prepared).
6. Microcentrifuge.
7. Vortex mixer.
8. Heated-lid thermocycler.
9. Agilent 2100 Bioanalyzer system (Agilent Technologies).
10. Agencourt AMPure XP beads (Beckman Coulter).
11. Turbo DNase-freeTM kit (ThermoFisher Scientific).
12. Agilent Bioanalyzer RNA 6000 Nano/Pico Chip (Agilent Technologies).
13. TruSeq RNA sample preparation V2 kit (Illumina).

14. NEBNext rRNA Depletion Kit (New England BioLabs).
15. High sensitivity DNA Kit (Agilent Technologies).
16. NextSeq 2000 system (Illumina).

3 Methods

This protocol describes the preparation of untargeted RNA-seq libraries. Bench, pipettes, and racks used should be thoroughly cleaned with DNA decontamination solution (RNase Away), and the use of a PCR cabinet with a UV light operation is strongly suggested. We retrospectively analyzed total nucleic acids (stored at −80 °C) extracted from pharyngeal swabs of patients with MeV infection (Fig. 1).

3.1 Purification of Total RNA

It is recommended to perform purification of total RNA in order to significantly reduce genomic DNA contamination and ensure the purified RNA is suitable for sensitive downstream applications.

1. Add 0.1 volume 10× TURBO DNase Buffer and 1 μL TURBO DNase to the RNA, and mix gently (*see* **Note 1**).

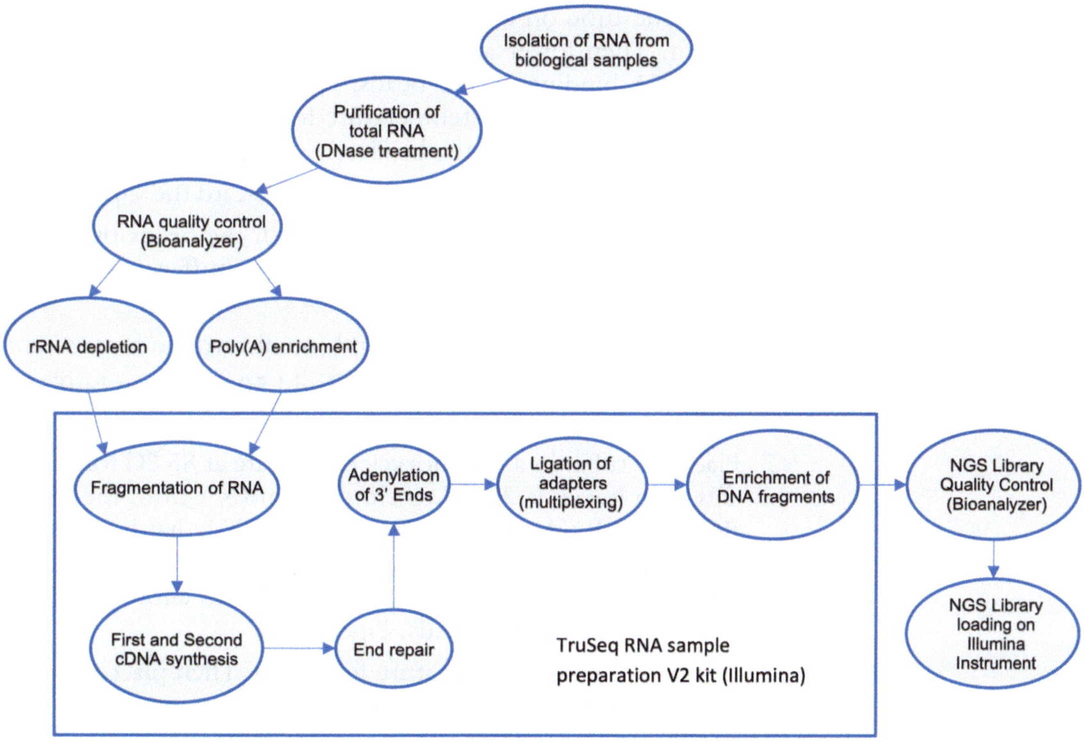

Fig. 1 Dual RNA-seq schematic protocol. Viral RNA genomes and host transcriptomes are captured using rRNA depletion and poly(A) enrichment, respectively. Equimolecular dilutions are pooled and undergo multiplexed library preparation

2. Incubate at 37 °C for 20–30 min.
3. Add resuspended DNase Inactivation Reagent (typically 0.1 volume), and mix well (*see* **Note 2**).
4. Incubate for 5 min at room temperature, mixing occasionally by flicking.
5. Centrifuge at 10,000 × g for 1.5 min, and transfer the supernatant RNA to a fresh tube.

3.2 Assessment of RNA Quality and Quantity

Run the RNA sample on an Agilent Bioanalyzer RNA 6000 Nano/ Pico Chip (Agilent Technologies, Santa Clara, CA, USA) to determine the RNA quantity and Integrity Number (RIN). RIN values range from 10 (intact) to 1 (totally degraded), whereas high-quality RNA will contain an RIN of at least *8*. Both intact and degraded RNA can be used in the subsequent depletion protocol. However, the use of degraded RNA in the Poly(A) purification protocol can result in low yield, overrepresentation of the 3′ ends of the RNA molecules, or failure of the protocol.

3.3 Enrichment of Specific RNA Species

3.3.1 Poly(A) Selection of mRNA

1. Pipette 50 μL of fully resuspended oligo-dT beads into a 0.2 μL PCR tube. Add at least 100 ng of total RNA diluted with nuclease-free ultrapure water to a final volume of 50 μL and pipette mix.
2. Place the tube on a thermocycler, and incubate at 65 °C for 5 min. Then lower at 4 °C for RNA denaturation and poly (A) RNA binding to the beads. Remove from thermocycler, and incubate at room temperature for 5 min to further allow the mRNA to bind to the beads.
3. Place the tube on a magnetic stand and discard the supernatant.
4. Remove from magnetic stand, and wash the beads by pipette mixing with 200 μL bead washing buffer to remove unbound RNA.
5. Place the tube on a magnetic stand and discard the supernatant.
6. Remove from magnetic stand, and add 50 μl elution buffer by pipette mixing.
7. Place the tube on a thermocycler, incubate at 85 °C for 2 min, and then lower at 25 °C to elute the mRNA from the beads. Both the mRNA and any contaminant rRNA that have bound the beads nonspecifically are released. Remove from thermocycler and add 50 μL bead binding buffer to allow mRNA to specifically rebind the beads. Pipette mix.
8. Incubate at room temperature for 5 min. Then place the tube on a magnetic stand, and discard the supernatant.
9. Remove from magnetic stand, and wash the beads by pipette mixing with 200 μL bead washing buffer.

10. Place the tube on a magnetic stand, and discard the supernatant that contains residual rRNA and other contaminants that were released in the first elution and did not rebind the beads.
11. Remove from magnetic stand, and add 19.5 μL Elute, Prime, Fragment Mix (*see* **Note 3**).

3.3.2 rRNA Depletion (See Note 4)

1. Dilute 5 ng–1 μg of total RNA with nuclease-free water to a final volume of 12 μL in a PCR tube. Keep on ice and add 1 μL rRNA depletion solution and 2 μL probe hybridization buffer. Pipette mix.
2. Place the tube on a thermocycler, incubate at 95 °C for 2 min, and then ramp down to 22 °C (0.1 °C/s). Hold for 5 min.
3. Remove from thermocycler, and place on ice while adding 2 μL RNAse H reaction buffer, 2 μL NEBNExt RNase H, and 1 μL nuclease-free water. Pipette mix.
4. Place the tube on a preheated thermocycler for 30 min at 37 °C.
5. Add 110 μL of fully resuspended beads, and incubate on ice for 15 min (*see* **Note 5**).
6. Place the tube on a magnetic stand and discard the supernatant.
7. Wash the beads by adding 200 μL of 80% freshly prepared ethanol. Incubate at room temperature for 30 s and discard supernatant. Repeat washing step.
8. Air-dry the beads for up to 5 min with the lid open.
9. Remove from magnetic stand, and add 18 μL Elute, Prime, Fragment Mix (*see* **Note 3**).

3.4 Library Preparation

In either case of the above RNA enrichment methods, heat the RNA in Elute, Prime, Fragment Mix at 94 °C for 8 min. Then hold the reaction at 4 °C.

Proceed with First and Second Strand cDNA Synthesis, End Repair, Adenylation of 3′ Ends, Ligation of Adapters and Enrichment of DNA Fragments (*see* **Note 6**).

3.5 Library Quality Control and Quantification

Perform quality control on the final library. Run 1 μL of library on the Agilent 2100 Bioanalyzer using the High sensitivity DNA Kit (Agilent Technologies, Santa Clara, CA, USA). Barcoded libraries can be pooled in equimolar ratios according to Illumina's pooling protocols and procedures for sequencing on the Illumina NextSeq platform.

3.6 Sequencing and MeV Genome Reconstruction

Sequence individual libraries or library pools using the Illumina platform, such as NextSeq, according to the manufacturer's protocol. Once sequencing data is obtained in the form of raw reads, quality control and sequence filtering need to be carried. This is crucial because next-generation sequencing data may contain unexpected artifacts and poor quality reads due to low complexity or high GC content regions and sequencing errors. The presence of these low-quality sequences will further affect downstream analysis leading to inaccuracies in overall RNA-seq data interpretation.

1. Generate the consensus genomes using mapping alignment of the reads to the reference strain MeVs/London.GBR/3.14 (genotype B3-KT732219.1) using Bowtie2 [5].
2. Use Samtools for all file transformations and for calling the variations from the mapping alignments [6].
3. Conduct quality confirmation of the genotyping results by visual inspection of the mapping alignments using IGV [7].
4. Perform RNA-seq analysis using Kallisto [8] and the human reference transcriptome v.GRCh38.rel79, in order to calculate the abundances of the transcripts.
5. Use Sleuth package [9] to perform differential expression analysis and R-base functions to interpret and visualize the RNA-seq analysis results.

4 Notes

1. There are separate DNase digestion conditions depending on the amount of contaminating DNA and the nucleic acid concentration of the sample. Consult the Turbo DNase-freeTM kit protocol for Routine or Rigorous DNase treatment (Thermo-Fisher Scientific, Waltham, MA, USA).
2. Always use at least 2 μL of DNase Inactivation Reagent, even if it is more than 0.1 volume.
3. Elute, Prime, Fragment Mix contains random hexamers for RT priming and serves as the first strand cDNA synthesis reaction buffer.
4. Follow the rRNA depletion protocol as an alternative to the Poly(A) Selection of mRNA. Comparative results of these two enrichment methods revealed a better genome coverage when using the rRNA depletion protocol (Fig. 2). On the other hand, Zhao et al. [10] showed that while rRNA depletion could record more unique characteristics of the transcriptome, poly(A) selection was more accurate in terms of gene quantification.
5. Use either Agencourt RNAClean XP Beads or NEBNext RNA Sample Purification Beads.
6. Follow instructions of the TruSeq RNA sample preparation V2 kit (Illumina).

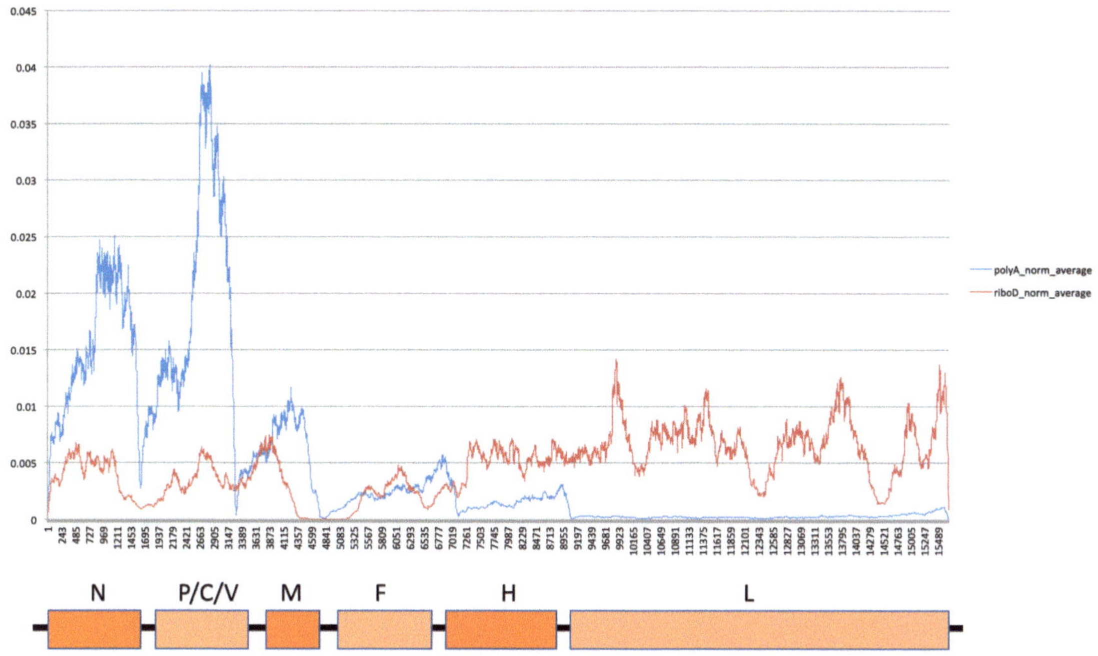

Fig. 2 Comparative read coverage plot. Normalized read counts – averaged from 6 MeV(+) clinical samples – across MeV genome map, as a result after poly-A enrichment (blue line) or after ribodepletion (red line)

References

1. Garcıa-Sastre A, Sansonetti PJ (2010) Host–pathogen interactions. Curr Opin Immunol 2010(22):425–427. https://doi.org/10.1016/j.coi.2010.06.005
2. Westermann AJ, Barquist L, Vogel J (2017) Resolving host–pathogen interactions by dual RNA-seq. PLoS Pathog 13(2):e1006033. https://doi.org/10.1371/journal.ppat.1006033
3. Saliba A-E, Santos SC, Vogel J (2017) New RNA-seq approaches for the study of bacterial pathogens. Curr Opin Microbiol 35:78–87. https://doi.org/10.1016/j.mib.2017.01.001
4. Karamitros T, Pogka V, Papadopoulou G, Tsitsilonis O, Evangelidou M, Sympardi S, Mentis A (2021) Dual RNA-Seq Enables Full-Genome Assembly of Measles Virus and Characterization of Host-Pathogen Interactions. Microorganisms 9(7):1538. https://doi.org/10.3390/microorganisms9071538
5. Langmead B, Salzberg S (2012) Fast gapped-read alignment with Bowtie2. Nat Methods 9:357–359. https://doi.org/10.1038/nmeth.1923
6. Li H, Handsaker B, Wysoker A, Fennell T, Ruan J, Homer N, Marth G, Abecasis G, Durbin R (2009) The sequence alignment/map format and SAMtools. Bioinformatics 25:2078–2079. https://doi.org/10.1093/bioinformatics/btp352
7. Thorvaldsdσttir H, Robinson JT, Mesirov JP (2013) Integrative genomics viewer (IGV): high-performance genomics data visualization and exploration. Brief Bioinform 14:178–192. https://doi.org/10.1093/bib/bbs017
8. Church DM, Schneider VA, Steinberg KM, Schatz MC, Quinlan AR, Chin CS, Kitts PA, Aken B, Marth GT, Hoffman MM et al (2015) Extending reference assembly models. Genome Biol 16:13. https://doi.org/10.1186/s13059-015-0587-3
9. Bray NL, Pimentel H, Melsted P, Pachter L (2016) Near-optimal probabilistic RNA-seq quantification. Nat Biotechnol 34:525–527. https://doi.org/10.1038/nbt.3519
10. Zhao S, Zhang Y, Gamini R, Zhang B, von Schack D (2018) Evaluation of two main RNA-seq approaches for gene quantification in clinical RNA sequencing: polyA+ selection versus rRNA depletion. Sci Rep 8(1):4781

Chapter 10

Investigating Liquid-Liquid Phase Separation in Virus-Generated Inclusion Bodies Using Fluorescence Recovery After Photobleaching of Fluorescently Labeled Host Proteins

Matthew B. T. Asuelime-Smith and Dzwokai Z. Ma

Abstract

Many negative-sense single-stranded RNA viruses within the order *Mononegavirales* harm humans. A common feature shared among cells infected by these viruses is the formation of subcellular membraneless structures called biomolecular condensates, also known as inclusion bodies (IBs), that form through a process called liquid-liquid phase separation (LLPS). Like many other membraneless organelles, viral IBs enrich a specific subset of viral and host proteins involved in the formation of viral particles. Elucidation of the properties and regulation of these IBs as they mature throughout the viral replication process are important for our understanding of viral replication, which may also lead to the development of alternative antiviral treatments. The protocol outlined in this chapter aims to characterize the intrinsic properties of LLPS within the measles virus (MeV, a member of *Mononegavirales*) IBs by using an imaging approach that fluorescently tags an IB-associated host protein. This method uses common laboratory techniques and is generalizable to any host factors as well as other viral systems.

Key words Fluorescence recovery after photobleaching, FRAP, Host factors, Innate immune response, Liquid-liquid phase separation, LLPS, Measles virus, *Mononegavirales*, *Paramyxovirus*, Viral inclusion bodies

1 Introduction

Many single-stranded RNA viruses of the order *Mononegavirales* are prominent human pathogens, and they form subcellular viral inclusion bodies (IBs) during their replication cycle [1–3]. These IBs have physical properties similar to membraneless organelles such as P granules, chromatin, nucleolus, Cajal bodies, PML bodies, nuclear speckles, P bodies, stress granules, and spindle apparatus [4–16]. These organelles form via liquid-liquid phase separation (LLPS), a process that arises when molecular components with overlapping multivalent interactions aggregate to form a

Dzwokai Z. Ma and Christian K. Pfaller (eds.), *Measles and Related Morbilliviruses: Methods and Protocols*, Methods in Molecular Biology, vol. 2808, https://doi.org/10.1007/978-1-0716-3870-5_10,

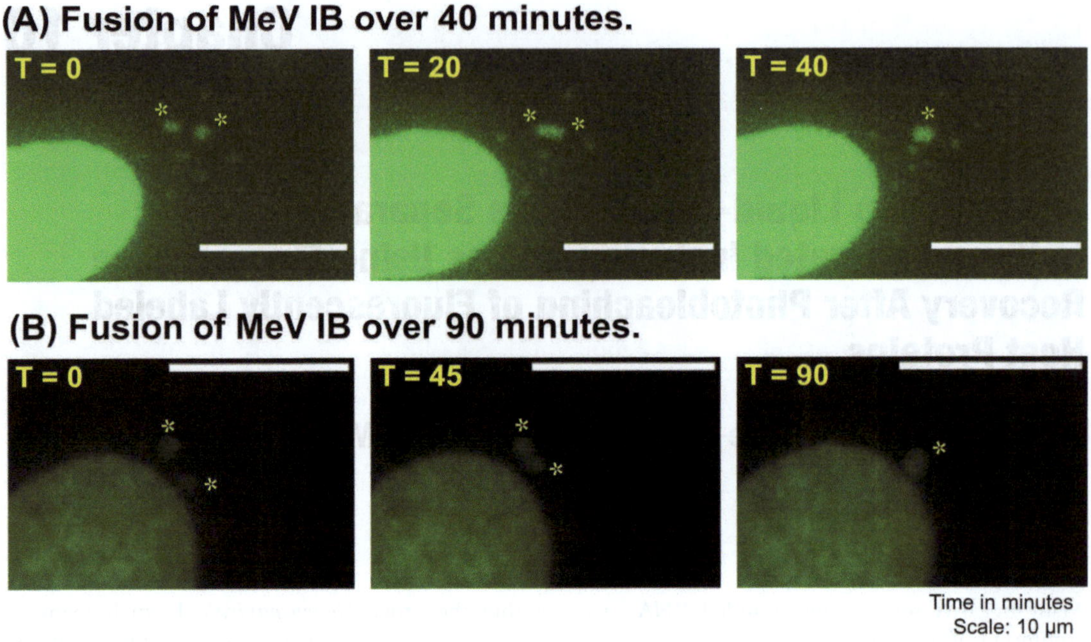

Fig. 1 Fusion and relaxation of inclusion bodies in MeV-infected HeLa cells stably expressing an EGFP fusion of WDR5. Live-cell imaging was performed 14 h postinfection (MOI = 0.2) with time-lapse microscopy to track the fusion and relaxation of IBs (designated by asterisks) using WDR5-EGFP as the marker. Snapshots at different timepoints show the fusion and subsequent relaxation between two IBs in MeV-infected cells. Fusion of two IBs spanning (**a**) 40 min and (**b**) 90 min. Scale bar, 10 μm

biomolecular condensate [17–22]. In some RNA viruses, only one to three viral components are needed for viral molecular condensates to form via LLPS [22–25]. For example, during measles virus (MeV) replication, the nucleation and enlargement of the MeV IBs are driven by the multivalent interactions, including those within the intrinsically disordered domains of the viral N and P proteins [22]. Although the formation and maturation of MeV IB only require two viral protein components, it also enriches specific host proteins for certain reasons [26, 27]. Identifying additional IB-associated host proteins and investigating their roles in viral replication represent an important and active research direction in the field of virus-host interactions.

Three criteria are used to determine if a subcellular structure is formed via LLPS: it maintains a spherical shape (Fig. 1), fusion occurs between compartments upon contact (Fig. 1), and there is an exchange of molecules with the surrounding environment (Fig. 2) [28]. These criteria are routinely examined using time-lapse live-cell microscopy. However, there are provisions when determining each criterion. Spherical inclusion bodies are not always observed due to intracellular crowding, especially for IBs of large size. Fusion is a stochastic event that may require many time-lapse images to capture. Fluorescence recovery after photobleaching (FRAP) is the "gold standard" for characterizing "liquid-

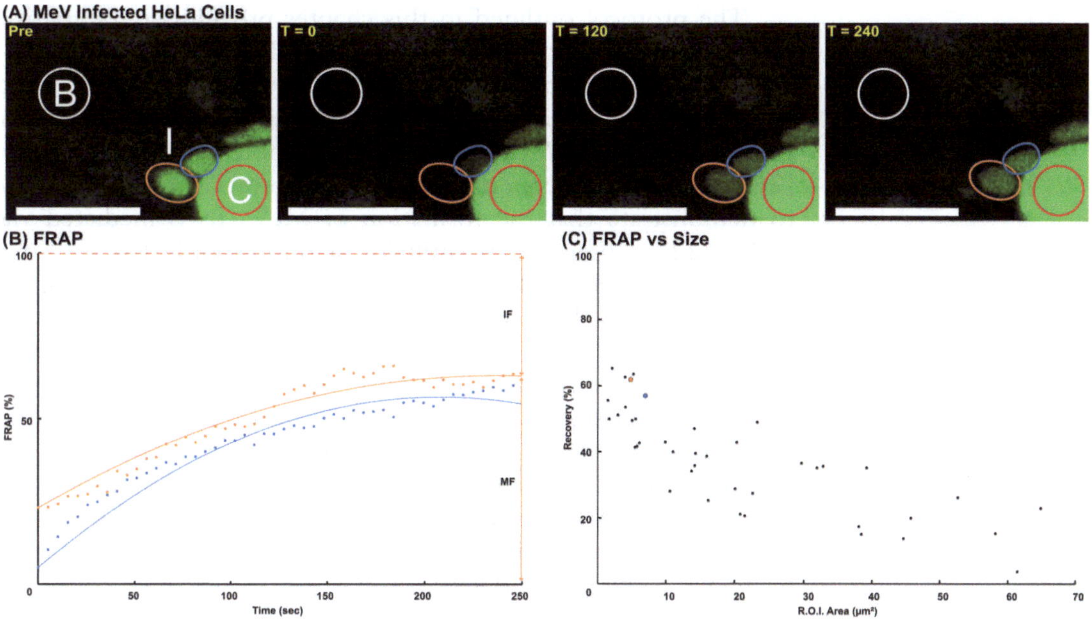

Fig. 2 FRAP measurements of inclusion body-associated WDR5-EGFP in MeV-infected HeLa cells. (**a**) Representative images of two inclusion bodies at 24 h postinfection (MOI = 0.2) in a HeLa cell before (Pre) and at 0, 120, and 240 s after photobleaching. Scale bars = 10 μm. Regions of inclusion body (I; orange and blue circles), background (**b**; white circle), and control (**c**; red circle). (**b**) An inverse correlation between inclusion body size and percent fluorescence recovery within MeV-infected cells. n = 42. (**c**) FRAP curve of WDR5-EGFP in two inclusion bodies with a high rate of recovery. Dots of FRAP curve show recovery rate every 5 s over 4-min period. Immobile fraction (IF) and the mobile fraction (MF)

like" subcellular compartments, but the size of the region of interest (ROI) and the model used to fit the data may need to be adjusted depending on the timepoint and targeted molecule undergoing LLPS [29, 30].

Studying the dynamics of virus-related LLPS typically uses the fluorescent tagging of one of the minimal viral components required for forming biomolecular condensates as a molecular probe during time-lapse live-cell microscopy. There are three methods for doing this: transfecting a tagged viral protein or an expression plasmid for the tagged viral protein into the cell during viral infection [22, 23, 31–41], co-transfecting the known minimal LLPS components in addition to the tagged viral protein [22, 23, 25, 32–34], or generating a transgenic virus natively expressing the fluorescently tagged viral protein [23, 33–35]. These methods are excellent for examining the de novo formation of nascent IBs and their subsequent maturation. On the other hand, some caveats do exist. The first two methods may not accurately model viral IBs if a viral protein is overrepresented. Whereas the third method uses native levels of viral proteins and is likely the most optimal for studying the formation and maturation of viral IBs, labeling a core viral component essential for the IB formation may perturb the inherent properties of LLPS within the viral IB.

The protocol outlined in this chapter provides an alternative approach by tagging a known IB-associated host protein to monitor the IB dynamics. We have previously shown that the host WD40 repeat-containing protein 5 (WDR5) localizes to IBs in cells infected by MeV, where it promotes viral replication [26]. We have subsequently utilized a fluorescent tagged WDR5 to demonstrate that the formation of MeV IBs is mediated via LLPS [22]. Compared to utilizing a fluorescently labeled core viral protein as the LLPS probe, a tagged host protein is less likely to interfere with the intrinsic properties of the viral IBs. However, one should keep in mind that this approach has its own limits. For example, the IB association of a host protein may be dependent on the maturation stage of viral IBs and/or cell type-specific factors. Moreover, one cannot rule out the possibility that overexpression of an IB-associated host protein can modulate the LLPS process of viral IBs. If the latter scenario becomes a concern, one should conduct the investigation in cells expressing the fluorescent host protein at a physiological level.

2 Materials

2.1 Equipment

1. Leica TCS SP8 X confocal laser resonant scanning microscopy platform using a Leica DMi8 inverted microscope with a high-contrast (HC) PL Apo 63× objective (NA, 1.3 glycerol; Leica) with a stage top incubator system (Bold Line H301 T; Okolab) that maintains an environment at 37 °C and 5% CO_2. An Oko-Touch (Okolab) maintains the climate within the stage top incubator by regulating the perfusion of CO_2 and by controlling the circulating bath (Lauda ECO RE 415 S; Lauda-Brinkmann) that adjusts heating/cooling. Alternate imaging systems can be used.
2. Class II Biological safety cabinet (BSC II).
3. Low-speed benchtop centrifuge with a swing-out rotor that is compatible with 15 mL conical tubes.
4. Phase contrast inverted microscope.
5. Water bath set to 37 °C.
6. Tissue culture incubator set to 37 °C, 5% CO_2, and 95% relative humidity.

2.2 Consumables

2.2.1 General

1. 70% isopropanol or ethanol.
2. 10% bleach.
3. 15 mL conical tubes.
4. 1.5 mL microfuge tubes.

2.2.2 Tissue Culture

1. HeLa cells (*see* **Note 1**).
2. Measles virus (MVvac2(mCherryNLS)H; *see* **Note 2**).
3. Advanced Dulbecco's Modified Eagle Medium (DMEM) supplemented with 4% fetal bovine serum (FBS), 1× GlutaMAX, 1× penicillin-streptomycin (P/S), and 3 μg/mL puromycin dihydrochloride.
4. Opti-MEM, serum-free.
5. Matrigel.
6. 1× phosphate buffered saline (PBS), without Ca^{2+} and Mg^{2+}.
7. 0.25% Trypsin-0.1% EDTA solution.
8. T75 cm^2 tissue culture flask.
9. Trypan Blue solution.

2.3 Live-Cell Imaging

1. 0.17 mM glass-bottomed delta T dish (Bioptechs; delta T dish; *see* **Note 3**).
2. Hemocytometer.

3 Methods

3.1 Tissue Culture (See Note 4)

3.1.1 Prepare a Delta T Dish for Tissue Culture

1. Begin preparing a 1:500 master mix of Matrigel/Opti-MEM for your number of samples (N) by dispensing X μL of Opti-MEM into a 15 mL conical tube:

$$X = N \times 499 \times 1.05$$

(*see* **Notes 5** and **6**)

2. Overnight at 4 °C, thaw the Matrigel and prechill the Opti-MEM from **step 1** on ice, and prechill N delta T dishes and pipet tips for a p10 and p1000 (*see* **Note** 7).
3. The next day, keeping everything on ice, use the prechilled pipet tips to dispense, and mix M μL of the Matrigel into the prechilled Opti-MEM, and then replace on ice.

$$M = N \times 1 \times 1.05$$

4. Evenly dispense 500 μL of the Matrigel mixture onto the surface of each a prechilled delta T dish, and then agitate gently until the surface is completely covered.
5. Place the delta T dish with Matrigel in the tissue culture incubator overnight.
6. The next day, aspirate the Matrigel mixture, replace with PBS, and return to the incubator.

3.1.2 Preparing HeLa Cells for Infection

1. Cells are maintained in a T75 flask in a tissue culture incubator and are passaged to a confluency between 15% and 20% every 2–3 days.
2. Tissue culture media is completely aspirated, and a 5 mL PBS wash is performed.
3. The PBS is completely aspirated, and 1.5 mL of Trypsin-EDTA is added and gently agitated until the surface is completely covered.
4. The flask is placed in the incubator until the cells begin to round and there is greater than 90% detachment, and at that time, 4.5 mL of culture media is added to quench the activity of Trypsin-EDTA.
5. The dissociated cells are triturated and transferred to a 15 mL conical tube.
6. Transfer 20 μL of the homogenous cell suspension to a microfuge tube to be counted.
7. Mix 10 μL of the cell suspension with 10 μL Trypan Blue solution (*see* **Note 8**), and count using a hemocytometer or cell counter.

Quadrant I	Quadrant II	Quadrant III	Quadrant IV	Cell Density

(*see* **Note 9**)

$$\text{Cell Density}\ (\text{Cells}/\text{mL}) = \frac{(\text{Total Number of Cells}) \times (\text{Dilution Factor}) \times 10^4}{(\text{Number of Quadrants Counted})}$$

Use culture media to dilute cells to a density of 4.2×10^5 cells/mL in a volume (mL) equal to the number of delta T dishes plus one (*see* **Notes 10** and **11**).

Seed 4.2×10^5 cells into the delta T dish by dispensing 1 mL of the mixture of diluted cells, and place into the incubator to attach overnight (*see* **Note 12**).

3.2 Infection

1. Infections are performed approximately 16 h after seeding or once the cells reach 100% confluency (*see* **Note 13**).
2. For each delta T dish at 100% confluence, virus infections are performed using a TCID50 at an MOI of 0.2 in a total volume

of 180 μL of Opti-MEM. For example, to calculate the amount of virus needed to achieve an MOI of 0.2 at 100% confluency using a viral titer of 1.0×10^7 TCID50/mL,

$$\mathrm{MOI} = \mathrm{TCID50/number\ of\ cells}$$

$$\mathrm{MOI} = (\mathrm{mL} \times (\mathrm{TCID50/mL}))/\mathrm{number\ of\ cells}$$

$$\mathrm{mL} = (\mathrm{MOI} \times \mathrm{number\ of\ cells})/(\mathrm{TCID50/mL})$$

There are an estimated 6.72×10^5 cells in a delta T dish at 100% confluence.

$$\mathrm{mL} = \left(0.2 \times 6.72 \times 10^5\right)/\left(1.0 \times 10^7\right)$$

$$\mathrm{mL} = \left(1.344 \times 10^5\right)/\left(1.0 \times 10^7\right)$$

$$\mathrm{mL} = 0.0134 = 13.4\ \mu L\ \mathrm{virus}$$

A total volume of 180 μL of diluted virus stock is needed for each delta T dish.

$\mu L = 180 - 13.4 = 166.6\ \mu L$ Opti-MEM.

3. Incubate the infection in a tissue culture incubator for 2 h, and gently rock back and forth every 15 min (*see* **Note 14**).
4. At the end of the 2-h incubation, the viral supernatant is aspirated and replaced with 1 mL of culture media.
5. The now-infected cells are placed in the tissue culture incubator for 23 h (*see* **Note 15**).

3.3 Observation Fusion and Relaxation of Spherical Inclusion Bodies Using Confocal Microscopy and Manual Image Analysis

1. Live-cell confocal microscopy experiments are performed on HeLa cells at 24 h postinfection, and images are captured every 20–45 min for 14 h (*see* Fig. 1, **Notes 16** and **17**).
2. Images for EGFP and mCherry are taken using laser excitation (488 nM and 578 nM) at 10% power for 50 ms, and the emissions (520.5 nM and 610 nM) from the sample are filtered (558 nM and 593 nM) before being captured.
3. Use a 63× objective to capture a field of view with a high number of infected cells (mCherry; *see* **Note 2**) and at a Z-plane where the cells have the largest area.
4. Preprocess image by adjusting the image to improve its clarity and contrast.
5. Find cells infected by measles virus by identifying nuclei stained with mCherry (*see* **Note 18**).
6. For each infected cell within a field of view, scan through each frame of the timelapse, and look for inclusion bodies that fuse (Fig. 1).

$$[\text{Initial Confluency}] = \frac{[\text{Final Confluency}]}{2^{(\text{Incubation Time/Doubling Time})}}$$

$$[\text{Initial Confluency}] = \frac{1}{2^{(16/24)}} = 0.63$$

13. Monitor the cell density by checking periodically a few hours before the estimated infection time. This is especially important if a transfection is being performed after infection (*see* **Note 1**).
14. Rocking is performed to keep the cells in the middle of the dish from drying out.
15. The timing for a 24-h postinfection timepoint begins 1 h after adding the virus to the cells.
16. The time postinfection at which the interval at which timepoints are collected should be optimized. Collecting multiple fields of interest is important for capturing the highly stochastic fusion of two inclusion bodies.
17. When selecting for each field of interest, focus on regions with a diverse collection of small, medium, and large IBs, as the physical properties of IBs at different sizes may change. Additionally, the capture of more fields may be required when detecting the fusion events or FRAP of small IBs as their movement is highly dynamic and often shifts out of the visual plane.
18. Viral mCherry has a nuclear localization signal, and the EGFP tagged WDR5 protein localizes to the nucleus and MeV IBs.

References

1. Novoa RR, Calderita G, Arranz R, Fontana J, Granzow H, Risco C (2005) Virus factories: associations of cell organelles for viral replication and morphogenesis. Biol Cell 97:147–172. https://doi.org/10.1042/BC20040058
2. Su JM, Wilson MZ, Samuel CE, Ma D (2021) Formation and function of liquid-like viral factories in negative-sense single-stranded RNA virus infections. Viruses 13:126. https://doi.org/10.3390/v13010126
3. Etibor TA, Yamauchi Y, Amorim MJ (2021) Liquid biomolecular condensates and viral lifecycles: review and perspectives. Viruses 13:366. https://doi.org/10.3390/v13030366
4. Jiang H, Wang S, Huang Y, He X, Cui H, Zhu X, Zheng Y (2015) Phase transition of spindle-associated protein regulate spindle apparatus assembly. Cell 163:108–122. https://doi.org/10.1016/j.cell.2015.08.010
5. Brangwynne CP, Eckmann CR, Courson DS, Rybarska A, Hoege C, Gharakhani J, Jülicher F, Hyman AA (2009) Germline P granules are liquid droplets that localize by controlled dissolution/condensation. Science 324:1729–1732. https://doi.org/10.1126/science.1172046
6. Larson AG, Elnatan D, Keenen MM, Trnka MJ, Johnston JB, Burlingame AL, Agard DA, Redding S, Narlikar GJ (2017) Liquid droplet formation by HP1α suggests a role for phase separation in heterochromatin. Nature 547:236–240. https://doi.org/10.1038/nature22822
7. Strom AR, Emelyanov AV, Mir M, Fyodorov DV, Darzacq X, Karpen GH (2017) Phase separation drives heterochromatin domain formation. Nature 547:241–245. https://doi.org/10.1038/nature22989
8. Gibson BA, Doolittle LK, Schneider MWG, Jensen LE, Gamarra N, Henry L, Gerlich DW, Redding S, Rosen MK (2019) Organization of chromatin by intrinsic and regulated phase separation. Cell 179:470–484.e21. https://doi.org/10.1016/j.cell.2019.08.037

9. Brangwynne CP, Mitchison TJ, Hyman AA (2011) Active liquid-like behavior of nucleoli determines their size and shape in Xenopus laevis oocytes. Proc Natl Acad Sci USA 108: 4334–4339. https://doi.org/10.1073/pnas.1017150108
10. Feric M, Vaidya N, Harmon TS, Mitrea DM, Zhu L, Richardson TM, Kriwacki RW, Pappu RV, Brangwynne CP (2016) Coexisting liquid phases underlie nucleolar subcompartments. Cell 165:1686–1697. https://doi.org/10.1016/j.cell.2016.04.047
11. Handwerger KE, Cordero JA, Gall JG (2005) Cajal bodies, nucleoli, and speckles in the Xenopus oocyte nucleus have a low-density, sponge-like structure. Mol Biol Cell 16:202–211. https://doi.org/10.1091/mbc.e04-08-0742
12. Corpet A, Kleijwegt C, Roubille S, Juillard F, Jacquet K, Texier P, Lomonte P (2020) PML nuclear bodies and chromatin dynamics: catch me if you can! Nucleic Acids Res 48:11890–11912. https://doi.org/10.1093/nar/gkaa828
13. Marzahn MR, Marada S, Lee J, Nourse A, Kenrick S, Zhao H, Ben-Nissan G, Kolaitis R-M, Peters JL, Pounds S, Errington WJ, Privé GG, Taylor JP, Sharon M, Schuck P, Ogden SK, Mittag T (2016) Higher-order oligomerization promotes localization of SPOP to liquid nuclear speckles. EMBO J 35: 1254–1275. https://doi.org/10.15252/embj.201593169
14. Kroschwald S, Maharana S, Mateju D, Malinovska L, Nüske E, Poser I, Richter D, Alberti S (2015) Promiscuous interactions and protein disaggregases determine the material state of stress-inducible RNP granules. elife 4:e06807. https://doi.org/10.7554/eLife.06807
15. Molliex A, Temirov J, Lee J, Coughlin M, Kanagaraj AP, Kim HJ, Mittag T, Taylor JP (2015) Phase separation by low complexity domains promotes stress granule assembly and drives pathological fibrillization. Cell 163:123–133. https://doi.org/10.1016/j.cell.2015.09.015
16. Patel A, Lee HO, Jawerth L, Maharana S, Jahnel M, Hein MY, Stoynov S, Mahamid J, Saha S, Franzmann TM, Pozniakovski A, Poser I, Maghelli N, Royer LA, Weigert M, Myers EW, Grill S, Drechsel D, Hyman AA, Alberti S (2015) A liquid-to-solid phase transition of the ALS protein FUS accelerated by disease mutation. Cell 162:1066–1077. https://doi.org/10.1016/j.cell.2015.07.047
17. Shin Y, Brangwynne CP (2017) Liquid phase condensation in cell physiology and disease. Science 357:eaaf4382. https://doi.org/10.1126/science.aaf4382
18. Banani SF, Lee HO, Hyman AA, Rosen MK (2017) Biomolecular condensates: organizers of cellular biochemistry. Nat Rev Mol Cell Biol 18:285–298. https://doi.org/10.1038/nrm.2017.7
19. Boeynaems S, Alberti S, Fawzi NL, Mittag T, Polymenidou M, Rousseau F, Schymkowitz J, Shorter J, Wolozin B, Van Den Bosch L, Tompa P, Fuxreiter M (2018) Protein phase separation: a new phase in cell biology. Trends Cell Biol 28:420–435. https://doi.org/10.1016/j.tcb.2018.02.004
20. Gomes E, Shorter J (2019) The molecular language of membraneless organelles. J Biol Chem 294:7115–7127. https://doi.org/10.1074/jbc.TM118.001192
21. Alberti S, Dormann D (2019) Liquid-liquid phase separation in disease. Annu Rev Genet 53:171–194. https://doi.org/10.1146/annurev-genet-112618-043527
22. Zhou Y, Su JM, Samuel CE, Ma D (2019) Measles virus forms inclusion bodies with properties of liquid organelles. J Virol 93: e00948–e00919. https://doi.org/10.1128/JVI.00948-19
23. Nikolic J, Le Bars R, Lama Z, Scrima N, Lagaudrière-Gesbert C, Gaudin Y, Blondel D (2017) Negri bodies are viral factories with properties of liquid organelles. Nat Commun 8:58. https://doi.org/10.1038/s41467-017-00102-9
24. Galloux M, Risso-Ballester J, Richard C-A, Fix J, Rameix-Welti M-A, Eléouët J-F (2020) Minimal elements required for the formation of respiratory syncytial virus cytoplasmic inclusion bodies in vivo and in vitro. MBio 11: e01202–e01220. https://doi.org/10.1128/mBio.01202-20
25. Heinrich BS, Maliga Z, Stein DA, Hyman AA, Whelan SPJ (2018) Phase transitions drive the formation of vesicular stomatitis virus replication compartments. MBio 9:e02290–e02217. https://doi.org/10.1128/mBio.02290-17
26. Ma D, George CX, Nomburg JL, Pfaller CK, Cattaneo R, Samuel CE (2018) Upon infection, cellular WD repeat-containing protein 5 (WDR5) localizes to cytoplasmic inclusion bodies and enhances measles virus replication. J Virol 92:e01726–e01717. https://doi.org/10.1128/JVI.01726-17
27. BenDavid E, Pfaller CK, Pan Y, Samuel CE, Ma D (2022) Host 5′-3′ exoribonuclease XRN1 acts as a proviral factor for measles virus replication by downregulating the dsRNA-activated kinase PKR. J Virol 96:e0131922. https://doi.org/10.1128/jvi.01319-22
28. Alberti S, Gladfelter A, Mittag T (2019) Considerations and challenges in studying liquid-liquid phase separation and biomolecular

condensates. Cell 176:419–434. https://doi.org/10.1016/j.cell.2018.12.035

29. McSwiggen DT, Mir M, Darzacq X, Tjian R (2019) Evaluating phase separation in live cells: diagnosis, caveats, and functional consequences. Genes Dev 33:1619–1634. https://doi.org/10.1101/gad.331520.119
30. Taylor NO, Wei M-T, Stone HA, Brangwynne CP (2019) Quantifying dynamics in phase-separated condensates using fluorescence recovery after photobleaching. Biophys J 117: 1285–1300. https://doi.org/10.1016/j.bpj.2019.08.030
31. Heinrich BS, Cureton DK, Rahmeh AA, Whelan SPJ (2010) Protein expression redirects vesicular stomatitis virus RNA synthesis to cytoplasmic inclusions. PLoS Pathog 6: e1000958. https://doi.org/10.1371/journal.ppat.1000958
32. Guseva S, Milles S, Jensen MR, Salvi N, Kleman J-P, Maurin D, Ruigrok RWH, Blackledge M (2020) Measles virus nucleo- and phosphoproteins form liquid-like phase-separated compartments that promote nucleocapsid assembly. Sci Adv 6:eaaz7095. https://doi.org/10.1126/sciadv.aaz7095
33. Alenquer M, Vale-Costa S, Etibor TA, Ferreira F, Sousa AL, Amorim MJ (2019) Influenza A virus ribonucleoproteins form liquid organelles at endoplasmic reticulum exit sites. Nat Commun 10:1629. https://doi.org/10.1038/s41467-019-09549-4
34. Monette A, Niu M, Chen L, Rao S, Gorelick RJ, Mouland AJ (2020) Pan-retroviral nucleocapsid-mediated phase separation regulates genomic RNA positioning and trafficking. Cell Rep 31:107520. https://doi.org/10.1016/j.celrep.2020.03.084
35. Sengupta P, Seo AY, Pasolli HA, Song YE, Johnson MC, Lippincott-Schwartz J (2019) A lipid-based partitioning mechanism for selective incorporation of proteins into membranes of HIV particles. Nat Cell Biol 21:452–461. https://doi.org/10.1038/s41556-019-0300-y
36. Savastano A, Ibáñez de Opakua A, Rankovic M, Zweckstetter M (2020) Nucleocapsid protein of SARS-CoV-2 phase separates into RNA-rich polymerase-containing condensates. Nat Commun 11:6041. https://doi.org/10.1038/s41467-020-19843-1
37. Carlson CR, Asfaha JB, Ghent CM, Howard CJ, Hartooni N, Safari M, Frankel AD, Morgan DO (2020) Phosphoregulation of phase separation by the SARS-CoV-2 N protein suggests a biophysical basis for its dual functions. Mol Cell 80:1092–1103.e4. https://doi.org/10.1016/j.molcel.2020.11.025
38. Cascarina SM, Ross ED (2020) A proposed role for the SARS-CoV-2 nucleocapsid protein in the formation and regulation of biomolecular condensates. FASEB J 34:9832–9842. https://doi.org/10.1096/fj.202001351
39. Chen H, Cui Y, Han X, Hu W, Sun M, Zhang Y, Wang P-H, Song G, Chen W, Lou J (2020) Liquid-liquid phase separation by SARS-CoV-2 nucleocapsid protein and RNA. Cell Res 30:1143–1145. https://doi.org/10.1038/s41422-020-00408-2
40. Iserman C, Roden C, Boerneke M, Sealfon R, McLaughlin G, Jungreis I, Park C, Boppana A, Fritch E, Hou YJ, Theesfeld C, Troyanskaya OG, Baric RS, Sheahan TP, Weeks K, Gladfelter AS (2020) Specific viral RNA drives the SARS CoV-2 nucleocapsid to phase separate. bioRxiv 2020.06.11.147199. https://doi.org/10.1101/2020.06.11.147199
41. Perdikari TM, Murthy AC, Ryan VH, Watters S, Naik MT, Fawzi NL (2020) SARS-CoV-2 nucleocapsid protein undergoes liquid-liquid phase separation stimulated by RNA and partitions into phases of human ribonucleoproteins. bioRxiv 2020.06.09.141101. https://doi.org/10.1101/2020.06.09.141101
42. Xu Z, Gong Q, Xia B, Groves B, Zimmermann M, Mugler C, Mu D, Matsumoto B, Seaman M, Ma D (2009) A role of histone H3 lysine 4 methyltransferase components in endosomal trafficking. J Cell Biol 186:343–353. https://doi.org/10.1083/jcb.200902146
43. Bailey JK, Fields AT, Cheng K, Lee A, Wagenaar E, Lagrois R, Schmidt B, Xia B, Ma D (2015) WD repeat-containing protein 5 (WDR5) localizes to the midbody and regulates abscission. J Biol Chem 290:8987–9001. https://doi.org/10.1074/jbc.M114.623611
44. Devaux P, von Messling V, Songsungthong W, Springfeld C, Cattaneo R (2007) Tyrosine 110 in the measles virus phosphoprotein is required to block STAT1 phosphorylation. Virology 360:72–83. https://doi.org/10.1016/j.virol.2006.09.049
45. Pfaller CK, Mastorakos GM, Matchett WE, Ma X, Samuel CE, Cattaneo R (2015) Measles virus defective interfering RNAs are generated frequently and early in the absence of C protein and can be destabilized by adenosine deaminase acting on RNA-1-like hypermutations. J Virol 89:7735–7747. https://doi.org/10.1128/JVI.01017-15
46. Langereis MA, Rabouw HH, Holwerda M, Visser LJ, van Kuppeveld FJM (2015) Knockout of cGAS and STING rescues virus infection of plasmid DNA-transfected cells. J Virol 89: 11169–11173. https://doi.org/10.1128/JVI.01781-15

Chapter 11

Culturing Immortalized Human Airway Epithelial Cells at an Air-Liquid Interface for Measles Virus Infection

Justin W. Kaufman, Brajesh K. Singh, Ni Li, and Patrick L. Sinn

Abstract

Measles virus (MeV) infection of airway surface epithelial cells provides a site for final amplification before being released back into the environment via coughing and sneezing. Multiple cell lines have served as models of polarized epithelia for MeV infection, such as Caco2 cells (intestinal derived human epithelia) or MDCK cells (kidney derived canine epithelia). In this chapter, we describe the materials and air-liquid interface (ALI) culture conditions for maintaining four different cell lines derived from human airway epithelial cells: 16HBE14o-, Calu-3, H358, and NuLi-1. We provide methods for confirming transepithelial electrical resistance (TER) and preparing samples for microscopy as well as expected results from apical or basolateral MeV delivery. Polarized human airway derived cells serve as tissue culture models for investigating targeted questions about how MeV exits a human host. In addition, these methods are generalizable to studies of other respiratory viruses or the biology of ALI airway epithelial cells.

Key words Paramyxovirus, Respiratory cells, Lung, Polarize, Differentiate, Respiratory virus

1 Introduction

Humans are the only natural reservoir for measles virus (MeV). As such, a challenge for the study of MeV is the identification of model systems that accurately reflect how MeV behaves in humans. Immortalized cell lines are easily accessible and relatively inexpensive to maintain. When properly maintained, cell lines can provide robust and reproducible results.

In organisms, many cell types show morphological and spatial directionality referred to as *polarity*. Within lungs, the columnar surface epithelial cells that line the airway lumen have defined apical and basolateral membranes. Tight junctions adhere epithelial cells together and define the boundary between the apical and basolateral surfaces; in addition, organelles, cytoskeletal features, and nuclei are reproducibly sorted to defined cellular loci [1]. Cell *differentiation* is a process of maturation from a progenitor cell

Dzwokai Z. Ma and Christian K. Pfaller (eds.), *Measles and Related Morbilliviruses: Methods and Protocols*, Methods in Molecular Biology, vol. 2808, https://doi.org/10.1007/978-1-0716-3870-5_11,

type to a more specialized cell type [2]. Immortalized lung derived epithelial cells grown in submerged cultures are typically neither polarized nor differentiated. However, many immortalized airway cells can be grown on semipermeable cell culture support inserts at an air-liquid interface (ALI). As discussed in this chapter, such models may polarize; however, differentiation into new cell types is uncommon. Typically, both polarization and differentiation processes are only observed in primary cells or iPSCs.

Submerged cultures of immortalized epithelial cells have greatly contributed to our understanding of MeV infection and replication. We and others use Vero E6 cells that stably express human Signaling Lymphocytic Activation Molecule Family Member 1 (hSLAM; aka CD150) to propagate and titer MeV for laboratory studies [3]. In addition to submerged cultures, multiple groups have used polarized human colon carcinoma cells (Caco2) and Madin-Darby canine kidney (MDCK) cells grown at an ALI to study MeV entry and release [4–8]. However, not all polarized epithelia behave the same, and experimental questions should be paired with an appropriate cell culture model.

Hundreds of lung derived epithelial cell lines are available for study, and many of these cell lines can establish transepithelial electrical resistance when grown at an ALI on semipermeable membranes. Here we focus on four cell lines: 16HBE14o-, H358, Calu-3, and NuLi-1. Gruenert and colleagues transformed human bronchial epithelial cells using SV40 large T-antigen to create 16HBE14o- cells [9]. When grown at an ALI, these cells are commonly used for electrophysiological studies due to their suitable regulation of tight junctions, transepithelial resistance, ion channels, and transporters [10]. Although these cells are a standard model of epithelial cell electrophysiology, they have not been widely used as a model of MeV infection. H358 cells are derived from a human non-small cell lung carcinoma [11]. These cells are well known to be permissive to MeV and were used to help identify nectin-4 as the cellular receptor in epithelial cells [12–14]. Calu-3 cells were first isolated and described in the mid-1970s by Fogh and colleagues [15]. Calu-3 cells can be maintained in ALI cultures and will establish relatively high TERs as compared to other ALI cultures [16, 17]. As a result, Calu-3 cells are commonly used in drug delivery and toxicology research relevant to the integrity of the airway epithelial cell layer. Prior to the identification of nectin-4 as the epithelial receptor for MeV, Yanagi and colleagues [12, 18] demonstrated that MeV infects many epithelial derived cell lines, including Calu-3 and H358, independent of SLAM or CD46. Finally, NuLi-1 cells were generated using hTERT and human papillomavirus type 16 E6 and E7 genes to transform primary cultures of human airway epithelial cells [19]. Of the cell lines described in this chapter, Nuli-1 cells most closely morphologically resemble primary airway epithelial cells. These

Table 1
Cell lines discussed in this chapter

Cell line	Tissue origin	Age/sex	Source	Steady state TER
16HBE14o-	Lung	1 M	SCC150 (Millipore)	100 ± 36
Calu-3	Lung	25 M	HTB-55 (ATCC)	150 ± 84
H358	Lung	M	CRL-5807 (ATCC)	25 ± 14
NuLi-1	Lung	36 M	CRL-4011 (ATCC)	420 ± 91

The steady state transepithelial electrical resistance (TER, ohms × cm^2) result from three cultures averaged over 6 days, beginning 24 h post-seeding

cells differentially express ciliated and goblet cell markers; however, differentiation seems to wane with increasing passage number. Of the cell lines discussed, NuLi-1 cells have the highest transepithelial resistance. This report is the first to describe MeV infection of NuLi-1 cells.

In this chapter, we summarize the method for polarizing human lung derived epithelial cell lines (Table 1) that can be used to study MeV infection and spread. These methods do not require unusual equipment or materials and can be easily adapted to other respiratory viruses. These models may provide a compromise between submerged epithelial cells and well-differentiated primary cells or iPSCs.

2 Materials

2.1 Human Placental Collagen Type IV Solution

1. Human Placental Collagen Type IV (Sigma C7521-50MG).
2. ddH_2O.
3. Glacial acetic acid.
4. Stir bar.
5. 250 mL flask.
6. Parafilm.
7. Evaporating dish.
8. Hot plate.
9. 1 L graduated cylinder.
10. 500 mL 0.22 μm filter.
11. Make Human Placental Collagen Type IV Solution:
    (a) Place stir bar in 250 mL flask, and add 83 mL water, 166.6 μL glacial acetic acid, and 50 mg collagen. Cover flask with parafilm.

(b) Fill glass evaporating dish half full with tap water. Place on hot plate. Set flask of collagen into evaporating dish. Heat collagen solution. Spin until collagen is dissolved.
(c) Pour dissolved collagen mix into a 1 L graduated cylinder. Fill to 830 mL with distilled deionized water.
(d) Filter sterilize with 0.22 μm filter.
(e) Final concentration of collagen is 0.06 mg/mL.
(f) Solution can be stored at 4 °C for at least 1 week.

2.2 Collagen Coating Transwells

1. 0.06 mg/mL Human Placental Collagen Type IV Solution.
2. Costar Transwell Permeable Support; 24-well plate, 6.5 mM insert, 0.4 μm polycarbonate membrane, tissue culture treated (Corning 3413).
3. Clear filters for microscopy (Corning 3470) (*see* **Note 1**).
4. 1× PBS (without calcium or magnesium) solution.

2.3 Seeding Cells onto Transwells

1. Culture vessel containing cells.
2. 1× PBS solution.
3. TrypLE Express (Gibco 12604-013).
4. Hemocytometer.
5. 0.4% Trypan blue solution.
6. Collagen coated polycarbonate transwells.
7. Complete growth medium, pre-warmed to 37 °C. *See* **Note 2**.
8. Sterile 50 mL conical tube.
9. 37 °C incubator with a humidified atmosphere of 5% CO_2.
10. Swinging bucket centrifuge.

2.4 Measuring Transepithelial Electrical Resistance (TER)

1. 24-well plate, 6.5 mM insert (0.33 cm^2 area), 0.4 μm polycarbonate membrane, tissue culture treated (Corning 3413).
2. EVOM^2 Epithelial Volt/Ohm (TER) Meter (World Precision Instruments EVOM2).
3. Chopstick Electrode Set for EVOM2 (World Precision Instruments STX2).
4. 70% ethanol.
5. Sterile 1× PBS.

2.5 Infecting Transwells with Measles Virus

1. Transwells containing cells to be infected.
2. Measles virus. The MeV used for the examples in this chapter was a recombinant MeV derived from the wild-type IC-323 strain and expressing eGFP upstream of the nucleocapsid (N) gene as previously described [20, 21].

3. Opti-MEM Media (Gibco 31985-070).
4. Complete growth medium, pre-warmed to 37 °C.
5. 37 °C incubator with a humidified atmosphere of 5% CO_2.
6. Sterile 1.5 mL reaction tube.
7. Sterile 15 cm tissue culture dish.
8. Parafilm.

2.6 Preparing Samples for Microscopy

1. 2% paraformaldehyde (PFA) in PBS.
2. ddH_2O.
3. Sterile 1× PBS.
4. 0.2% Triton X-100 in SuperBlock Blocking Buffer in PBS (Thermo 37515).
5. Desired 1° and 2° antibodies.
6. Razor blade.
7. Superfrost Plus Microscope Slides (Fisher 12-550-15).
8. Surgipath Cover Glass 22 × 50 (Leica 3800120).
9. Vectashield mounting medium for fluorescence with DAPI (Vector Laboratories H-1200).
10. Clear fingernail polish.
11. Fluorescent or confocal microscope.

3 Methods

3.1 Collagen Coating Transwells

1. Place 150 μL of human placental collagen type IV solution onto the apical surface of each transwell.
2. Incubate the transwells containing collagen at room temperature overnight.
3. Aspirate the human placental collagen type IV solution from the transwells.
4. Wash the transwells three times with 150 μL 1× PBS, use the transwells immediately, or air-dry for 3 h to overnight.

3.2 Seeding Cells Onto Transwells

1. Prior to seeding on transwells, cells can be maintained in submerged cultures in culture vessels such as 75 cm^2 flasks or 100 mM plates. *See* **Note 2** for submerged media recipes for 16HBE14o-, H358, Calu-3, and NuLi-1 cells.
2. Remove and discard the spent cell culture media from the culture vessel.
3. Wash cells using PBS.

4. Remove and discard the wash solution from the culture vessel.
5. Add the TrypLE express to the culture vessel; use enough reagent to cover the cell layer.
6. Incubate the culture vessel at 37 °C for approximately 5 min.
7. When >90% of the cells have detached, transfer the cells to a 50 mL conical tube and centrifuge at 300 × g for 5 min.
8. Resuspend the cell pellet in a minimal volume of pre-warmed ALI medium, and remove a sample for counting.

   See **Note 3** ALI media recipes for 16HBE14o-, H358, Calu-3, and NuLi-1 cells.
9. Count the cells and determine the percent viability using a hemocytometer and Trypan blue exclusion. Start with a 1: 100 cell dilution, and adjust if too many or too few cells are observed.
10. Dilute the cells to 1×10^6/mL using pre-warmed complete ALI medium.
11. Place 125 μL (125,000 cells) of the cell suspension onto the apical surface of each transwell.
12. Add 500 μL of ALI media to the basolateral compartment of each transwell.
13. Place into a 37 °C incubator with a humidified atmosphere of 5% CO_2.
14. After 24 h, gently aspirate media from the apical surface, and incubate at an air-liquid interface changing the basolateral media as needed.

3.3 Measuring Transepithelial Electrical Resistance

1. Submerge the $EVOM^2$ epithelial volt ohmmeter (WPI EVOM2) chopsticks probe (WPI STX2) into 70% EtOH to sterilize.
2. Allow the probe to air-dry and rinse the probe with sterile media or buffer.
3. For cultures incubated at air-liquid interface, add 150 μL of media to the apical compartment.
4. Insert the probe into the transwell so that the longer portion of the probe touches the bottom of the basolateral compartment and the shorter probe arm is submerged in the apical compartment, but not touching the membrane (Fig. 1a).
5. Record the net resistance measured by the instrument.
6. Re-sterilize the probe in 70% EtOH and let dry between sample sets.
7. Calculate the TER by subtracting the resistance of a blank transwell from the recorded net resistance, and then multiply

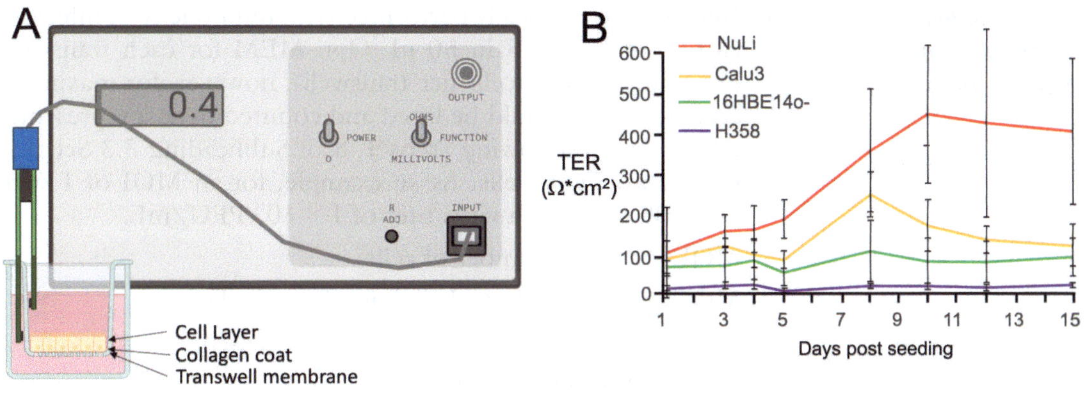

Fig. 1 Transepithelial electrical resistance (TER). (**a**) A volt ohmmeter is used to determine the TER (ohms × cm^2) as shown. (**b**) For each indicated cell line grown at ALI, the TER was measured for 15 days, beginning 24 h post-seeding. $n = 3$

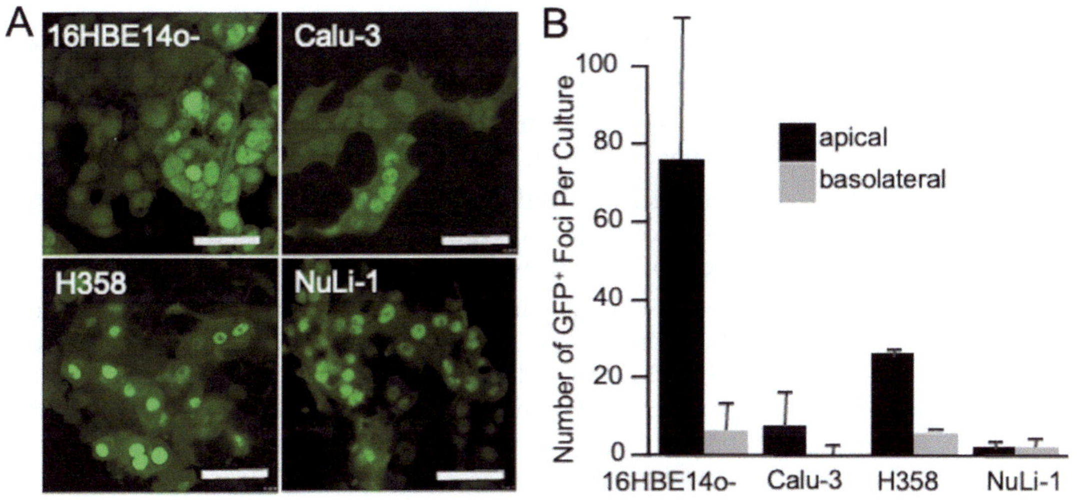

Fig. 2 Infection of ALI cell lines. (**a**) Individual MeV foci of infection for each indicated cell line are shown. Scale bar = 50 μm. (**b**) For each indicated cell line, the number of GFP+ foci was counted ~48 h following apical or basolateral MeV delivery. Total foci per culture are shown. $n = 3$

that difference by the surface area of the transwell, for example, measured net resistance = 1000 Ω in a 24-well, 0.33 cm^2 transwell with a blank resistance of 100 Ω. (1000 Ω −100 Ω) * 0.33 cm^2 = 297 Ω*cm^2.

8. Expected TER measurements from days 1 to 15 post-seeding for 16HBE14o-, H358, Calu-3, and NuLi-1 cells are shown (Fig. 2b).

3.4 Infecting Transwells with Measles Virus

1. Infections are typically performed with multiplicity of infection (MOI) of 0.1 to 5 in 50 μL Opti-MEM for each transwell. Estimate 300,000 cells per transwells; however, for maximum accuracy, cells should be lifted and counted from a representative transwell following **steps 1–8** of Subheading 3.3 Seeding Cells onto Transwells. As an example, for an MOI of 1 with 300,000 cells and a virus titer of 1×10^8 PFU/mL.

   MOI = PFU / number of cells.

   MOI = (ml * (PFU/mL)) / number of cells.

   ml = (MOI * number of cells) / (PFU/mL).

   ml = $(1 * 300{,}000) / 1 \times 10^8$.

   ml = $300{,}000 / 1 \times 10^8$.

   ml = 0.003.
2. To achieve an MOI of 1 with 300,000 cells and a virus titer of 1×10^8 PFU/ml, we need 3 μL of measles virus per transwell culture.
3. Place 47 μL (per transwell) of pre-warmed Opti-MEM into a sterile 1.5 mL reaction tube.
4. Add 3 μL (per transwell) of measles virus, lightly mix, and briefly centrifuge.
5. For apical infected samples cultured at air-liquid interface, add 50 μL of the virus/Opti-MEM mixture to the apical compartment of the transwell.
6. For basolaterally infected samples, remove the transwell from the culture plate, and place it upside down into a sterile 150 mM dish. Add 50 μL of the virus to the basolateral side of the transwell, and cover the 150 mM dish with a lid (the lid is propped up a little with parafilm to prevent contact with the inoculum).
7. Carefully place the samples into a 37 °C incubator with a humidified atmosphere of 5% CO_2, and allow the infection to proceed for 4 h.
8. Following apical infection, aspirate off the inoculation media from the apical surface, and leave dry.
9. Following basolateral infection, return the transwells upright into the culture plate.
10. Allow the infection to proceed as desired, and change the culture media as needed (~1×/week).
11. At the desired time points, prepare the samples for microscopy.

3.5 Preparing Samples for Microscopy

1. For each step below (fixing, blocking, antibody solution, washing, etc.), simultaneous application of a volume of 100 μL directly on the apical surface and 200 μL in the well (contacting

the basolateral surface) is sufficient for coverage and will not evaporate within the incubation times; however, the exact volumes are not critical.

2. Aspirate media from cells and fix cells by applying 2% PFA for 15 min at room temperature.
3. Wash 3× in PBS.
4. Block and permeabilize by adding SuperBlock Blocking Buffer containing 0.2% Triton X-100 for 1 h at room temperature (or overnight at 4 °C).
5. Wash 3× in PBS.
6. Stain cells with desired antibodies or counterstains.
7. Wash the cells 3× in PBS between consecutive antibody incubations.
8. Remove the filters containing the cells by cutting around the edges of the membrane with a razor blade.
9. Being careful to keep the apical surface facing up, mount the membranes on a slide using Vectashield with DAPI, coverslip, and clear nail polish.
10. Visualize and photograph the cells using a fluorescence and/or confocal microscope (Fig. 2a). GFP+ cells can be quantified by visual counting (Fig. 2b). For these studies, MeV conferred GFP expression indicates MeV infected cells (*see* **Note 4**).

4 Notes

1. *Choice of Transwell Supports.*

   Here we propose two alternatives for transwell permeable supports. Although either will support the growth and polarization of airway cells, the polycarbonate supports are often better for longevity of the cells, and the clear supports result in sharper microscopy images.

2. *Media Solutions for Propagation in Submerged Cultures.*

   16HBE14o- cultures are grown in serum-free PneumaCult^{tm}Ex Plus Media (Stem Cell Technologies 05040) supplemented with 500 U/mL penicillin and 50 μg/mL streptomycin at 37 °C in 5% CO_2. Cells are split 1:5 when they reach 80% confluency.

   H358 cells are cultured in RPMI-1640 media supplemented with 10% fetal calf serum, 500 U/mL penicillin, and 50 μg/mL streptomycin at 37 °C in 5% CO_2. Cells are split 1:5 when they reach 80% confluency.

   Calu-3 cells are cultured at 37 °C and 5% CO_2 in a 50%/50% mixture of Dulbecco's Modified Eagle's Medium/Ham's

F-12 (DMEM/F12) supplemented with 15% fetal calf serum (FCS), 500 U/mL penicillin, and 50 μg/mL streptomycin. Cells are split 1:5 when they reach 80% confluency.

NuLi-1 Cells are grown in serum-free PneumaCult[tm]Ex Plus Media (Stem Cell Technologies 05040) supplemented with 500 U/mL penicillin and 50 μg/mL streptomycin at 37 °C in 5% CO_2. Cells are split 1:5 when they reach 80% confluency.

3. *Media Solutions for ALI Cultures.*

16HBE14o- cells are seeded onto collagen coated transwells as described in Subheadings 3.2 and 3.3 and maintained in serum-free PneumaCult[tm]-ALI Media (Stem Cell Technologies 05001). Twenty-four hours after seeding, the apical media is removed, and the cells are cultured at an air-liquid interface.

H358 cells are seeded onto collagen coated transwells as described in Subheadings 3.2 and 3.3 and maintained in RPMI-1640 media supplemented with 10% fetal calf serum, 500 U/mL penicillin, and 50 μg/mL streptomycin at 37 °C in 5% CO_2. Twenty-four hours after seeding, the apical media is removed, and the cells are cultured at an air-liquid interface.

Calu-3 cells are seeded onto collagen coated transwells as described in Subheadings 3.2 and 3.3 and maintained in a 50%/50% mixture of Dulbecco's Modified Eagle's Medium/Ham's F-12 (DMEM/F12) supplemented with 15% fetal calf serum (FCS), 500 U/mL penicillin, and 50 μg/mL streptomycin. Twenty-four hours after seeding, the apical media is removed, and the cells are cultured at an air-liquid interface.

NuLi-1 cells are seeded onto collagen coated transwells as described in Subheadings 3.2 and 3.3 and maintained in serum-free PneumaCult[tm]-ALI Media (Stem Cell Technologies 05001). Twenty-four hours after seeding, the apical media is removed, and the cells are cultured at an air-liquid interface.

4. *Apical Versus Basolateral MeV Entry.*

In all four cell lines tested here, MeV preferentially entered at the apical surface and caused membrane fusion resulting in syncytia formation. In contrast, we previously reported that MeV infects at the basolateral surface of well-differentiated, unpassaged primary cultures of airway epithelial cells [8, 13]. As such, polarized immortalized airway cells may not be optimal for answering questions about polarity of viral entry.

Acknowledgments

We thank Cami Hippee and Ashley Cooney for their editorial comments. This work was supported by the National Institutes of Health R01 AI132402 and R01 HL105821.

References

1. Hasan S, Sebo P, Osicka R (2018) A guide to polarized airway epithelial models for studies of host-pathogen interactions. FEBS J 285(23): 4343–4358. https://doi.org/10.1111/febs.14582
2. Karp PH, Moninger TO, Weber SP, Nesselhauf TS, Launspach JL, Zabner J et al (2002) An in vitro model of differentiated human airway epithelia. Methods for establishing primary cultures. Methods Mol Biol 188:115–137. https://doi.org/10.1385/1-59259-185-X:115
3. Hippee CE, Singh BK, Thurman AL, Cooney AL, Pezzulo AA, Cattaneo R et al (2021) Measles virus exits human airway epithelia within dislodged metabolically active infectious centers. PLoS Pathog 17(8):e1009458. https://doi.org/10.1371/journal.ppat.1009458
4. Blau DM, Compans DW (1995) Entry and release of measles virus are polarized in epithelial cells. Virology 210:91–99. https://doi.org/10.1006/viro.1995.1320
5. Blau DM, Compans RW (1997) Adaptation of measles virus to polarized epithelial cells: alterations in virus entry and release. Virology 231:281–289. https://doi.org/10.1006/viro.1997.8520
6. Maisner A, Klenk H, Herrler G (1998) Polarized budding of measles virus is not determined by viral surface glycoproteins. J Virol 72(6):5276–5278. https://doi.org/10.1128/jvi.72.6.5276-5278.1998
7. Naim HY, Ehler E, Billeter MA (2000) Measles virus matrix protein specifies apical virus release and glycoprotein sorting in epithelial cells. EMBO J 19(14):3576–3585. https://doi.org/10.1093/emboj/19.14.3576
8. Sinn PL, Williams G, Vongpunsawad S, Cattaneo R, McCray PB Jr (2002) Measles virus preferentially transduces the basolateral surface of well-differentiated human airway epithelia. J Virol 76(5):2403–2409. https://doi.org/10.1128/jvi.76.5.2403-2409.2002
9. Cozens AL, Yezzi MJ, Kunzelmann K, Ohrui T, Chin L, Eng K et al (1994) CFTR expression and chloride secretion in polarized immortal human bronchial epithelial cells. Am J Respir Cell Mol Biol 10:38–47. https://doi.org/10.1165/ajrcmb.10.1.7507342
10. Wan H, Winton HL, Soeller C, Stewart GA, Thompson PJ, Gruenert DC et al (2000) Tight junction properties of the immortalized human bronchial epithelial cell lines Calu-3 and 16HBE14o. Eur Respir J 15(6): 1058–1068. https://doi.org/10.1034/j.1399-3003.2000.01514.x
11. Brower M, Carney DN, Oie HK, Gazdar AF, Minna JD (1986) Growth of cell lines and clinical specimens of human non-small cell lung cancer in a serum-free defined medium. Cancer Res 46(2):798–806
12. Takeda M, Tahara M, Hashiguchi T, Sato TA, Jinnouchi F, Ueki S et al (2007) A human lung carcinoma cell line supports efficient measles virus growth and syncytium formation via a SLAM- and CD46-independent mechanism. J Virol 81(21):12091–12096. https://doi.org/10.1128/jvi.01264-07
13. Leonard VH, Sinn PL, Hodge G, Miest T, Devaux P, Oezguen N et al (2008) Measles virus blind to its epithelial cell receptor remains virulent in rhesus monkeys but cannot cross the airway epithelium and is not shed. J Clin Invest 118(7):2448–2458. https://doi.org/10.1172/JCI35454
14. Muhlebach MD, Mateo M, Sinn PL, Prufer S, Uhlig KM, Leonard VH et al (2011) Adherens junction protein nectin-4 is the epithelial receptor for measles virus. Nature 480(7378): 530–533. https://doi.org/10.1038/nature10639
15. Fogh J, Fogh JM, Orfeo T (1977) One hundred and twenty-seven cultured human tumor cell lines producing tumors in nude mice. J Natl Cancer Inst 59(1):221–226. https://doi.org/10.1093/jnci/59.1.221
16. Grainger CI, Greenwell LL, Lockley DJ, Martin GP, Forbes B (2006) Culture of Calu-3 cells at the air interface provides a representative model of the airway epithelial barrier. Pharm Res 23(7):1482–1490. https://doi.org/10.1007/s11095-006-0255-0

17. Harcourt JL, Haynes LM (2013) Establishing a liquid-covered culture of polarized human airway epithelial Calu-3 cells to study host cell response to respiratory pathogens in vitro. J Vis Exp 72:e50157. https://doi.org/10.3791/50157
18. Tahara M, Takeda M, Shirogane Y, Hashiguchi T, Ohno S, Yanagi Y (2008) Measles virus infects both polarized epithelial and immune cells by using distinctive receptor-binding sites on its hemagglutinin. J Virol 82(9):4630–4637. https://doi.org/10.1128/jvi.02691-07
19. Zabner J, Karp P, Seiler M, Phillips SL, Mitchell CJ, Saavedra M et al (2003) Development of cystic fibrosis and noncystic fibrosis airway cell lines. Am J Physiol Lung Cell Mol Physiol 284(5):L844–L854. https://doi.org/10.1152/ajplung.00355.2002
20. Singh BK, Hornick AL, Krishnamurthy S, Locke AC, Mendoza CA, Mateo M et al (2015) The Nectin-4/Afadin protein complex and intercellular membrane pores contribute to rapid spread of measles virus in primary human airway epithelia. J Virol 89(14):7089–7096. https://doi.org/10.1128/jvi.00821-15
21. Singh BK, Pfaller CK, Cattaneo R, Sinn PL (2019) Measles virus ribonucleoprotein complexes rapidly spread across well-differentiated primary human airway epithelial cells along F-actin rings. MBio 10(6):e02434-19. https://doi.org/10.1128/mbio.02434-19

Chapter 12

In Vitro Modeling of Feline Morbillivirus Infections Using Primary Feline Kidney Cells

Michael Sieg, Johannes Busch, Denny Böttcher, and Thomas W. Vahlenkamp

Abstract

Domestic cats are the natural host of feline morbilliviruses (FeMV). Although other species can also be infected (such as dogs and opossums), no laboratory animal infection model is established so far. In vitro models for studying the molecular pathogenesis are therefore needed. For this purpose, propagation and titration of FeMV are key techniques. Unlike other morbilliviruses, such as canine distemper virus (CDV) or measles virus (MV), FeMV is a slow growing virus in cell culture and is difficult to titrate using classical plaque techniques. Here we describe methods for the efficient isolation of FeMV from natural sources (e.g., urine), the propagation of viral stocks, and their titration. In addition, we establish the generation of a three-dimensional infection model mimicking the feline tubular epithelium.

Key words Feline morbillivirus, Virus isolation, Virus quantification, Primary feline kidney cells, Differentiated epithelium

1 Introduction

Feline morbilliviruses (FeMV) represent a diverse group of enveloped, single stranded RNA viruses within the genus *Morbillivirus*, family *Paramyxoviridae* [1]. After its first description in 2012 in Hong Kong [2], FeMV was subsequently found in domestic cat populations in Asia and Europe as well as in North and South America (reviewed in [3]). FeMV was associated with tubulointerstitial nephritis (TIN) in the initial retrospective case-control study from stray cats [2]. Subsequent studies tried to elucidate the impact of FeMV to feline kidney disease, but results have not yet finally answered this question. Some authors found a correlation of FeMV to chronic kidney diseases of domestic cats [4–8], while others did not [9–12]. The situation is further complicated due to the presence of at least two different genotypes of FeMV. In addition, it was

Dzwokai Z. Ma and Christian K. Pfaller (eds.), *Measles and Related Morbilliviruses: Methods and Protocols*, Methods in Molecular Biology, vol. 2808, https://doi.org/10.1007/978-1-0716-3870-5_12,

shown by in vitro [13] and in vivo [7, 8] infection experiments that FeMV can infect cells of other organs than the kidney, e.g., immune cells, epithelial cells of the lung and the urinary bladder, as well as neuronal cells. The susceptibility of different organs and cells were also confirmed by postmortem analyses of naturally infected cats [14]. Furthermore, FeMV seems not to be restricted to domestic cats as infections were detected in black leopards [15], guignas [16], dogs, [17] and opossums [18]. To date, there is no laboratory animal model available to investigate FeMV pathogenesis under standardized conditions.

Primary feline proximal tubular epithelial cells can be used as an in vitro model for studying epithelial-to-mesenchymal transition (EMT), a key feature of chronic kidney disease [19]. As these cells can be successfully isolated from cadaverous feline kidney tissue, this method presents a good alternative to animal experiments. The protocols described below are suitable for the investigation of molecular pathomechanisms of FeMV infections in feline kidney cells. They may also be used as tools for elucidating the innate immune response against FeMV or for investigation of antiviral therapies.

2 Materials

2.1 Equipment and Consumables

1. Class II laminar flow hood.
2. Ultra-deep freezer (−86 °C).
3. Liquid nitrogen storage tank.
4. Incubator at 37 °C, 5% CO_2, and 90% relative humidity.
5. Rotary shaker.
6. Inverted microscope.
7. Water bath (37 °C and 95 °C).
8. Cell strainer, 100 μm.
9. Corning™ Primaria™ Tissue Culture Flasks 75 cm^2 (T75) and 6-well plates.
10. Polyethylene terephthalate (PET) hanging cell culture inserts, pore size 0.4 μm for 6-well plates.
11. Serological pipets, 5 mL and 10 mL, sterile.
12. Sterile centrifuge tubes, 2, 15, and 50 mL.
13. Tissue culture plates, 6-well and 96-well format.
14. Multichannel (8- or 12-channel) pipette.
15. Micropipettes and aerosol-resistant, sterile pipette tips.
16. Sterile scalpels and forceps.

17. Neubauer hemocytometer.
18. Cryovials.
19. Cell freezing apparatus, cooling rate ~ −1 °C/min (such as Nalgene® Mr. Frosty®).

2.2 Patient Sample Material

1. Urine from FeMV-excreting cats (*see* **Note 1**).
2. Kidney(s) from specific pathogen-free (SPF) healthy domestic cats (*see* **Note 2**) in ice-cold sterile HBSS (Subheading 2.3).

2.3 Media and Reagents for Tissue Culture

2.3.1 Isolation and Titration of FeMV

1. LLC-MK2 cells (ATCC #CCL-7) transfected with pDisplay™ vector (Invitrogen) expressing the ectodomain of feline SLAM-F1 receptor (*see* **Note 3**).
2. Sterile 1× phosphate buffered saline (PBS) cell culture grade.
3. Fetal bovine serum (FBS): heat-inactivate FBS for 30 min at 56 °C. Store at 4 °C and use within 3 months.
4. Basis medium: Dulbecco's Modified Eagle's Medium (DMEM, high glucose, GlutaMax™, Gibco), 1% (v/v) nonessential amino acids (Gibco), 1% (v/v) sodium pyruvate (Gibco), 100 U/mL penicillin, and 100 μg/mL streptomycin.
5. Sterile trypsin (0.05%)—EDTA (0.02%) solution in PBS. This working solution can be used for approximately 3 months when stored at 4 °C.
6. Cultivation medium: basis medium supplemented with 5% (v/v) FBS.
7. Infection medium: basis medium supplemented with 2% (v/v) FBS.
8. IFA blocking medium: 5% (w/v) bovine serum albumin (cell culture grade, Sigma-Aldrich) in PBS. Filter-sterilize through 0.45 μm filter and store at 4 °C.
9. 80% (v/v) acetone (research grade) in distilled water.
10. Primary antibody: α-FeMV nucleoprotein polyclonal rabbit serum (see supplementary file in Ref. [20]) diluted 1:1000 in IFA blocking buffer.
11. Secondary antibody: fluorochrome-coupled α-Rabbit IgG antibody (e.g., Goat anti-Rabbit IgG (H + L) Cross-Adsorbed Secondary Antibody, Alexa Fluor™ 546) diluted 1:1000 in IFA blocking medium as stated by the manufacturer.

2.3.2 Culture of Primary Feline Kidney Cells

1. Sterile 1× HBSS (Hank's balanced salt solution) with calcium, magnesium, and phenol red.
2. Collagenase solution: dissolve lyophilized collagenase type II (Gibco, produced from *Clostridium histolyticum*) in HBSS to a final concentration of 1 mg/mL, and pass through a 0.22 μm filter. Store at 4 °C, and use within 1 week.

3. Basal kidney epithelial cell medium (basal-KECM): mix equal amounts of Dulbecco's Modified Eagle's Medium (DMEM, high glucose, Gibco) and Ham's F12 Nutrient Medium (Gibco). Add Insulin-Transferrin-Selenium (ITS-G, Gibco, 100×) to a final concentration of 1×, and supplement with 100 U/mL penicillin and 100 μg/mL streptomycin.
4. Complete kidney epithelial cell medium (c-KECM): to basal-KECM, add 10 ng/mL human epidermal growth factor and 5×10^{-8} M hydrocortisone.
5. Soybean trypsin inhibitor (Sigma-Aldrich).

2.4 Equipment, Consumables, and Reagents for Histology

1. Sliding or rotary microtome.
2. 4% (w/v) neutral buffered (PBS) formaldehyde solution.
3. Xylenes (mixed isomers).
4. Ethanol, 100% (v/v).
5. Ethanol, 95% (v/v).
6. Ethanol, 70% (v/v).
7. Butyl acetate.
8. Isopropyl alcohol (2-propanol).
9. Paraffin wax (suitable for histology).
10. Superfrost Plus™ slides (Thermo Fisher Scientific).
11. Hemalum staining solution.
12. Eosin staining solution.
13. IHC antigen retrieval solution: dissolve 0.05% (v/v) citraconic anhydride by vigorous stirring in distilled water; adjust pH to 7.2–7.6 with 1 N sodium hydroxide solution. Prepare immediately before use.
14. IHC blocking solution: SuperBlock™ T20 (PBS) Blocking Buffer (Thermo Fisher Scientific).
15. Primary antibody: affinity purified anti-FeMV nucleoprotein polyclonal rabbit IgG diluted to 1 μg/mL in 0.1% (v/v) IHC blocking solution.
16. Primary antibody: mouse antihuman CD10 (clone F-4, Santa Cruz Biotechmology, verified cross-reactive with feline CD10).
17. Secondary antibody: Goat anti-Rabbit IgG (H + L) Cross-Adsorbed Secondary Antibody, Alexa Fluor™ 546, diluted in 0.1% (v/v) IHC blocking medium as stated by the manufacturer.
18. Secondary antibody: Goat anti-mouse IgG (H + L) Cross-Adsorbed Secondary Antibody, Alexa Fluor™ 488, diluted in 0.1% (v/v) IHC blocking medium as stated by the manufacturer.
19. DAPI nuclear staining solution.
20. Mounting medium for fluorescent dye-stained tissues.

3 Methods

3.1 Isolation of FeMV from Urine Samples

1. Seed LLC-MK2-CatSLAM cells in T75 cell culture flask to an appropriate density so that a confluency of approximately 80% will be achieved next day. Normally, 2×10^6 cells per flask are suitable for this purpose. Add 10 mL cultivation medium and incubate overnight at 37 °C and 5% CO_2 (*see* **Note 4**).
2. Wash cells twice with warmed PBS and add about 5 mL of diluted urine sample (*see* **Note 5**). Incubate at 37 °C for 2 h gently rocking the flask every 15 min to ensure proper sample distribution. Include a noninfected (MOCK) flask as a negative control for monitoring the normal cell morphology during the whole virus isolation process.
3. Gently wash cell monolayer twice with warm PBS to remove inoculum and cell debris. Add 10 mL of infection medium per flask, incubate for 7 days at 37 °C and 5% CO_2, and observe daily for the presence of a cytopathic effect (*see* Fig. 1a). Normally, only few (if any) syncytia are observed after this time, reflecting low viral load in the inoculum. Therefore, prolonged incubation times are needed to isolate FeMV strains.
4. Remove the infection medium from the cell monolayer and gently wash once with warmed PBS. Add 10 mL of fresh prewarmed infection medium, and incubate for additional 7 days at 37 °C and 5% CO_2. After this period, typical syncytia should be observed in the infected flask (*see* Fig. 1b). If no cytopathic effect can be seen, viral isolation is likely to have failed and flasks can be discarded.
5. Shock-freeze virus-positive flasks at −80 °C for several hours or overnight, and slowly thaw on ice. Repeat freeze-thaw cycle once. After second thaw, scrape down cell lysates and transfer into prechilled sterile conical 15 mL tubes.
6. Remove the cellular debris by centrifugation at 500× *g* and 4 °C for 10 min.

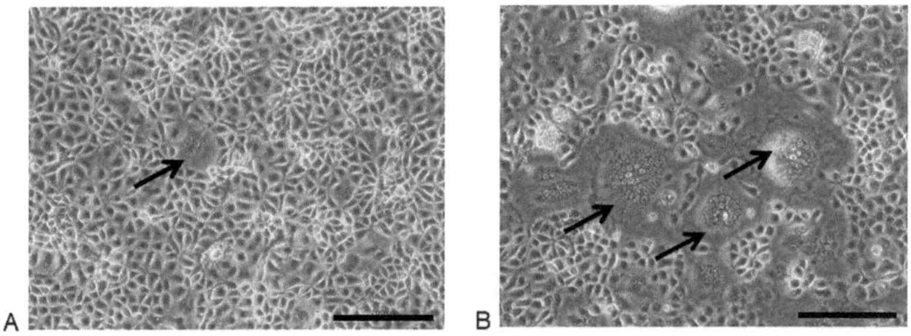

Fig. 1 Cytopathic effects of FeMV on LLC-MK2-CatSLAM cells at passage no. 1 after 7 (**a**) or 7 + 7 days (**b**) of incubation. Arrows show the formation of syncytia due to viral replication. Scale bars represent 200 μm

7. Discard debris and aliquot supernatant into prechilled, labeled 1.5 mL sterile cryovials. Store aliquots at −80 °C (or below) until needed. This is passage no. 1 of the viral isolate (*see* **Note 6**).

3.2 Titration of FeMV Virus Stock

1. Prepare a 96-well plate with LLC-MK2-CatSLAM cells the day before viral titration. For this purpose, seed 2×10^4 cells per well in 100 μL cultivation medium, and incubate overnight at 37 °C and 5% CO_2. Cells should be nearly 100% confluent for viral titration the next day.
2. Wash cells twice with warmed PBS.
3. Prepare eight 2.0 mL sterile tubes each containing 900 μL of pure DMEM. Store on ice until further use.
4. Add 100 μL of virus stock to the first tube and mix by inverting several times (= 10^{-1} dilution). Repeat 1:10 dilution process until 10^{-8}. Store diluted virus on ice and use immediately.
5. Remove the PBS from the cells and add 100 μL of pure viral stock to row A. To the wells of row B–I, add 100 μL of the prepared virus dilutions (10^{-1}–10^{-8}). Negative controls are included at row J by adding 100 μL of pure DMEM. Each dilution should be tested in six or eight replicates.
6. Incubate 96-well plate for 2 h at 37 °C and 5% CO_2.
7. Add 100 μL of infection medium to each well without discharging the viral dilutions. Incubate for 5 days at 37 °C and 5% CO_2.
8. Remove the medium from the cells and wash once with ice-cold PBS. Discard the PBS and add 100 μL of 80% acetone (prechilled at −20 °C) to each well, and incubate for 10 min at −20 °C.
9. Remove the acetone and wash cells twice with ice-cold PBS and discard.
10. Add 100 μL IFA blocking medium to each well and incubate for 30 min at 37 °C.
11. Discard the blocking solution and replace with 100 μL of primary antibody solution. Incubate for 1–2 h at 37 °C.
12. Remove the primary antibody and wash cells three times with PBS.
13. Add 100 μL of secondary antibody solution to each well, and incubate for 1 h at 37 °C.
14. Discard secondary antibody from the cells, wash twice with PBS, and analyze cells using a fluorescence microscope. Calculate $TCID_{50}$ titer by the Spearman-Karber method (*see* **Note 7**).

3.3 Isolation and Infection of Primary Feline Kidney Epithelial Cells

1. Use sterile techniques for all steps in the following protocol to ensure that primary cells will not become contaminated with bacteria or molds. For best practices, work in a Class II laminar flow hood.
2. Quickly remove the kidney from a euthanized cat/carcass. Discharge the fatty tissue, dissect the kidney longitudinally into two halves, and remove the capsule on a sterile Petri dish (*see* Fig. 1a). Immediately transfer the organ material into ice-cold 1× HBSS.
3. Dissect 5–10 mm pieces from the renal cortex, and be sure not to prepare the medullary part as this will lead to marked fibrocyte contamination of the primary cells. Always completely emerge the kidney and tissue pieces in ice-cold HBSS during preparation!
4. Using two sterile scalpels, mince cortex pieces into small parts of approximately 1 mm^3 size. Transfer the minced tissue into a 50 mL centrifuge tube, and cover with fresh ice-cold HBSS (*see* Fig. 1b).
5. After allowing the minced tissue pieces to settle, discharge most of the upper HBSS medium and refill with fresh ice-cold HBSS. Invert several times, and let settle down the tissue pieces again. Repeat this washing step until the upper HBSS medium is clear and most of the erythrocytes and cellular debris are washed out (*see* Fig. 1c).
6. Aliquot the minced cortical tissue into new conical 50 mL centrifuge tubes, and gently discharge the supernatant leaving about 15 mL tissue pellet per tube.
7. Add 25 mL prewarmed (37 °C) collagenase solution per tube, and incubate on a rotary shaker for 20–30 min at 37 °C.
8. Allow the minced tissue to settle down and then discharge the supernatant. Add 25 mL of new prewarmed (37 °C) collagenase solution, and repeat incubation on the rotary shaker for 30 min.
9. The resulting suspension is filtered through a 100 μm mesh, and the flow-through is collected in a sterile 50 mL centrifuge tube (*see* Fig. 1d). The remaining tissue can be retreated with collagenase as described in **step 8** for two additional times.
10. Centrifuge the filtered cell suspension at 400× *g* for 5 min at 4 °C to pellet the primary cells. Discharge to supernatant, and wash cell pellet with 50 mL ice-cold HBSS for two additional times to remove the collagenase (Fig. 2).
11. After the final centrifugation step, resuspend the cell pellet from one 50 mL tube in 10 mL of prewarmed (37 °C) c-KECM, and transfer the suspension into a T75 Primaria™ cell culture flask.

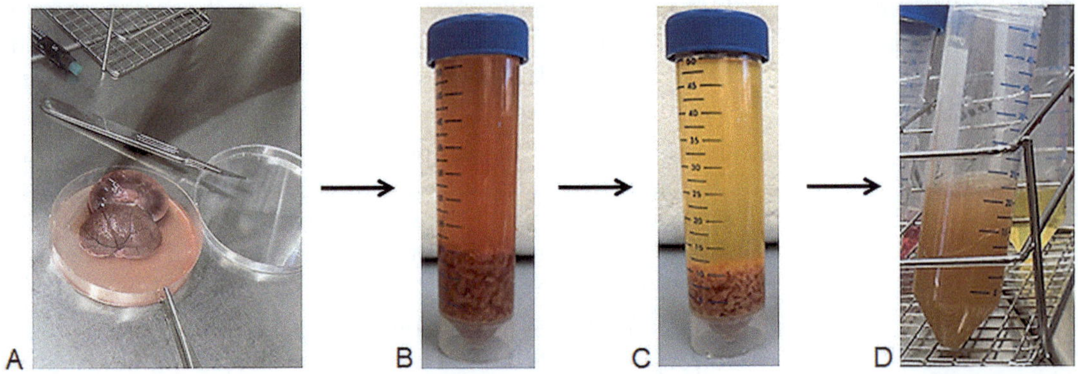

Fig. 2 Preparation of feline primary kidney cells. Kidneys are first bisected (**a**) and minced and then washed vigorously with HBSS (**b**) until most of the debris and erythrocytes are removed (**c**). Collagenase digestion is used to obtain single cell suspensions of primary kidney cells (**d**)

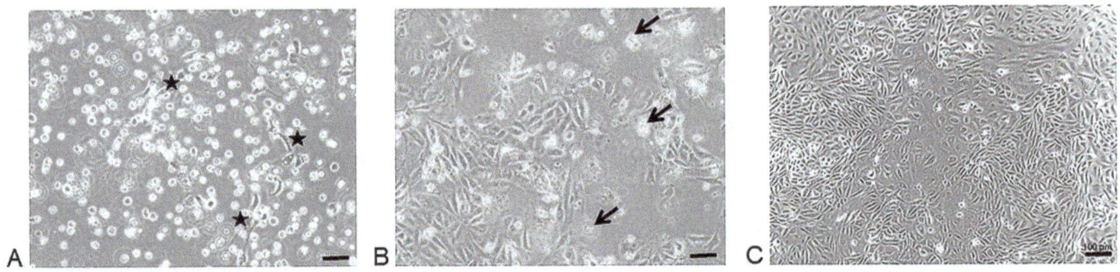

Fig. 3 Cultivation of feline primary kidney cells. One day after seeding, only a minor portion of the cells are attached to the flask (**a**). After 48 h, viable epithelial cells begin to divide, forming cell clusters (**b**). Five to seven days post seeding, cells form a confluent monolayer (**c**). Asterisks highlight epithelial cells, and arrows show dead cells unable to attach to the flask. Scale bar represents 100 μm

12. Cultivate cells for 2 days at 37 °C, 5% CO_2, and 90% humidity.
13. After 2 days, remove the medium from the attached primary cells including all debris and nonadherent cell clumps, wash once with c-KECM, and add 10 mL of fresh prewarmed (37 °C) c-KECM.
14. Cultivate until cells become 80–90% confluent and renew medium every second day (*see* Fig. 3).
15. Cells are now ready for 2D infection experiments with FeMV and can be expanded by standard splitting techniques using trypsin-ETDA (see below). Primary feline kidney cells normally can be passaged approximately five times before passing into senescence.
16. Wash cells (low passage no.) with warmed PBS twice before infecting them with FeMV strains. MOI of 0.01–1.0 should be used. Viral stock should be diluted in basal-KCEM to avoid toxic effects to the cells.

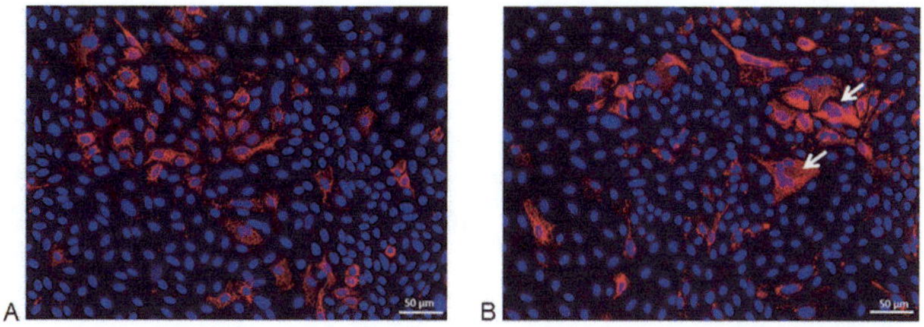

Fig. 4 Immunofluorescence staining of FeMV-infected feline primary kidney cells 5 days postinfection. A MOI of 1.0 was used for infection. Both FeMV-1 (**a**) and FeMV-2 (**b**) are capable of infecting feline kidney cells. Arrows show syncytia formation. Scale bar represents 50 μm

17. Incubate for 1–2 h at 37 °C with frequent agitation to ensure even distribution of the inoculum.
18. Wash cells twice with prewarmed PBS and replace with prewarmed basal-KCEM. Incubate at 37 °C and 5% CO_2 for intendent periods.
19. After various incubation times, infected cells can be used for immunofluorescence staining (*see* Subheading 3.2; Fig. 4), flow cytometry, qPCR, Western blot, etc.

3.4 Infection of Polarized Feline Primary Epithelial Cells (3D Model)

1. Gently rinse monolayer of low passage primary feline kidney epithelial cells with prewarmed PBS once. Add an appropriate volume of trypsin-EDTA to cover the whole monolayer.
2. Incubate at 37 °C until all cells have detached from the flask.
3. Add an equal volume of soybean trypsin inhibitor to stop enzyme activity. Centrifuge at 300× *g* for 5 min at room temperature. Discard supernatant.
4. Suspend cell pellet in prewarmed c-KECM.
5. Seed cells on the top of Millicell® inserts in a density of approximately 0.5 million cells per insert. Do not use more than 250 μL of volume. Be careful not to damage the membrane.
6. Add 1 mL of c-KECM to the basal compartment of the transwell in a 6-well cell culture dish (*see* Fig. 5a).
7. Incubate for 2 days at 37 °C and 5% CO_2 without disturbance to allow adherence of cells.
8. Change medium at the basal compartment every second day. Incubate until three-dimensional structure (so-called cell dome) developed. This will take approximately 7–10 days (*see* Fig. 5b).
9. For infection of cells, wash twice with prewarmed PBS to remove cellular debris.

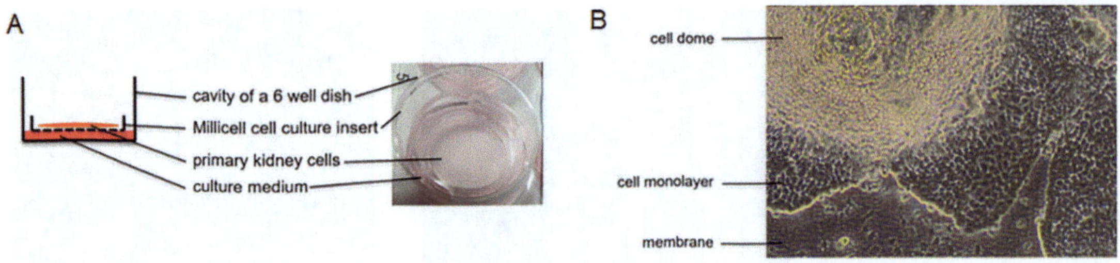

Fig. 5 Transwell model of feline kidney epithelium. Schematic diagram of transwell (**a**) and 3D cell structures ("cell dome") after 10 days of incubation (**b**)

10. Calculate the volume of virus stock to achieve the desired MOI. Dilute virus in basal-KECM if needed. Add virus directly to the top of the cultured cells on the transwell membrane. Be sure not to destroy the cell layer or the membrane. Do not use more than 250 μL of volume for infection.
11. Incubate for 2 h at 37 °C and 5% CO_2.
12. Carefully wash cells twice with prewarmed basal-KECM, and incubate infected transwells for additional 5–7 days at 37 °C and 5% CO_2. Change basal medium every second day.

After various incubation times, infected cells can be used for immunofluorescence staining (see below), flow cytometry, qPCR, Western blot, etc. For immune staining, the following protocol may be used.

3.5 Immunostaining of Infected Polarized Feline Primary Epithelial Cells

1. Wash transwells twice with ice-cold PBS.
2. Fix cells cultured on Millicell® inserts in 4% neutral buffered formaldehyde solution for 24–48 h at room temperature.
3. Dehydrate inserts in ascending sequence of ethanol: 70% for 30 min, 80% for 30 min, and two times in 96% ethanol for 30 min.
4. Incubate inserts in 100% isopropanol for 2 × 30 min.
5. Cut the membrane out of its plastic frame using a scalpel blade. To keep it flat and strained during the next step, the membrane should be bisected, and the cell-free sides of both halves should be attached to cork using fixing pins.
6. Incubate membranes in butyl acetate for 10 min at room temperature followed by embedding in liquid paraplast perpendicular to the cutting edge.
7. The firm paraffin blocks are then ready for sectioning into 3–4-μm-thick slices using a sliding or rotary microtome.
8. Mount paraffin sections on Superfrost Plus™ slides, dry at 37 °C for 1–2 h, and store desiccated in the dark at room temperature until further processing. Slides should be used within 1 week after sectioning.

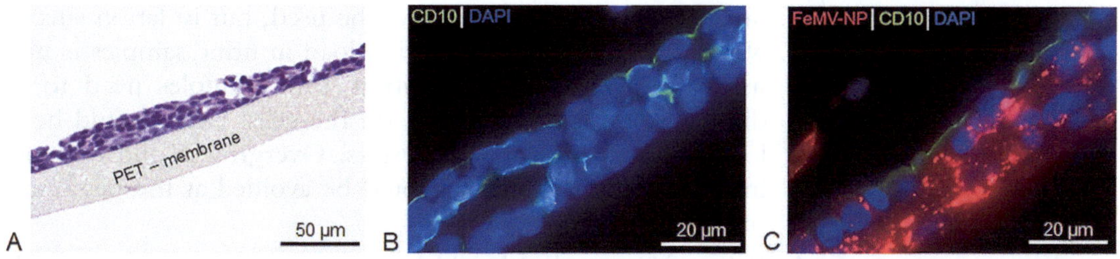

Fig. 6 Three-dimensional model of feline kidney epithelial cells using transwell systems. HE-staining (**a**) shows multilayered kidney epithelial cells on the surface of the membrane. Immunofluorescence staining of MOCK (**b**) and FeMV (**c**) infected transwells shows polarized epithelial cells (CD10 marker staining, shown in green) which can be infected with FeMV (shown in red). Cell nuclei were stained with DAPI (shown in blue). Scale bars represent 50 μm (**a**) and 20 μm (**b**, **c**), respectively

9. Stain slides with hemalum-eosin using standard protocols [21]. A representative result is shown in Fig. 6a.
10. For immunofluorescence, standard histological methods can be used. A representative result, staining for FeMV antigen and CD10 (apical surface marker of polarized kidney tubule cells), is shown in Fig. 6b, c.

4 Notes

1. Investigate urine samples via qPCR for the presence of FeMV (samples from persistently infected animals work best due to relative high and robust viral titers excreted with the urine). Spontaneous urine or urine collected via cystocentesis can be used. Urine should be cooled on ice directly after collection and should be immediately processed for virus isolation. Alternatively, urine samples can be diluted 1:3 in viral transport medium (e.g., BD universal viral transport system™) and stored for up to 2 days at 4 °C or at −80 °C for long term storage.
2. Refer to your country-specific animal testing regulations. Kidney samples should be bisected and decapsulated to achieve proper diffusion of cell culture medium into the tissue. Organ samples should be processed as soon as possible although overnight storage in HBSS at 4 °C is possible without significant effects on viability and isolation success of primary epithelial cells.
3. Other recombinant cell lines expressing the feline SLAM-F1 receptor can be used, for instance, Vero-CatSLAM [7] or CrFK-CatSLAM [8] cell lines as described previously. In our hands, LLC-MK2-CatSLAM cell line was superior for the isolation of FeMV due to higher viral titers compared to recombinant Vero or CrFK cell lines.

4. Smaller or larger flask sizes can be used, but isolation success was highest in T75 flask as viral load in urine samples is normally relatively low. In addition, urine samples need to be diluted to avoid toxic effects on the cells. Cells should be in logarithmic stage for best results. Overgrowth and entrance into the stationary phase should be avoided at the beginning of the infection process.
5. Pure urine samples should be diluted 1:5 to 1:10 in DMEM before adding to the cell monolayer. This will avoid toxic effects of the urine and will ensure physiologic pH of the medium. Urine samples which were collected in viral transport medium can be diluted 1:2 to 1:3. If marked bacteriuria is observed in the urine samples, they may be passed through a 0.45 μm filter before infecting the cell monolayer. On the other hand, this will reduce overall viral load of the inoculum and isolation success. As an alternative, additional antibiotics (e.g., gentamicin) can be supplemented to the medium to avoid bacterial overgrowth.
6. Normally, viral titers are low (approximately 10^4 $TCID_{50}$/mL) in this first passage of the virus. Therefore, a second or third passage is needed to achieve proper viral titers. For this purpose, repeat viral isolation process as described using passage no. 1 at a MOI of 0.01 to avoid the accumulation of defective interfering particles.
7. Calculation of TCID50 was done after the following formula:

$$\log_{10} 50\%\text{end point dilution} = -\left(x_0 - d/2 + d\sum r_i/n_i\right)$$

where

$x_0 = \log_{10}$ of the reciprocal of the highest dilution at which all wells are positive.

$d = \log_{10}$ of the dilution factor.

n_i = number of replicates used in each individual dilution.

r_i = number of positive wells (out of ni).

References

1. Rima B, Balkema-Buschmann A, Dundon WG et al (2019) ICTV virus taxonomy profile: paramyxoviridae. J Gen Virol 100(12): 1593–1594. https://doi.org/10.1099/jgv.0.001328
2. Woo PCY, Lau SKP, Wong BHL et al (2012) Feline morbillivirus, a previously undescribed paramyxovirus associated with tubulointerstitial nephritis in domestic cats. Proc Natl Acad Sci USA 109(14):5435–5440. https://doi.org/10.1073/pnas.1119972109
3. de Luca E, Sautto GA, Crisi PE et al (2021) Feline morbillivirus infection in domestic cats: what have we learned so far? Viruses 13(4): 683. https://doi.org/10.3390/v13040683
4. Busch J, Heilmann RM, Vahlenkamp TW et al (2021) Seroprevalence of infection with feline morbilliviruses is associated with FLUTD and increased blood creatinine concentrations in domestic cats. Viruses 13(4):578. https://doi.org/10.3390/v13040578

5. Sieg M, Heenemann K, Rückner A et al (2015) Discovery of new feline paramyxoviruses in domestic cats with chronic kidney disease. Virus Genes 51(2):294–297. https://doi.org/10.1007/s11262-015-1232-7
6. Sutummaporn K, Suzuki K, Machida N et al (2019) Association of feline morbillivirus infection with defined pathological changes in cat kidney tissues. Vet Microbiol 228:12–19. https://doi.org/10.1016/j.vetmic.2018.11.005
7. Nikolin V, Hatsue Sobreda Doi L, Sieg M et al (2022) In vitro growth, receptor usage and pathogenesis of feline morbillivirus in the natural host. Viruses 14(7):1503. https://doi.org/10.3390/v14071503
8. Nambulli S, Rennick LJ, Acciardo AS et al (2022) FeMV is a cathepsin-dependent unique morbillivirus infecting the kidneys of domestic cats. Proc Natl Acad Sci USA 119(43): e2209405119. https://doi.org/10.1073/pnas.2209405119
9. Park E-S, Suzuki M, Kimura M et al (2016) Epidemiological and pathological study of feline morbillivirus infection in domestic cats in Japan. BMC Vet Res 12(1):228. https://doi.org/10.1186/s12917-016-0853-y
10. McCallum KE, Stubbs S, Hope N et al (2018) Detection and seroprevalence of morbillivirus and other paramyxoviruses in geriatric cats with and without evidence of azotemic chronic kidney disease. J Vet Intern Med 32(3): 1100–1108. https://doi.org/10.1111/jvim.15097
11. Stranieri A, Lauzi S, Dallari A et al (2019) Feline morbillivirus in Northern Italy: prevalence in urine and kidneys with and without renal disease. Vet Microbiol 233:133–139. https://doi.org/10.1016/j.vetmic.2019.04.027
12. Yilmaz H, Tekelioglu BK, Gurel A et al (2017) Frequency, clinicopathological features and phylogenetic analysis of feline morbillivirus in cats in Istanbul, Turkey. J Feline Med Surg 19(12):1206–1214. https://doi.org/10.1177/1098612X16686728
13. Sieg M, Busch J, Eschke M et al (2019) A new genotype of feline morbillivirus infects primary cells of the lung, kidney. Brain Periph Blood Viruses 11(2):146. https://doi.org/10.3390/v11020146
14. de Luca E, Crisi PE, Marcacci M et al (2020) Epidemiology, pathological aspects and genome heterogeneity of feline morbillivirus in Italy. Vet Microbiol 240:108484. https://doi.org/10.1016/j.vetmic.2019.108484
15. Piewbang C, Chaiyasak S, Kongmakee P et al (2020) Feline morbillivirus infection associated with tubulointerstitial nephritis in black leopards (Panthera pardus). Vet Pathol 57(6): 871–879. https://doi.org/10.1177/0300985820948820
16. Sieg M, Sacristán I, Busch J et al (2020) Identification of novel feline paramyxoviruses in guignas (Leopardus guigna) from Chile. Viruses 12(12):1397. https://doi.org/10.3390/v12121397
17. Piewbang C, Wardhani SW, Dankaona W et al (2022) Feline morbillivirus-1 in dogs with respiratory diseases. Transbound Emerg Dis 69(4):e175–e184. https://doi.org/10.1111/tbed.14278
18. Lavorente FLP, de Matos AMRN, Lorenzetti E et al (2022) First detection of Feline morbillivirus infection in white-eared opossums (Didelphis albiventris, Lund, 1840), a non-feline host. Transbound Emerg Dis 69(3):1426–1437. https://doi.org/10.1111/tbed.14109
19. Lawson JS, Liu H-H, Syme HM et al (2018) The cat as a naturally occurring model of renal interstitial fibrosis: Characterisation of primary feline proximal tubular epithelial cells and comparative pro-fibrotic effects of TGF-β1. PLoS One 13(8):e0202577. https://doi.org/10.1371/journal.pone.0202577
20. Busch J, Sacristán I, Cevidanes A et al (2021) High seroprevalence of feline morbilliviruses in free-roaming domestic cats in Chile. Arch Virol 166(1):281–285. https://doi.org/10.1007/s00705-020-04882-2
21. Mulisch M, Welsch U (eds) (2015) Romeis – Mikroskopische Technik. Springer, Berlin/Heidelberg

Chapter 13

Setup of Aerosol Delivery System to Prevent Measles Virus Infection in Nonhuman Primate Model

Olivier Reynard, Jérôme Montharu, Mathieu Iampietro, Claudia Gonzalez, Sandrine Le Guellec, Branka Horvat, and Laurent Vecellio

Abstract

Measles virus is one of the most contagious airborne human viruses which keeps causing outbreaks in numerous countries over the world despite the existence of an efficient vaccine. Fusion inhibitory lipopeptides were shown to inhibit viral entry into target cells, and their adequate administration into the respiratory tract may provide a novel preventive approach against airborne infections. Aerosol delivery presents the best administration route to deliver such preventive compounds to the upper and lower respiratory tract. This approach offers a conceptually new strategy to protect the population at risk against infection by respiratory viruses, including measles. It is a noninvasive needle-free approach, which may be used when antiviral protection is required, without any medical assistance. In this chapter, we describe the nebulization approach of lipopeptide compounds in nonhuman primates and the subsequent measles virus challenge.

Key words Paramyxovirus, Measles, Nebulization, Nonhuman primate, Fusion inhibitor, Lipopeptide

1 Introduction

Measles virus (MeV), a member of the *Paramyxoviridae* family of single-stranded negative sense RNA viruses, is one of the most infectious microorganisms worldwide [1]. MeV is an airborne pathogen, transmitted by inhalation of respiratory droplets and smaller aerosol [2]. MeV uses several different receptors to infect target cells, and after binding to the receptor, it promotes conformational refolding of the fusion protein, leading to the formation of a fusion pore [3]. Fusion inhibitor lipopeptides are short peptides derived from the C-terminal heptad repeat domain of the

Olivier Reynard and Jérôme Montharu have equally contributed to this chapter.

Dzwokai Z. Ma and Christian K. Pfaller (eds.), *Measles and Related Morbilliviruses: Methods and Protocols*, Methods in Molecular Biology, vol. 2808, https://doi.org/10.1007/978-1-0716-3870-5_13,

fusion protein that are capable to efficiently prevent virus-cell fusion. Several studies have demonstrated their efficacy in the prevention of infection by different paramyxoviruses [3, 5, 8]. Efficient administration of those compounds into the respiratory tract by nebulization requires several optimizations (*see* **Note 1**), to ensure that generated aerosols reach the targeted portion of the respiratory tract. A mean diameter below 5 μm is required for correct delivery to the deep lung area; additionally, high enough concentration of treatment molecules per droplet is required to ensure efficacy. Adequate dosing and distribution of the compounds into the respiratory tract is of critical importance to efficiently prevent infection. Several drug delivery technologies have been implemented for local or systemic drug administration, among them, metered dose inhaler, dry powder inhaler, jet nebulizer, ultrasonic nebulizer, and vibrating mesh nebulizers [7]. The latest has proven its flexibility and usefulness through its ability to distribute a wide variety of molecule suspensions (antibiotics, proteins, and peptides), with a low residual volume and a high delivery efficacy. Vibrating mesh nebulizers consist of a three-part device composed serially of a piezoelectric probe, a reservoir chamber, and a mesh containing thousands of holes of few micrometers. Under voltage, the piezoelectric part induces a vibration of the mesh in contact with the liquid. This action extrudes fluid through the pores in the mesh, where surface tension and hydrodynamic effects result in breaking the extruded fluid into a stream of precisely controlled droplets. The size of the droplets is controlled primarily by the exit diameter of the apertures. This technology allows the production of precise particle sizes, with residual volume as low as microliters [4].

Development of the peptide nebulization approach requires the initial tests in a nonhuman primate model. We have used cynomolgus macaques, a well-characterized model that recapitulates measles infection in humans [8]. Drugs deposited in the lungs will undergo loss at different levels during the nebulization process, and the quantity deposited in the lungs will not correspond to the quantity of the drug loaded in the nebulizer reservoir (Fig. 1). The initial loss of the drug is due to the residual dose in the nebulizer and interface. A second presents the loss in a form of aerosol during the expiratory phase because the nebulizer continues to generate aerosol during the expiratory phase. The final loss is provoked by the deposition on the animal's face due to aerosol velocity and macaque fur. When aerosol is inhaled, a part is deposited in the upper airways (nose, mouth), a second is deposited in the lung, and a third is exhaled by the macaque. Therefore, only a small fraction (from 0.1% to 10%) of the drug loaded in the nebulizer is deposited in the lungs depending on the nebulizer's efficiency, macaque ventilation, and anatomy. It is then important to take into account these losses when there is an evaluation of drug efficiency in macaque lungs. To assess activity of the aerosolized fusion

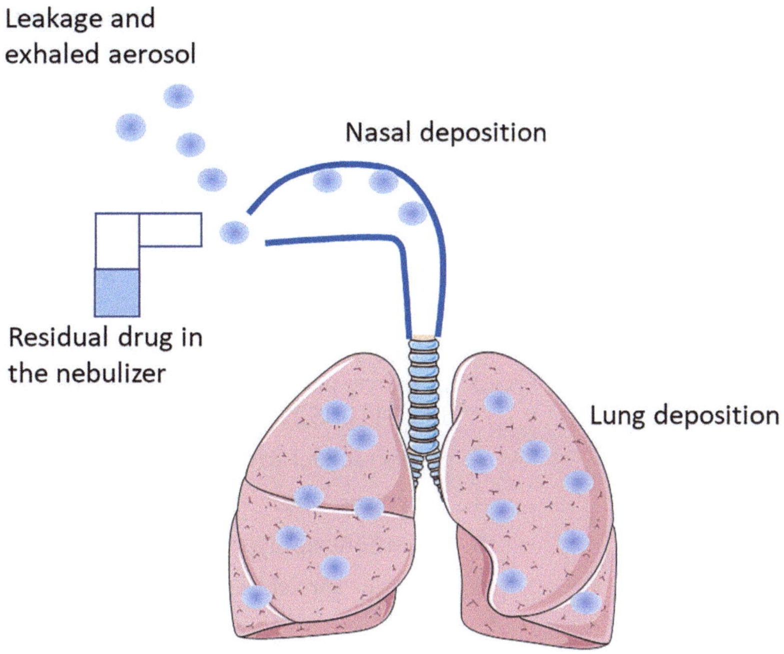

Fig. 1 Drug distribution in airways and drug losses after delivery by nebulization. Assessment of aerosol droplet size can be performed by laser diffraction or cascade impactor to ensure that aerosol size is optimum for a deep delivery [6]. Deposition of the nebulized compound in the different compartments can be assessed by the nebulization of a radioactive tracer and the subsequent scintigraphy imaging

inhibitory lipopeptides against MeV, we have nebulized three times (−24 h, −6 h, +24 h) cynomolgus macaques with 12 mg (3 mL of 4 mg/mL concentrated product) of peptides. Animals were challenged with the respective viruses 6 h after the second nebulization. The following methods describe precisely the procedures for primate anesthesia, nebulization, and challenge.

2 Materials

2.1 Animals and Fusion Inhibitory Lipopeptides

1. Measles vaccine naive cynomolgus macaques (different breeding colonies may be needed to contact to identify animals not receiving the MeV vaccine).
2. Fusion inhibitory peptide HRC4 [9] diluted in water (milliQ) or in water/DMSO (5%) at 4 mg/mL.

2.2 Anesthesia

1. Domitor (Orion Corporation, Espo, Finland).
2. Alzane (Laboratorios Syva, Leon, Spain).
3. Imalgene 1000 (Boehringer Ingelheim Animal Health, Lyon, France).

2.3 Nebulization

1. Face mask: modified commercial face mask (economy, anesthetic face mask, infant, Intersurgical, Wokingham, UK).
2. Bath baby seat.
3. Aerosol generator (mesh prototype nebulizer, DTF Medical, Saint Etienne, France).
4. Radical-7® Pulse CO-Oximeter (Masimo, Irvine, USA).

2.4 Viral Challenge

1. 10^4 pfu of recombinant MeVIC323 expressing the gene encoding eGFP (MeV-IC323-eGFP, GenBank accession number: LC420351.1).
2. Pediatric nasogastric tube (ø 2.0 mM-L. 40 cm Securimed, ref. 11214100522)
3. Endotracheal tube (COVETO ref 697000)
4. Laryngoscope.
5. Anesthetic spray (xylocaine spray 5% AstraZeneca).
6. Syringe for viral delivery (5 mL volume syringe without needle).
7. Syringe for intramuscular anesthesia (1 mL volume syringe with a 25G 5/8"needle).
8. Saline solution (NaCl 9 g/L).
9. Viral suspension diluted in DMEM (Thermo).
10. Dräger Primus anesthesia machine.
11. FastGene Blue/Green LED Flashlight (Nippon Genetics).

3 Methods

3.1 Animal Anesthesia

1. Inject the animals with an intramuscular dose of ketamine (3 mg/kg) and medetomidine (125 μg/kg) in the leg (a volume smaller than 0.5 mL/kg is recommended).
2. Leave the accommodation room and observe the animal falling asleep through the glass door.
3. Once the animal is asleep, test the state of sleep by touching the animal's head between the bars of the cage. If no withdrawal reflex is observed, the door can be opened.
4. Take the animal out of the cage and place it in the baby seat.
5. The depth of anesthesia is observed and monitored continuously. Measure the oxygen saturation and heart rate using a clamp placed on the palm of the primate.
6. After verifying the absence of pupillary and withdrawal reflexes, the aerosol is administered.
7. Reversion is accomplished by injecting 625 μg/kg of atipamezole.

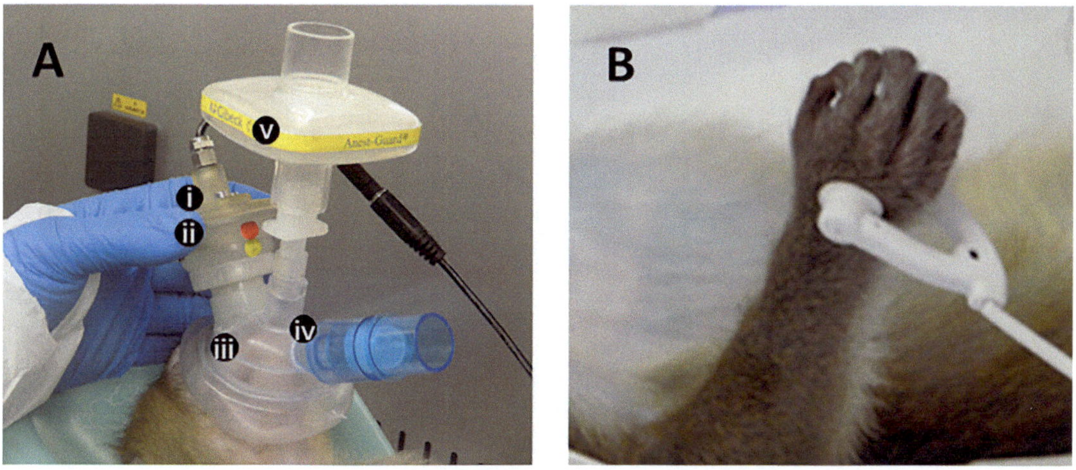

Fig. 2 Setup used for cynomolgus nebulization. (**a**) Sedated animals are placed under a biosafety cabinet, and mask is hand kept during the process to ensure capture of aerosols' leakage. The mesh nebulizer is composed of (i) a piezoelectric vibrator, (ii) a reservoir containing the mesh, (iii) a facemask, (iv) a one-way inspiratory valve, and (v) an absolute filter. (**b**) Oxygen saturation is continuously controlled to evaluate animal status

3.2 Animal Nebulization

1. Place animals for proper administration under a biosafety cabinet (BSC) as the third administration is given post-challenge.
2. Install animals in an in-house modified commercial bath baby seat covered with a draw sheet.
3. Align the whole body.
4. Place the head slightly back to open the airways; place eye gel on the eyes.
5. Place the face mask on the face of the animal covering both the nose and mouth.
6. Hold the mask manually on the face of the animal by the experimenter (Fig. 2).
7. The nebulization is carried out for 10 min or until the end of aerosol's generation.
8. Perform a visual inspection of animal's breathing during the nebulization, and control saturation along heartbeat with a monitor placed on the upper limb.
9. Ensure the production of aerosol and its inhalation by the animal by using a transparent mask.
10. At the end of nebulization, remove the mask from the face of the animal and clean it with a paper towel. Remove the animal afterward from the BSC to be awakened by anesthetic reversion.

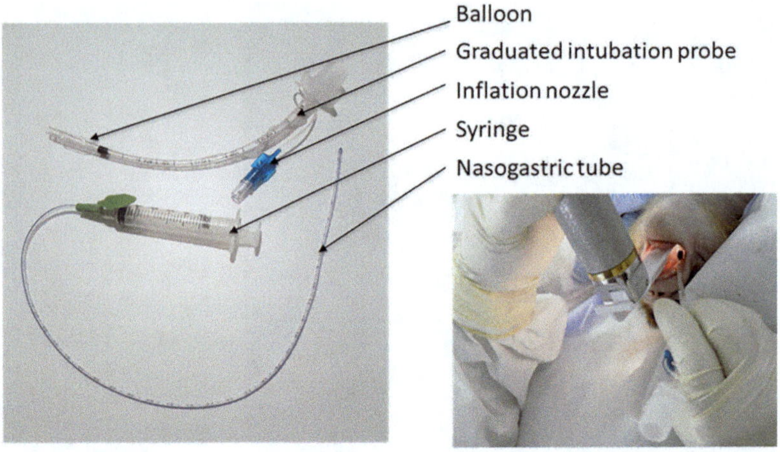

Fig. 3 Virus instillation setup. Before instillation, a spray of local anesthetic is applied to the vocal cords of the animal. During the whole process, oxygen saturation is evaluated to control animal status

3.3 Intratracheal Challenge

1. In order to mimic natural infection, inoculate the virus through intratracheal route under anesthesia. To ensure a good delivery of the liquid in the respiratory tract, without ambient air contamination, use an endotracheal tube (ETT) to reach the respiratory airway. Introduce a nasogastric tube into the endotracheal tube to deliver the pathogen in the respiratory tract (*see* **Note 2**).
2. Hold the macaque's head, open gently the mouth, and slid the laryngoscope blade into the mouth, and lift the vocal cords.
3. Introduce the 3-mM-diameter endotracheal tube into the trachea just after the carina, then inflate the balloon, and check its position using a stethoscope (Fig. 3).
4. Connect the syringe containing inoculum to the nasogastric tube, and administer the liquid (*see* **Note 3**).
5. Before removing the device, chase the dead volume by 2 mL of saline solution to ensure that all the volume is expelled into the lungs.
6. Then push 5 mL of air into the lung.
7. Remove the inoculation setup by taking off the nasogastric tube while aspirating at the same time, and then deflate the balloon of the ETT.
8. After this step, extract the ETT and monitor the state of the animal.
9. Awake the animal using a dose of atipamezole (625 μg/kg) intramuscularly. During the process, if sudden dyspnea appears, re-intubate animal and connect the ETT to the respirator to relieve the animal and avoid hypoxia.

10. Animals may have spasms or cough during the inoculation step and further shed viruses during infection; therefore, specific personal protective equipment are required when working with airborne viruses, and specific care should be applied to high-risk steps (*see* **Note 3**).

4 Notes

1. Key parameters for optimal and efficient delivery (Table 1).
2. Setup of the nasogastric tube: an endotracheal tube (ETT) containing a nasogastric tube is used to deliver the pathogen into the respiratory tract. To ensure a good delivery of the liquid in the respiratory tract, the depth of the nasogastric tube within the ETT should be primarily defined (Fig. 4 right panel). The dead volume is calculated by gravimetric method. A syringe is filled with water and weighed. It is connected to the nasogastric tube, and the liquid is pushed inside until a drop of water appears at the extremity of the tube. Then, the weight of the syringe is evaluated once again. The difference between the initial weight and the one post-push provides the volume of the nasogastric tube (based on the liquid density of 1). For the virus instillation step, the nasogastric tube is slid into the intubation endotracheal tube up to the previously annotated level (Fig. 4 comments). A piece of Parafilm M is placed on the top of the endotracheal tube to cover the open area, letting the nasogastric tube to pass through (Fig. 4 left panel). This important step prevents any liquid projection that could arise if coughing occurs during inoculation.

Table 1
Key delivery parameters and optimized values

Parameters	Values	Optimization
Diameter of the aerosols	Below 5 μm	Nebulization methods, mesh pore size in the case of mesh nebulizer
Duration of nebulization	Below 10 min	Optimal delivered dose O_2 administration
Compound concentration	Depending on expected volume to be delivered in biocompatible vehicle	Initial drug design for high solubility Vehicle optimization (pH, salt, chaotropic agent, osmolarity)
Liquid volume	Depends on dose target in the airways and inhaler	Mesh nebulizer with optimized facemask

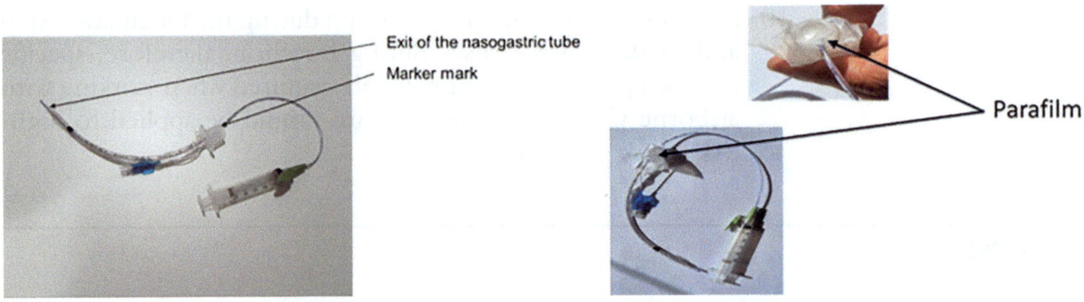

Fig. 4 Position and mark of the nasogastric tube. Left: the pediatric nasogastric tube is introduced into the endotracheal tube (ETT); when the end of the tube protrudes 1–2 mM from the EET, a mark is placed to identify the ideal position of the nasogastric tube. Then the nasogastric tube can be positioned ideally in the ETT when the pathogen will be delivered to the animal. Right: a piece of paraffin film is placed at the exterior extremity of the EET to prevent virus spillage if the animal coughs

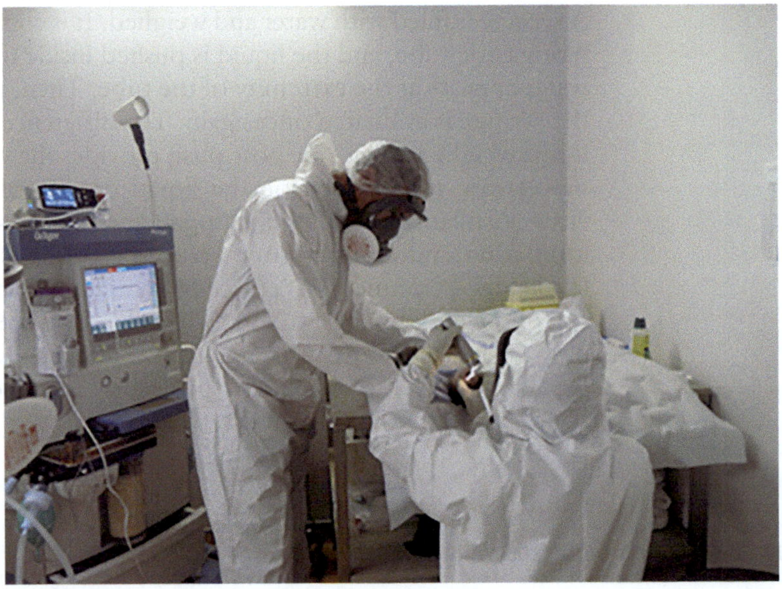

Fig. 5 Personnel equipped with PPE in BSL2. Animal practicians are equipped with PPE covering the full face to ensure complete protection against aerosolized viral particles

3. Key personal protective equipment (PPE) required for proper BSL2 intratracheal challenge with MeV (Fig. 5): the PPE listed below are not exhaustive and may vary depending on country-specific regulation:
   - Disposable full Tyvek suit.
   - Two pairs of gloves.
   - Disposable protective sleeves.

- Powered air system with hood or full-face piece respirators with high-efficiency particulate filters (P100).
- Disposable overshoes.

Acknowledgments

We thank the animal experimentation PST-A team of Tours University for the realization of the animal experiments. We acknowledge the Servier Medical Art (smart.servier.com) for providing the images used for the schemas presented in the article. The study was supported by Region AURA (Pack Ambition Recherche, project AerVirStop-BH).

References

1. Anderson RM, May RM (1982) Directly transmitted infections diseases: control by vaccination. Science 215:1053–1060. https://doi.org/10.1126/science.7063839
2. Figueira TN, Palermo LM, Veiga AS, Huey D, Alabi CA, Santos NC, Welsch JC, Mathieu C, Horvat B, Niewiesk S, Moscona A, Castanho MARB, Porotto M (2017) In vivo efficacy of measles virus fusion protein-derived peptides is modulated by the properties of self-assembly and membrane residence. J Virol 91:e01554–e01516. https://doi.org/10.1128/JVI.01554-16
3. Harrison SC (2008) Viral membrane fusion. Nat Struct Mol Biol 15:690–698. https://doi.org/10.1038/nsmb.1456
4. Martin AR, Finlay WH (2015) Nebulizers for drug delivery to the lungs. Expert Opin Drug Deliv 12:889–900. https://doi.org/10.1517/17425247.2015.995087
5. Mathieu C, Huey D, Jurgens E, Welsch JC, DeVito I, Talekar A, Horvat B, Niewiesk S, Moscona A, Porotto M (2015) Prevention of measles virus infection by intranasal delivery of fusion inhibitor peptides. J Virol 89:1143–1155. https://doi.org/10.1128/JVI.02417-14
6. Mitchell JP, Nagel MW, Nichols S, Nerbrink O (2006) Laser diffractometry as a technique for the rapid assessment of aerosol particle size from inhalers. J Aerosol Med 19:409–433. https://doi.org/10.1089/jam.2006.19.409
7. Pritchard JN, Hatley RH, Denyer J, Hollen D von (2018) Mesh nebulizers have become the first choice for new nebulized pharmaceutical drug developments. Ther Deliv 9:121–136. doi: https://doi.org/10.4155/tde-2017-0102
8. Reynard O, Gonzalez C, Dumont C, Iampietro M, Ferren M, Le Guellec S, Laurie L, Mathieu C, Carpentier G, Roseau G, Bovier FT, Zhu Y, Le Pennec D, Montharu J, Addetia A, Greninger AL, Alabi CA, Brisebard E, Moscona A, Vecellio L, Porotto M, Horvat B (2022) Nebulized fusion inhibitory peptide protects cynomolgus macaques from measles virus infection. Nat Commun 13:6439. https://doi.org/10.1038/s41467-022-33832-6
9. Welsch JC, Talekar A, Mathieu C, Pessi A, Moscona A, Horvat B, Porotto M (2013) Fatal measles virus infection prevented by brain-penetrant fusion inhibitors. J Virol 87:13785–13794. https://doi.org/10.1128/JVI.02436-13

Chapter 14

Histopathological Analysis of Brains from Dogs Infected with Canine Distemper Virus

José Manuel Verdes, Camila Larrañaga, Belén Varela, Victoria Iribarnegaray, Victoria Yozzi, Gimena Feijóo, and Kanji Yamasaki

Abstract

We describe the use of conventional histology and immunohistochemistry against canine distemper virus (CDV) to examine the brains of domestic dogs with a confirmed diagnosis of CDV infection. Histologically, to identify the main typical lesions, we used conventional H&E stain; to evaluate the progressive demyelination, we used Luxol Fast Blue stain; and to identify the presence of viral particles in these affected regions, we used immunohistochemistry against CDV. We confirm that the histopathological analysis of brains of distemper-infected dogs is a powerful tool to evaluate the typical brain lesions and could be used as an interesting natural model to continue studying the pathogenesis of canine distemper in different species and/or other morbillivirus infections, like measles.

Key words Animal model, Distemper, Dog, Histopathology, Measles, Comparative neuropathology

1 Introduction

Canine distemper is one the most infectious diseases of domestic dogs. It is also known as a highly prevalent viral infectious disease of various non-canine hosts, including nonhuman primates [3, 8]. Recently, we performed a systematic study of neuropathological changes to clarify the histological hallmarks in vaccinated dogs infected with CDV [4]. The main changes observed in these cases were demyelination, gliosis, nonsuppurative meningitis, lymphocytic perivascular cuffing, and inclusion bodies in the cerebrum and cerebellum [4]. Based on histopathological changes, all lesions were classified as subacute or chronic. Lesions suspected as being acute were not observed [4]. Previous reports propose canine distemper virus infection as an animal model to study pathogenesis of multiple sclerosis [14] or measles in humans [2]. Here, we describe the use of conventional histology and immunohistochemistry against CDV to study neuropathological changes in dogs

Dzwokai Z. Ma and Christian K. Pfaller (eds.), *Measles and Related Morbilliviruses: Methods and Protocols*, Methods in Molecular Biology, vol. 2808, https://doi.org/10.1007/978-1-0716-3870-5_14,

naturally infected with CDV. Both techniques offer an easy approach to study distemper pathogenesis in domestic dogs and could be interesting to study with similar methods in other species or in other morbillivirus infections, like measles infection in humans.

2 Materials

2.1 Histological Specimen Preparation

1. Central nervous system (CNS) specimens: cerebrum and cerebellum recovered during autopsy from naturally infected domestic dogs.
2. Buffered neutral formalin solution (pH 7.4): 3.7–4% formaldehyde, 33 mM sodium phosphate, monobasic, 46 mM sodium phosphate, dibasic (anhydrous).
3. 100% ethanol (*see* **Note 1**).
4. 95% ethanol (*see* **Note 1**).
5. 80% ethanol (*see* **Note 1**).
6. 70% ethanol (*see* **Note 1**).
7. Xylene (*see* **Note 1**).
8. Paraffin (*see* **Note 1**).
9. Distilled water.
10. Tap water.
11. Mayer's hematoxylin solution (*see* **Note 2**).
12. Eosin Y solution (*see* **Note 3**).
13. Luxol Fast Blue Staining Kit (e.g., Abcam® Luxol Fast Blue Staining Kit, cat. N° ab 150675) (*see* **Note 4**).
14. High adherent glass slides (StarFrost®, Germany).
15. Mounting medium (Canada balsam for microscopy, Sigma-Aldrich®, 60610, USA).
16. Coverslips.
17. Automatic tissue processor (*see* **Note 5**).
18. Laboratory oven.
19. Dispensing module.
20. 1 L Pyrex beaker.
21. 1 L stainless steel beaker.
22. Stainless steel specimen containers.
23. Rotary microtome.
24. Tissue flotation bath.
25. Carousel slide stainer.

2.2 Immunohistochemistry Against CDV Reagents

1. 10× phosphate buffered saline (PBS), pH 7.4: 80 g/L NaCl, 2 g/L KCl, 14.4 g/L Na_2HPO_4, 2.4 g/L KH_2PO_4. (*i*). Using a 1 L graduated cylinder, add 800 mL of distilled water into a 1 L Pyrex bottle. (*ii*). Add 80 g of NaCl. (*iii*). Add 2 g of KCl. (*iv*). Add 14.4 g of Na_2HPO_4. (*v*). Add 2.4 g of KH_2PO_4. (*vi*). Adjust the pH to 7.4 with HCl. (*vii*). Add distilled water to a total volume of 1 L. (*viii*). Store in refrigerator (4 °C) until use. (*ix*). To prepare 1× PBS, dispense 100 mL of 10× PBS solution in a 1 L graduated cylinder, adding 900 mL of distilled water to a total volume of 1 L. (*x*). Store in refrigerator (4 °C) until use.
2. pH meter.
3. Stir bar.
4. Hot plate magnetic stirrer.
5. 1 L Pyrex bottle.
6. 1 L graduated cylinder.
7. Blocking reagent (e.g., Biocare's Peroxidazed 1, Biocare Medical, USA).
8. 1x retrieval solution (e.g., Borg Decloaker RTU, Biocare Medical, USA).
9. Mouse anti-CDV monoclonal primary antibody (MCA 1893, dilution 1:250; BioRad, UK).
10. Secondary antibody (e.g., Mouse-on-Canine HRP-Polymer, Biocare Medical, USA).
11. 3,3′-Diaminobenzidine (DAB) buffer (e.g., Betazoid DAB Buffer, Biocare Medical, USA) (*see* **Note 6**).
12. Mayer's hematoxylin solution (*see* **Note 2**).
13. Steamer.
14. Wet incubation chamber (*see* **Note 7**).
15. Light microscope.

3 Methods

3.1 Brain Autopsy and Tissue Fixation

Domestic dogs with previous clinical diagnosis of CDV infection that had to be euthanized or had died after the presence of severe and uncontrolled neurological signs (*see* **Note 8** and Figs. 1 and 2) are autopsied.

1. During a complete routine autopsy, to properly recover the cerebrum and cerebellum, the disarticulation of atlanto-occipital joint must be carried out with a sharp thin-blade autopsy knife, to completely separate the head from the body.

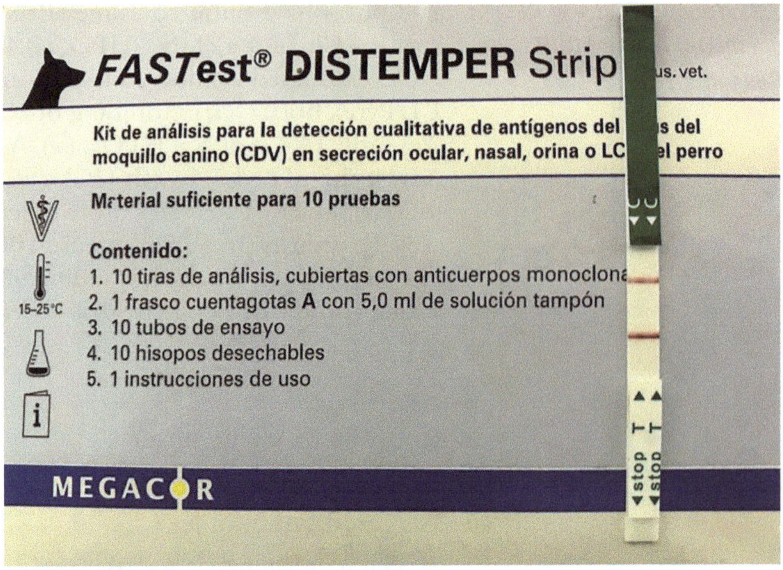

Fig. 1 A positive immunochromatographic assay to detect CDV antigen in ocular discharge. Two bands in the strip indicate positivity. (Courtesy MSc. Mary Gutierrez)

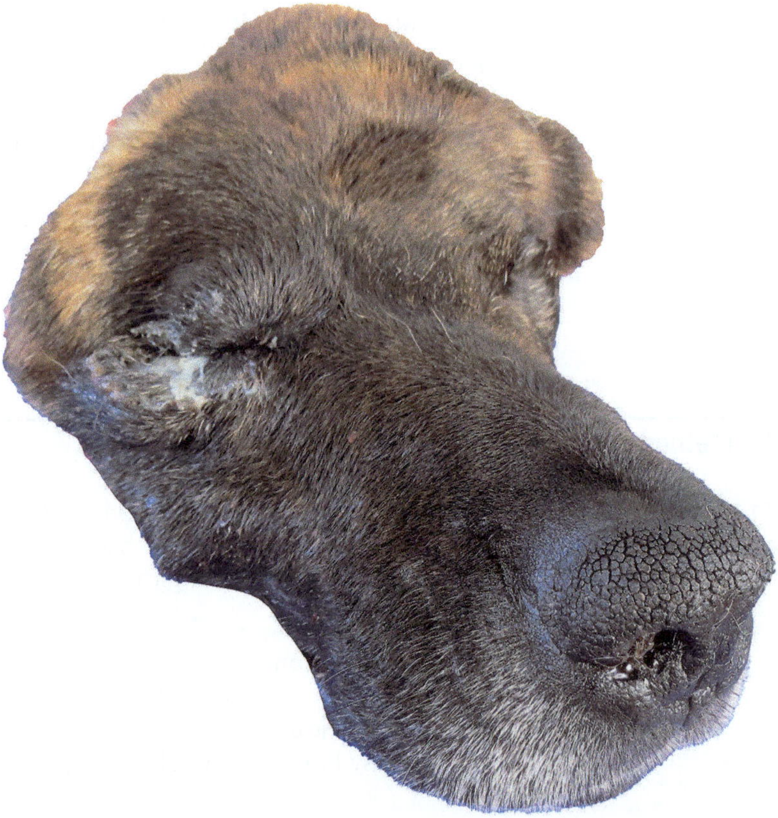

Fig. 2 Details of the head of an autopsied CDV infected dog, with previous immunochromatographic positive test. See the classical signs recognized in CDV infected dogs (nasal hyperkeratosis and the purulent ocular and nasal discharge)

2. The point of incision to the atlanto-occipital joint is the joint space between occipital bone and the first cervical vertebra (atlas) and can be identified by touch from the ventral face of the head and neck muscles.
3. At this point, cut carefully to separate the joint completely, cut the neck muscles, and cut the cervical spinal cord, to finally remove the head.
4. With the head completely separated, remove the head skin and cranial muscles, until a clean surface of the skull is obtained. It is crucial to remove most of cranial muscles to have a better cut surface along the skull bones.
5. To cut the skull bones, use a hacksaw to cut manually or, better, an electric autopsy saw.
6. The first cut is a transverse cut line above frontotemporal bone joint. The second and third cuts are placed to connect the right and the left ends of the transverse frontotemporal cut with the *foramen magnum* of the occipital bone, through the occipital condyles.
7. If the three cuts are properly done, you can easily remove the bone from the frontotemporal cut prying with a wire cutting plier.
8. Finally, remove the meninges (Fig. 3a), and cut the cranial nerve roots from the cranial base with the sharp tip of dressing scissors, until the brain is completely removed from the cranial base (Fig. 3b).
9. The complete right hemisphere, including the cerebrum (intermediate hypothalamus) and cerebellum, is fixed by placing it in 10% buffered neutral formalin solution (pH = 7.4) for 5 days and routinely processed for histologic examination (*see* **Note 9**).
10. The left hemisphere is kept in an ultra-freezer at −80 °C for future molecular analyses.

3.2 CNS Tissue Processing

After 5 days in 10% buffered neutral formalin solution, the selected CNS specimens (*see* **Note 10**) are being trimmed and processed with a microtome blade to macroscopically select tissue specimens before further analysis. For optimal fixation and subsequent high quality processing, the dimensions of tissue specimens are important. Ideally, the thickness should not exceed 4 mm, and the specimen should fit into a histology cassette without distortion [7] (Fig. 4).

1. The first transection level includes telencephalon (mainly frontal cerebral cortex and basal nuclei as gray matter and *corpus callosum* and internal capsule as white matter) and is represented as S1 in Fig. 4a or S1 in histologic cassettes in Figs. 4b–d.

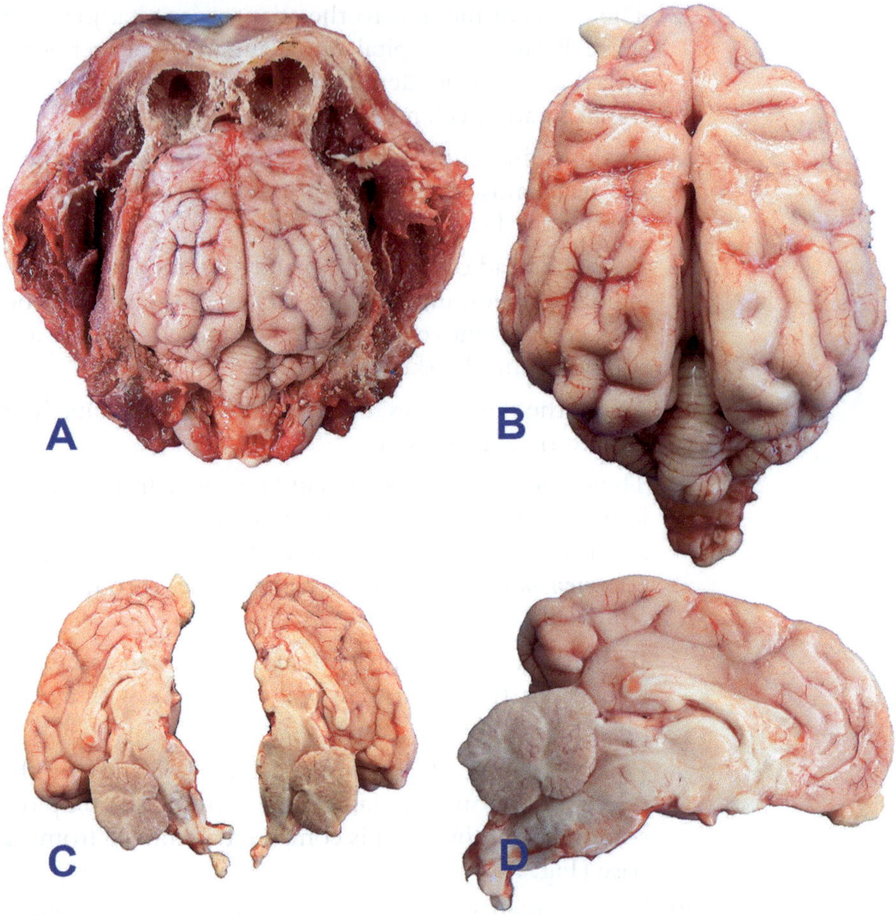

Fig. 3 Step-by-step brain removal of the same dog of Fig. 2. (**a**) Dorsal view of the brain after the removal of cut dorsal cranial bones and meninges (according to **steps 7** and **8** of Subheading 3.1). (**b**) Dorsal view of the brain removed from the cranial base (according to **step 8** of Subheading 3.1). (**c**) Sagittal section of the brain showing both brain hemispheres. (**d**) Detail of left hemisphere before to be frozen for future molecular studies (we recommend to keep some frozen samples of each autopsied animals)

2. The second transection includes telencephalon (mainly neocortex and *cingulate gyrus* as gray matter, *corpus callosum* and internal capsule as white matter, right lateral ventricle, and rhinencephalon including hippocampus) and diencephalon (mainly thalamus, hypothalamus, and third ventricle) and is represented as S4 in Fig. 4a or S4 in histologic cassettes in Figs. 4b–d.
3. The third transection includes telencephalon (mainly neocortex, cortical white matter, right lateral ventricle, and rhinencephalon) and mesencephalon (mainly rostral colliculus, part of mesencephalic aqueduct,and periaqueductal gray matter) and is represented as S8 in Fig. 4a or S8 in histologic cassettes in Figs. 4b–d.

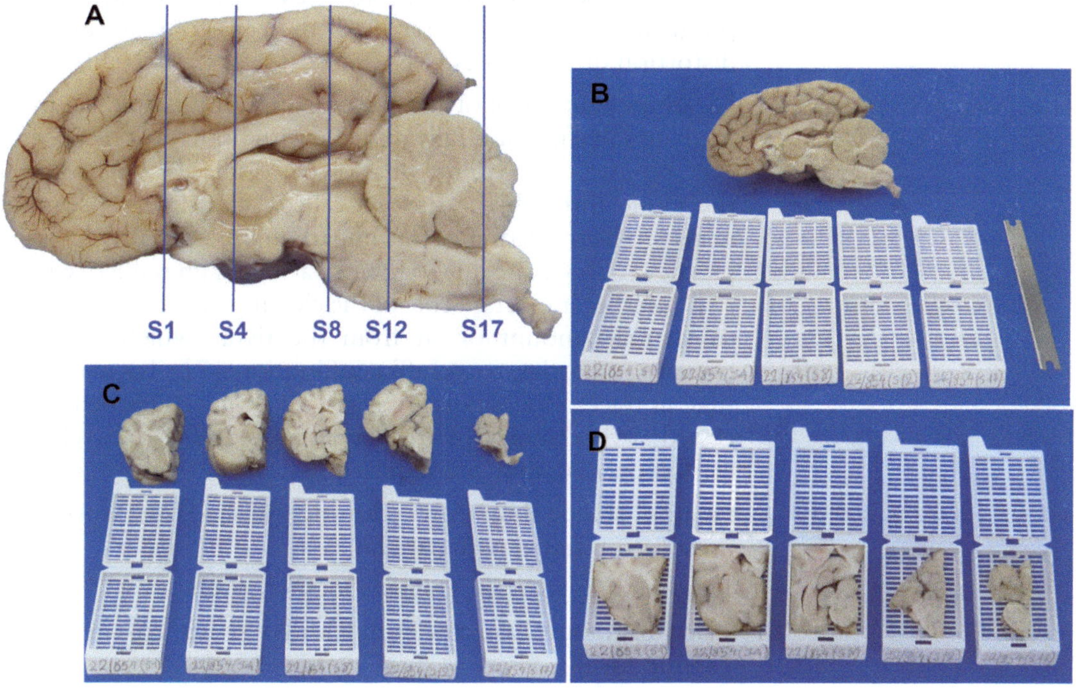

Fig. 4 (**a**) Right brain hemisphere of the same dog after formalin fixation for 5 days (see the color change of fixed CNS tissues, comparing with colors seen in Fig. 3d), showing the five recommended sections. (**c**) See the cut tissue section in parallel to the selected sections shown in **a** (S1, S4, S8, S12, and S17). (**b**–**d**) Step-by-step trimming and processing of the five mentioned specimens into each histological cassettes, to continue with the tissue processing for histopathological evaluation

4. The fourth transection includes telencephalon (neocortex), metencephalon (mainly rostral cerebellar vermis and lateral hemisphere), and pons (including fourth ventricle) and is represented as S12 in Fig. 4a or S12 in histologic cassettes in Figs. 4b–d.
5. The fifth transection includes metencephalon (mainly caudal cerebellar vermis and lateral hemisphere) and myelencephalon (including fourth ventricle and choroid plexus) and is represented as S17 in Fig. 4a or S17 in histologic cassettes in Figs. 4b–d.
6. These five transections are placed into their specific histologic cassettes and kept in a glass beaker containing 10% buffered neutral formalin solution, during 12 h to finishing the fixation step to continue with tissue dehydration.
7. Dehydration: Since paraffin is hydrophobic (not mixable with water), water inside the tissue must be removed before it can be infiltrated with paraffin. This process is carried out by immersing specimens in an increasing series of alcohols (begun with ethanol 70% and gradually increased concentrations, usually

80%, 95%, and 100%) (*see* **Note 11**). In this way, ethanol progressively replaces water in all the cells to avoid excessive distortion of the tissues [7].

8. Clearing: Since alcohol and paraffin are not miscible, an intermediate solvent that is fully miscible with both must be used. In this step, we used xylene; this solvent displaces the alcohol in the tissue through the process called "clearing." "Clearing" relates to how clearing agents impart an optical clarity or transparency to the tissue due to their relatively high refractive index [7]. Another important role of the clearing agent is to remove a substantial amount of fat from the tissue which otherwise presents a barrier to paraffin infiltration [7]. To make sure that all traces of alcohols are removed from tissues being processed, two changes of fresh xylene (*see* **Note 12**), clear of carried over ethanol, are required [7].
9. Paraffin embedding: After clearing finished, the specimens must be infiltrated with paraffin.
10. Heat paraffin to 61.5 °C in a laboratory oven.
11. Place histology cassettes with cleared specimens in melted paraffin for 3 h.
12. Place cassettes in fresh, melted paraffin for additional 3 h.
13. A final embedding step includes the specimen removal from the histologic cassette, to include in a stainless steel container, to fill again with new paraffin, using an automatic dispensing module or manually and topping with the specific marked part of the cassette.
14. Allow the specimens in the paraffin to cool to room temperature (RT), at which point the paraffin will solidify to a consistency that allows for sectioning. Since solidified paraffin holds the cells and intercellular structures in their proper relationship, it is also referred to as supporting medium [7].

3.3 Tissue Sectioning

1. Each paraffin-embedded tissue block is placed into the paraffin block holder of a rotary microtome.
2. Obtain five to ten consecutive sections of 5 μm, to form a continuous ribbon of these sections (*see* **Note 13**).
3. Place the ribbons on the warm water surface of a tissue flotation bath (*see* **Note 14**), to straighten them.
4. Once straightened, each tissue section of the ribbon is individually adhered to the surface of a high adherent glass slide.

3.4 Deparaffination and Rehydration

1. Slide drying, adhesion of specimens to the slides and deparaffinization: Keep brain slides overnight in a laboratory oven at 61.5 °C to dissolve paraffin.

2. After that, completely remove the paraffin from the slides by two consecutive passages in xylene (1 × 10 min, 1 × 15 min) (*see* **Note 15**).
3. Rehydration: To remove the xylene, pass the slides through several changes of ethanol (100% for 1 min, 95% for 1 min, 70% for 1 min), and then thoroughly rinse in tap water, with a final step in distilled water.

3.5 H&E Staining

1. Hematoxylin nuclear stain: Stain the rehydrated slides with a nuclear stain, which consists of a dye (Mayer's hematoxylin), for 10 min.
2. Rinse for 10 min in tap water (*see* **Note 16**).
3. Eosin counterstain: Stain the section during 3–4 min with a solution of eosin. Eosin colors many nonnuclear elements in different shades of pink.
4. Rinse, dehydrate, clear, and mount: After eosin staining, rinse the slides in distilled water for 1 min, followed by rinses with ethanol three times (95% for 2 min, 100% for 2 min, 100% for 2 min) to remove all traces of water. Next, rinse in two baths of xylene for 2 min each to "remove" the tissue and make it completely transparent [7].
5. Apply a thin layer of mounting medium and then a glass coverslip. If the staining and all subsequent steps were performed correctly, the slides will show all-important microscopic components and remain stable for many years [7] (*see* **Note 17**).

3.6 LFB Staining

1. In a new slide prepared according previous Subheadings 3.3 and 3.4, incubate in 100 mL Luxol Fast Blue ready-to-use solution for 24 h into a covered Coplin staining jar at room temperature or for 2 h at 61.5 °C in a laboratory oven (*see* **Note 18**).
2. Rinse in water.
3. Differentiate by submerging the slide in lithium carbonate ready-to-use solution for a few seconds.
4. Differentiate further by submerging in 70% ethanol reagent.
5. Rinse in water.
6. Incubate in Cresyl Echt Violet ready-to-use solution for 2–5 min.
7. Rinse, dehydrate, and clear: After Cresyl Echt Violet staining, rinse the slides in distilled water for 1 min, followed by rinses with ethanol three times (95% for 2 min, 100% for 2 min, 100% for 2 min) to remove all traces of water. Next, rinse in two baths of xylene for 2 min each to "remove" the tissue and make it completely transparent [7].

8. Apply a thin layer of mounting medium and then a glass coverslip. If the staining and all subsequent steps were performed correctly, the slides will show all-important microscopic components and remain stable for many years [7] (*see* **Note 17**).
9. If the staining and all subsequent steps were performed correctly, the slide will show myelinated fibers in white matter as intense blue stain, and the demyelinated regions will be recognized by the loss of characteristic bluish stain. Nissl substance and neurons must have an intense violet stain.

3.7 CDV Immunohistochemistry

1. In a new slide prepared according previous Subheadings 3.3 and 3.4, apply 150 μl of blocking reagent onto specimen surface for 5 min.
2. Rinse slides in distilled water for 5 min.
3. Place slides into a Coplin staining jar with 1x retrieval solution in a steamer, and heat to 95 °C for 30 min. After this time, keep the slides in the steamer for another 20 min at RT.
4. Remove slides and rinse in 1× PBS buffer solution (pH = 7.4) for 5 min.
5. Dry the contour of specimen with a tissue paper.
6. Place the slide into a wet incubation chamber (*see* **Note 7**), and apply mouse anti-CDV monoclonal primary antibody (dilution 1:250 in PBS) (or PBS buffer solution if "negative control" slides are required) overnight at 4 °C.
7. Rinse in PBS buffer solution for 5 min, followed by another wash with distilled water for 5 min.
8. Dry the contour of specimen with a tissue paper.
9. Apply secondary antibody (ready-to-use dilution) for 30 min at RT in a wet incubation chamber.
10. Rinse in PBS buffer solution for 5 min, followed by another wash with distilled water for 5 min.
11. Dry the contour of specimen with a tissue paper.
12. Mix one drop of 3, 3′-diaminobenzidine (DAB) solution with 1 mL of DAB buffer prior to use.
13. Apply 150 μL of DAB buffer for 4 min (*see* **Note 19**).
14. See under light microscope to confirm the HRP conjugation to the secondary antibody. Now, the antibody quickly reacts with DAB to produce the typical brown staining. Choose the best reaction time to stop it (e.g., 1 min).
15. Stop the DAB reaction, directly washing with tap water.
16. Counterstain with hematoxylin solution for 4 min.
17. Keep in tap water for 10 min.

18. Dehydrate and clear: Dehydrate by rinsing with ethanol solutions three times (95% for 2 min, 100% for 2 min, 100% for 2 min) to remove all traces of water. Next, rinse in two baths of xylene for 2 min each to "remove" the tissue, and make it completely transparent [7].
19. Apply a thin layer of mounting medium and then a glass coverslip. If the staining and all subsequent steps were performed correctly, the slides will show all-important microscopic components and remain stable for many years [7] (*see* **Note 17**).

If the immunohistochemical staining and all subsequent steps were performed correctly, the slide will show the CDV infected cells and their organelles, as characteristic intense brown colored particles in nucleus and cytoplasm.

3.8 Histopathological Analysis

1. Histopathological evaluation: Examine the H&E-stained slides under a light microscope for histological changes at 4×, 10×, and 40× magnifications.
2. Demyelination evaluation: Examine the LFB-stained slides under a light microscope for demyelination of white matter at 4×, 10×, and 40× magnifications.
3. CDV immunohistochemical evaluation: Examine the immunostained slides under a light microscope for immunoreactivity in the cerebrum and cerebellum at 4×, 10×, and 40× magnifications.
4. Characterize acute lesions by focal vacuolation, gliosis, intranuclear and intracytoplasmic inclusion bodies, and CDV-immunopositive cells (Fig. 5) [1, 4–6, 10–15].
5. Identify subacute lesions by demyelination, gliosis, necrosis, perivascular lymphocytic infiltration (cuffing) of two to three layers of thickness, intranuclear and intracytoplasmic inclusion bodies, and CDV-positive cells (Figs. 6 and 7) [1, 4–6, 10–15].
6. Chronic lesions are similar to the subacute stage, but perivascular infiltration is more prominent (at least four layers of thickness) (Fig. 8) [1, 4–6, 10–15].
7. The degree of demyelination is characterized by examining H&E and LFB staining and classified as mild (one-third demyelination of the whole slide), moderate (one-third to two-thirds demyelination of the whole slide), or severe (greater than two-thirds demyelination of the slide). In Fig. 9, we show the use of H&E, LFB, and IHC against CDV in simultaneous slides of section 12 (S12, representing regions of cerebellum, pons and floor of fourth ventricle), recognizing the presence of focal demyelination (subacute stage), one of the main lesions caused by CDV, in the same microscopic field of the same CDV infected dog.

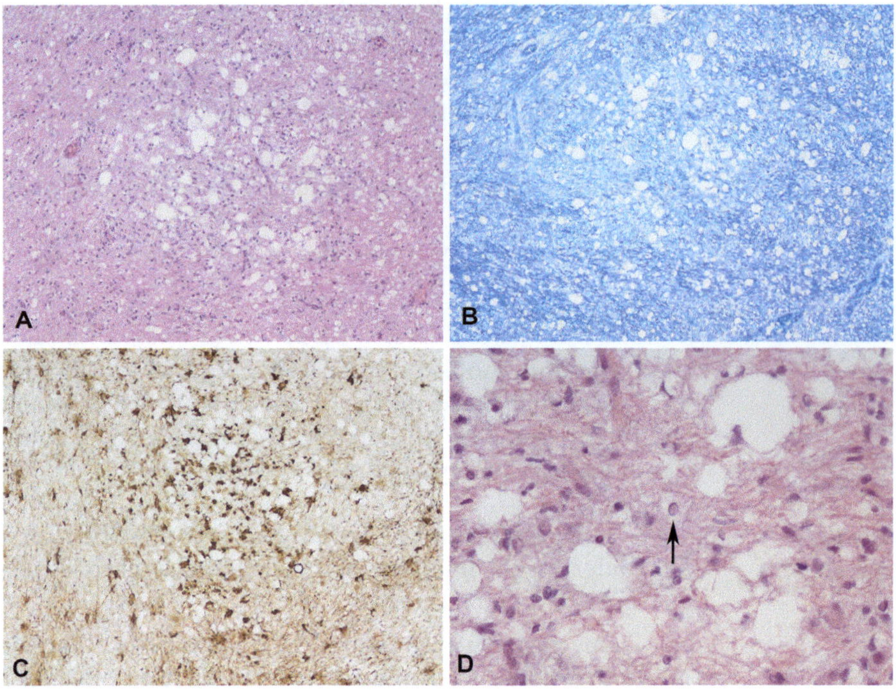

Fig. 5 Canine distemper cerebral acute lesions in non-vaccinated dogs (S4). (**a**) Demyelination in the cerebral parenchyma. H&E, original objective 10×. (**b**) Demyelination is apparent by loss of typical blue staining of myelin. Same part as in **a**. LFB, original objective 10×. (**c**) Canine distemper virus (CDV)-positive particles in the demyelinated area. Same part as in **a**. IHQ against CDV, original objective 10×. (**d**) High magnification of a. Inclusion bodies (arrow) and gliosis in the demyelinated area. H&E, original objective 40×

4 Notes

1. Technical grade is enough to these reagents.
2. To prepare Mayer's hematoxylin solution, in a glass beaker, dissolve 1 g hematoxylin with 1000 mL distilled water. After that, add 0.2 g of sodium iodate; immediately add 50 g of potassium alum, 1 g of citric acid, and 50 g of chloral hydrate; and dissolve with a magnetic stirrer. The working solution color is reddish violet; keep in hermetic bottle at room temperature.
3. To prepare Eosin stock solution, in a Pyrex glass beaker, smoothly dissolve 1 g of eosin Y with 20 mL of distilled water in a hot plate magnetic stirrer. Keep until the solution recover room temperature. Add 80 mL of 95% ethanol. Before use, mix 25 mL of stock solution with 75 mL of 80% ethanol; filter the working solution using filter paper in a funnel. Finally, add 0.5 mL of glacial acetic acid to the filtered solution to reinforce the red staining of the tissues.

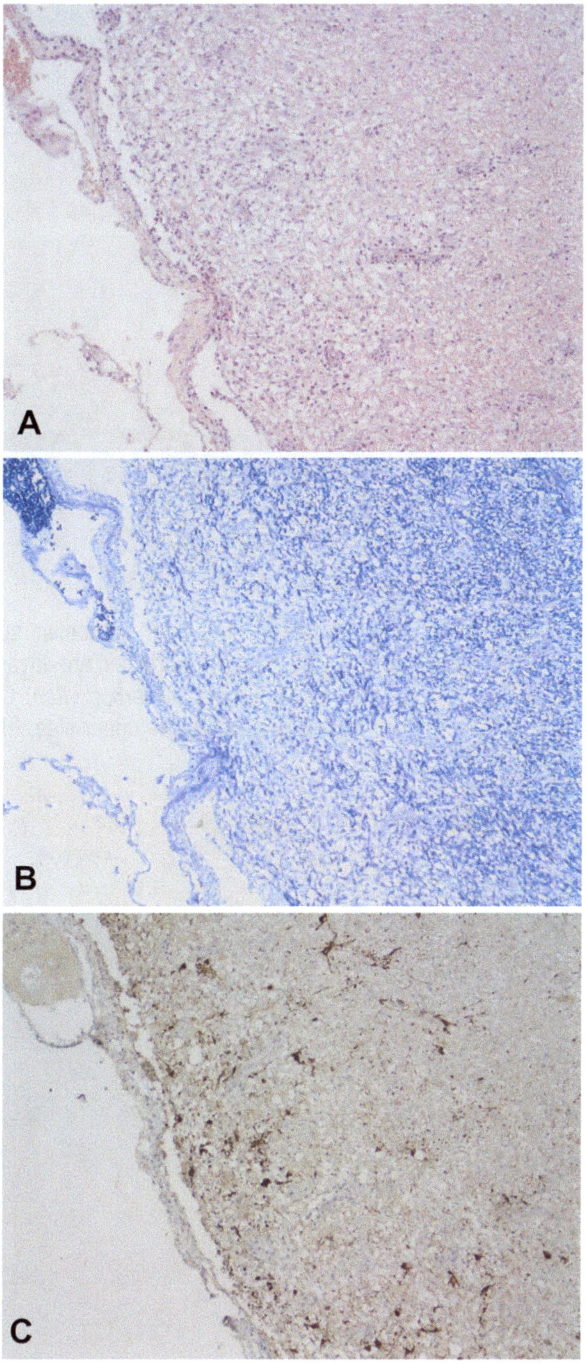

Fig. 6 Canine distemper cerebral subacute lesions in non-vaccinated dogs (S4). (**a**) Demyelination in the area beneath the lateral ventricle. H&E, original objective 10×. (**b**) Demyelination of the same part as in **a**. LFB, original objective 10×. (**c**) CDV-immunopositive particles in the same demyelinated area of **a**. IHQ against CDV, original objective 10×

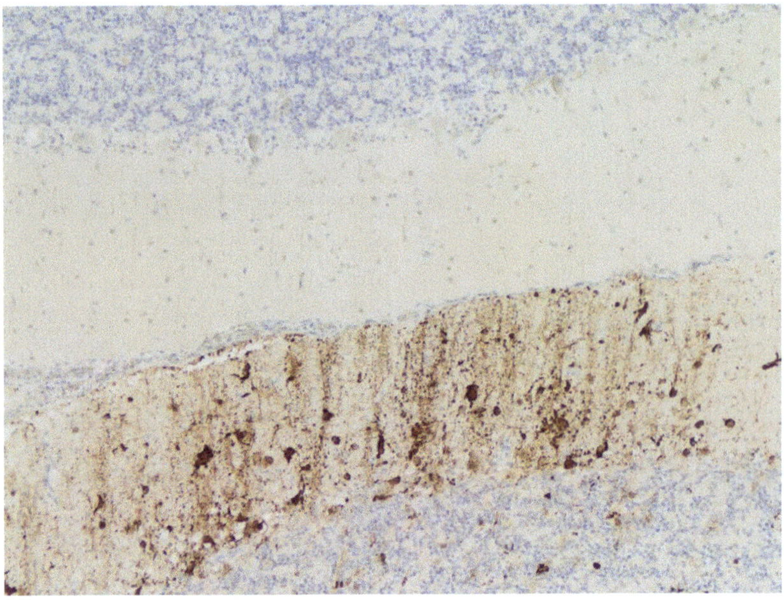

Fig. 7 Canine distemper cerebellar subacute lesions in non-vaccinated dog (S12). CDV-positive particles are invaded from the white matter to the pia mater in the cerebellar *arbor vitae*. Changes are seen in the cortex of the lower side, but not in the upper side. IHQ against CDV, original objective 10×

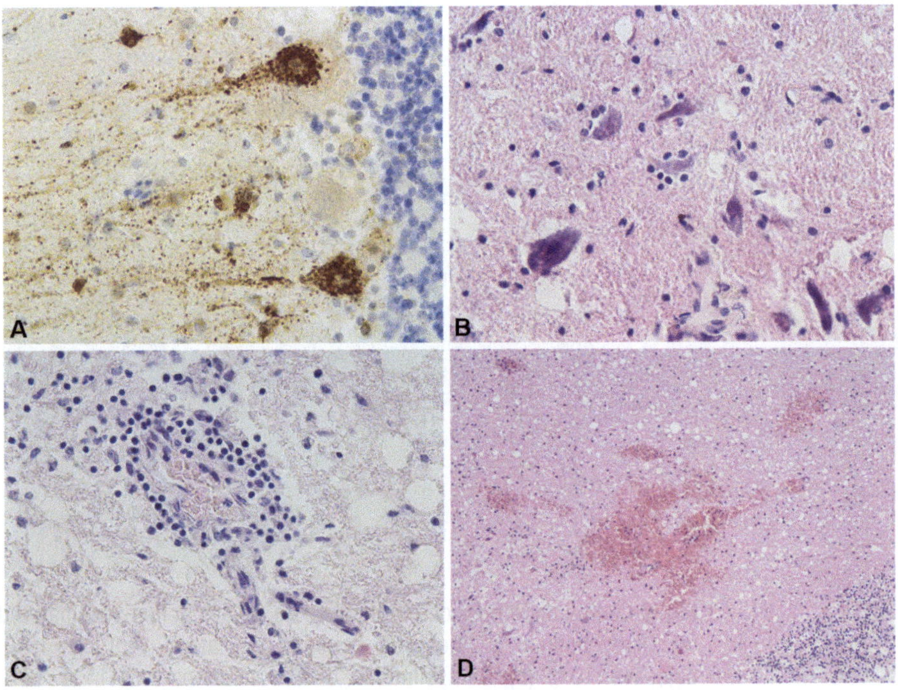

Fig. 8 Canine distemper cerebellar chronic lesions in non-vaccinated dogs (S12). (**a**) Purkinje cells contain many CDV-positive particles. IHQ against CDV, original objective 40×. (**b**) Satellitosis in the cerebellar parenchyma (arrow). H&E, original objective 40×. (**c**) Perivascular cuffing in the cerebellar parenchyma. H&E, original objective 40×. (**d**) Hemorrhages in the cerebellar parenchyma. H&E, original objective 10×

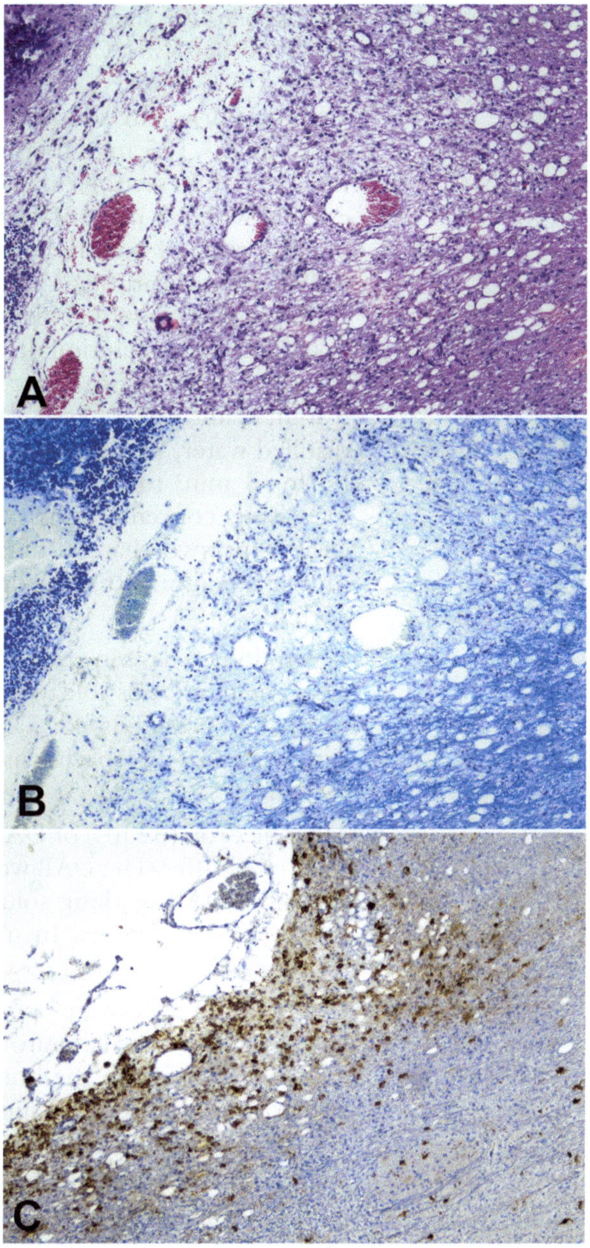

Fig. 9 Canine distemper cerebellar subacute lesions in non-vaccinated dogs (S12). (**a**) Focal demyelination in the floor of fourth ventricle. H&E, original objective 10×. (**b**) Demyelination is apparent by Luxol Fast Blue staining. Same part as Fig. **a**. LFB, original objective 10×. (**c**) Canine distemper virus (CDV)-positive particles in the demyelinated area. Same part as Fig. **a**. IHQ against CDV, original objective 10×

4. Luxol Fast Blue stain kit (myelin stain Abcam® commercial kit, cat N° ab 150675). Components for 100 tests: ethanol reagent (70%) 1 × 500 mL; Cresyl Echt Violet solution 1 × 125 mL; Lithium Carbonate solution 1 × 500 mL; and Luxol Fast Blue solution 1 × 125 mL. If you want to prepare your own solutions [9], stain sections in 0.1% Luxol Fast Blue diluted in 95% ethanol with 2.5 mL of 10% acetic acid at 56 °C for 24 h; rinse in 95% ethanol and then in distilled water; differentiate in 0.05% lithium carbonate until gray and white matters can be distinguished; rinse in 70% ethanol, two changes; counterstain in 0.1% aqueous solution of crystal violet with 5 mg oxalic acid for 10 min and at 56 °C for 2 min; rinse in distilled water for 5 min; stain sections in 0.2% Sirius Red for 10 min; rinse in distilled water, two changes; counterstain in Harris hematoxylin for 3 min; rinse in tap water for 3–5 min; dehydrate in increasing concentrations of ethanol (70%, 95%, 99%); and clear using two changes of xylene and mount using a hydrophobic medium. Let slides dry at room temperature for at least 15 min.
5. If you do not have in your laboratory an automatic processing equipment, it is possible to carry out all the steps manually. The manual processing is slow, and a more labor-intensive method since transfer of specimens or changing reagents is done by hand.
6. Mix one drop (32 μL) of DAB chromogen per 1.0 mL of DAB substrate buffer. The DAB working solution is stable for 5 days. Store the DAB working solution at 2–8 °C. Apply DAB mixture to tissue sections. Incubate for 2–5 min. Rinse in deionized water.
7. A common Tupperware® plastic container could be enough to substitute a specific laboratory wet chamber. In this case, place wet paper towels on the bottom of the box. After that, place two to four plastic slide support rods on top of wet towels, separated in parallel less than the length of each slide, to allow to place each slide on the top of parallel rods. Finally, apply the antibody and cover with the box lid.
8. Domestic dogs with previous clinical diagnosis of CDV infection confirmed by a commercial specific immunochromatographic assay to detect CDV antigen in nasal and ocular discharge (FASTest® Distemper Strip ad us. vet., Megacor Diagnostik, Germany) (Fig. 1). Clinical diagnoses were based on classical signs of CDV infection, such as ocular or nasal purulent discharge, diarrhea, fever, respiratory disturbance, and footpad and/or nasal hyperkeratosis (Fig. 2).

9. The ratio of fixative to tissue should never be less than ten parts of fixative to one-part tissue.
10. The better section to recognize specific lesions is the section 12 (see S12 section, in Fig. 4). More details of dog brain transection levels can be found in http://vanat.cvm.umn.edu/brainsect/levels.html?1.
11. To carry out dehydration, place the histology cassettes in 70% ethanol for 2 h; change and place the cassettes in 80% ethanol for 2 h; change and place the cassettes in 95% ethanol for 2 h; and, finally, change and place the cassettes in 100% ethanol for 2 h. After this final step, the specimens are completely dehydrated and ready for clearing.
12. To carry out clearing, place the cassettes in xylene for 2 h; change xylene and incubate in fresh xylene for 2 h to completely remove the ethanol.
13. Previously the cut surface of the paraffin block must be kept in contact with an ice bed in a freezer at −20 °C, and during the sectioning step, periodically rehydration of the paraffin block surface is needed (with a little piece of cotton imbibed with cold tap water); in this way, you could improve the paraffin section quality.
14. The water temperature into the tissue flotation bath is automatically controlled from 42 to 43 °C (or you could have a good manual control, using a laboratory thermometer in the water bath and adding progressively a little of hot tap water to keep constantly this temperature).
15. Deparaffination time could be reduced if keeping in the laboratory oven at 61.5 °C.
16. Another rinse in tap water for 10 min could improve complete withdraw of hematoxylin excess.
17. A very important recommendation here is to avoid the formation of air bubbles between the slide and the coverslip during this step and to remove the excess of mounting medium with a tissue paper, keeping the slide at room temperature for 6 h, until finishing completely the drying of the Canada balsam.
18. Coverage of Coplin staining jar is very important to avoid evaporation of solution.
19. To avoid personal or environmental contamination with DAB, after finishing the use of this hazardous substance, all the used laboratory surfaces and micropipette disposable tips must be cleaned with sodium hypochlorite before final disposal.

Acknowledgments

Victoria Iribarnegaray is PhD fellowship recipient of CAP-Udelar (Uruguay) and Camila Larrañaga and Belén Varela are MSc fellowship recipients of ANII and CAP-Udelar (Uruguay), respectively. José Manuel Verdes is Research Career Member of the National Research System (SNI-ANII), Uruguay.

Funding

This work was supported by the National Agency of Research and Innovation (ANII Fondo María Viñas 2019, Grant FMV-1-2019-1-155934), Committee of Scientific Research of the University of the Republic (CSIC, Udelar), and Programme for the Development of Basic Sciences (PEDECIBA), Uruguay.

References

1. Alldinger S, Fonfara S, Kremmer E et al (2000) Up-regulation of the hyaluronate receptor CD44 in canine distemper demyelinated plaques. Acta Neuropathol 99:138–146. https://doi.org/10.1007/pl00007417
2. Budaszewski RF, von Messling V (2016) Morbillivirus experimental animal models: measles virus pathogenesis insights from canine distemper virus. Viruses 8:274. https://doi.org/10.3390/v8100274
3. Cantile C, Youssef S (2016) Nervous System. In: Maxie MG (ed) Jubb, Kennedy, and Palmer's pathology of domestic animals, 6th edn, vol 1. Elsevier, St Louis, pp 250–406. ISBN: 978-0-7020-5317-7
4. Feijóo G, Yamasaki K, Delucchi L et al (2021) Central nervous system lesions caused by canine distemper virus in 4 vaccinated dogs. J Vet Diagn Invest 33:640–647. https://doi.org/10.1177/104063872110092
5. Gröters S, Alldinger S, Baumgärtner W (2005) Up-regulation of mRNA for matrix metalloproteinases-9 and -14 in advanced lesions of demyelinating canine distemper leukoencephalitis. Acta Neuropathol 110:369–382. https://doi.org/10.1007/s00401-005-1055-z
6. Lempp C, Spitzbarth I, Puff C et al (2014) New aspects of the pathogenesis of canine distemper leukoencephalitis. Viruses 6:2571–2601. https://doi.org/10.3390/v6072571
7. Lisowski A (2022) Science of tissue processing. Leica Biosystems. https://www.leicabiosystems.com/es/knowledge-pathway/science-of-tissue-processing/. Accessed 29 Sept 2022
8. Martinez-Gutierrez M, Ruiz-Saenz J (2016) Diversity of susceptible hosts in canine distemper virus infection: a systematic review and data synthesis. BMC Vet Res 12:78. https://doi.org/10.1186/s12917-016-0702-z
9. Sajadi E, Raoofi A, Abdi S et al (2020) The modified method of Luxol Fast Blue for paraffin-embedded myelin sheath staining. Int J Morphol 38:1197–1200. https://doi.org/10.4067/S0717-95022020000501197
10. Seehusen F, Orlando EA, Wewetzer K et al (2007) Vimentin-positive astrocytes in canine distemper: a target for canine distemper virus especially in chronic demyelinating lesions? Acta Neuropathol 114:597–608. https://doi.org/10.1007/s00401-007-0307-5
11. Spitzbarth I, Lempp C, Kegler K et al (2016) Immunohistochemical and transcriptome analyses indicate complex breakdown of axonal transport mechanisms in canine distemper leukoencephalitis. Brain Behav 3:e00472. https://doi.org/10.1002/brb3.472
12. Summers BA, Greisen HA, Appel MJ (1979) Early events in canine distemper demyelinating encephalomyelitis. Acta Neuropathol 46:1–10. https://doi.org/10.1007/BF00684797

13. Ulrich R, Puff C, Wewetzer K et al (2014) Transcriptional changes in canine distemper virus-induced demyelinating leukoencephalitis favor a biphasic mode of demyelination. PLoS One 9:e95917. https://doi.org/10.1371/journal.pone.0095917
14. Vandevelde M, Zurbriggen A (2005) Demyelination in canine distemper virus infection: a review. Acta Neuropathol 109:56–68. https://doi.org/10.1007/s00401-004-0958-4
15. Vandevelde M, Fankhauser R, Kristensen F et al (1981) Immunoglobulins in demyelinating lesions in canine distemper encephalitis. An immunohistological study. Acta Neuropathol 54:31–41. https://doi.org/10.1007/BF00691330

Chapter 15

Canine Distemper Virus Pathogenesis in the Ferret Model

Oliver Siering, Bevan Sawatsky, and Christian K. Pfaller

Abstract

Canine distemper virus (CDV) is a highly contagious pathogen within the morbillivirus genus infecting a wide range of different carnivore species. The virus shares most biological features with other closely related morbilliviruses, including clinical signs, tissue tropism, and replication cycle in the respective host organisms.

In the laboratory environment, experimental infections of ferrets with CDV were established as a potent surrogate model for the analysis of several aspects of the biology of the human morbillivirus, measles virus (MeV). The animals are naturally susceptible to CDV and display severe clinical signs resembling the disease seen in patients infected with MeV. As seen with MeV, CDV infects immune cells and is thus associated with a strong transient immunosuppression. Here we describe several methods to evaluate viral load and parameters of immunosuppression in blood-circulating immune cells isolated from CDV-infected animals.

Key words *Paramyxoviridae*, *Morbillivirus*, Pathogenesis, *Mustela putorius furo*, Animal model, Viremia, Surrogate

1 Introduction

The genus *Morbillivirus* within the family *Paramyxoviridae* includes several well-known and newly discovered pathogens. Measles virus (MeV), the human pathogenic prototype virus within the genus, is commonly known for its severe pathogenicity and immunosuppressive character in humans [1]. Besides MeV, the other known species within the genus are canine distemper virus (CDV), rinderpest virus, cetacean morbillivirus, phocine morbillivirus, peste-des-petits-ruminants virus, feline morbillivirus [2], and the most recently identified bat and porcine morbilliviruses [3, 4]. These viruses share many biological features, including their cell entry receptor usage. Their natural receptors are CD150, also known as signaling lymphocytic activation molecule (SLAM), which is expressed on immune cells [5, 6], as well as the adherens-junction molecule nectin-4, which is expressed on epithelial cells [7, 8]. Despite the common usage of these receptors,

Dzwokai Z. Ma and Christian K. Pfaller (eds.), *Measles and Related Morbilliviruses: Methods and Protocols*, Methods in Molecular Biology, vol. 2808, https://doi.org/10.1007/978-1-0716-3870-5_15,

morbilliviruses have very distinct host ranges [2]. MeV infection, for example, is restricted to humans and nonhuman primates [1, 9–12]. In contrast, CDV infects carnivores and is the cause of dog distemper, an often fatal disease in domestic dogs [13]. In addition, the virus infects other carnivore species such as red foxes [14], raccoons [15, 16], and mustelids such as ferrets (*mustela putorius furo*) and other marten species [17–19]. There are also reports stating susceptibility of other species outside of these families [20–23]. CDV was first isolated in 1905 by Henri Carré from an infected dog [24]. Earliest animal experiments were conducted by Green and colleagues who analyzed the disease in red foxes and termed it epizootic fox encephalitis [14]. Dunkin et al. performed experimental infections in dogs and showed that the underlying pathogen causing the known dog distemper was able to induce a similar disease in ferrets [13, 25]. With the technological progress toward the end of the twentieth century, the first genomic sequence of CDV was described for the Onderstepoort vaccine strain [26]. Since then, reverse genetics technology for generation of recombinant morbilliviruses including CDV [27, 28] has established CDV infections in ferrets as a reliable surrogate model for morbillivirus infection [17, 29, 30]. These studies provided molecular evidence for the similarities of transmission, infection, dissemination, and disease development between CDV and other morbilliviruses including MeV. Today, the ferret-CDV model is widely used to study the impact of mutations in the viral genome on pathogenesis [31–35] and the efficacy of antiviral substances [36, 37].

Ferrets are usually infected intranasally with CDV, or by aerosols [17, 25]. Pathogenic CDV strains, such as the ferret-adapted 5804P strain [17], cause fatal disease in ferrets (Fig. 1a). The first general clinical sign of disease is a short-lived mild fever peak within the first week of infection. The severe clinical disease usually starts after 1 week of infection and often consolidates around day 10 post-infection. At this stage, animals present with high fever, depression, coryza, and weight loss as well as the morbillivirus-specific rash that can develop either in localized spots or generalized over the whole body (Fig. 1b–d) [17, 25]. The occurrence of neurological complications is highly dependent on the viral strain and can vary from strains that do not seem to progress into the central nervous system at all, strains that do so after long persistence in the host organisms, and strains that cause neurological disease within 2 weeks after infection [17, 30, 38].

Another consequence of an infection with CDV is the typical morbillivirus-induced immunosuppression. The primary target cells of morbilliviruses are infiltrating CD150-positive immune cells in the airways of the host. After infection, these cells transport the virus through the lymphatic system to immune organs where the virus replicates to high titers and infected cells subsequently

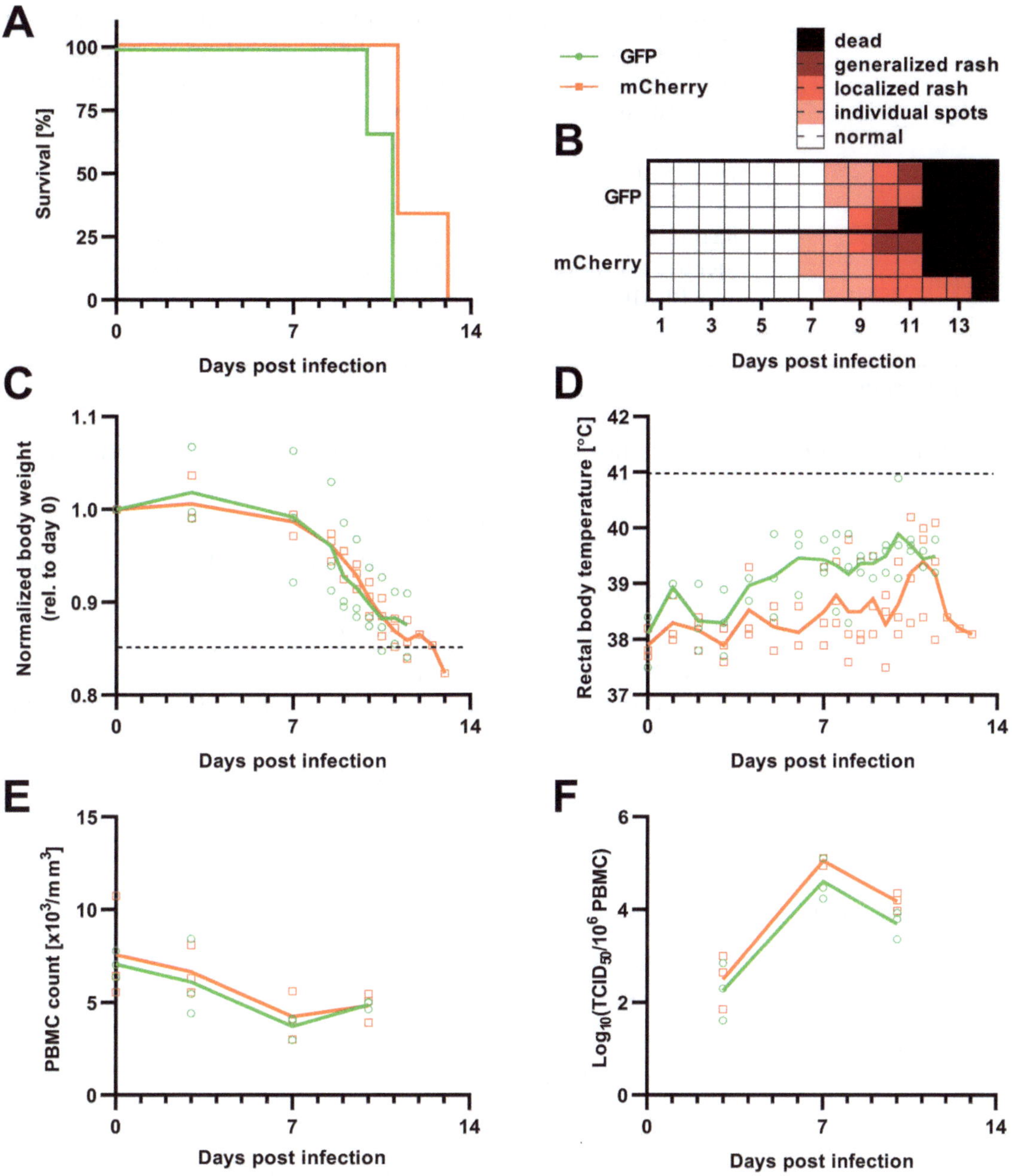

Fig. 1 In vivo pathogenesis of CDV-5804P-GFP and CDV-5804P-mCherry. Ferrets (n=3 per group) were infected intranasally with 2×10^5 $TCID_{50}$ of the respective viruses. (**a**) Survival curve, (**b**) individual development of rash, (**c**) body weight, (**d**) body temperature, (**e**) PBMC count, and (**f**) PBMC-associated viral blood titer per 10^6 PBMCs in the course of infection. CDV-5804P-GFP is depicted in green and CDV-5804P-mCherry in orange. (**b**) Severity of rash is indicated by darker red scales. The white boxes indicate the absence of rash, and the black boxes indicate the times when the animal was euthanized. (**c**–**f**) Lines indicate mean values from three groups, and symbols indicate individual values. (**c**) and (**d**) Dotted lines represent threshold values for humane endpoints. A weight loss of 15% or more in (**c**) and body temperatures of 41 °C or above in (**d**) are considered clinical endpoints

disseminate to the whole body. At this stage, virus-infected cells can also be detected among peripheral blood mononuclear cells (PBMCs), which eventually results in their depletion. Leucopenia (Fig. 1e) and viremia (Fig. 1f) usually occur even before the onset of other clinical symptoms and are an important measure of morbillivirus pathogenesis.

Here, we describe our standardized protocol for infection of ferrets by CDV and evaluation of disease development. We also include methods to determine viral load, white blood cell count, and PBMC proliferation activity.

2 Materials

2.1 Live Animal Work

1. Healthy male or female ferrets, preferably castrated or sterilized.
2. Veterinary thermometer.
3. Veterinary weighing scale (500–2500 g. Choose the weighing scale accordingly).
4. Insulin syringes (30G or 31G, 0.01–1 mL).
5. CDV stock solution with 5×10^5 to 1×10^6 tissue culture infectious dose 50 ($TCID_{50}$) in 100–200 μL volume per animal.
6. 1 mg/mL medetomidine hydrochloride solution.
7. 10% ketamine solution.
8. 5 mg/mL atipamezole hydrochloride solution.

2.2 Euthanasia of Moribund Animals

1. Insulin syringes (30G or 31G, 0.01–1 mL).
2. 1 mg/mL medetomidine hydrochloride solution.
3. 10% ketamine solution.
4. Blood collection needles (21G × 1–1/4″).
5. Blood collection tubes.
6. T61® euthanasia solution.

2.3 Blood Draw

1. Insulin syringes (30G or 31G, 0.01–1 mL).
2. 1 mg/mL medetomidine hydrochloride solution.
3. 5 mg/mL atipamezole hydrochloride solution.
4. Blood collection needles (21G × 1–1/4″).
5. 2 mL blood collection tubes coated with anticoagulant K_3EDTA.
6. 2 mL blood collection tubes coated with anticoagulant lithium heparin.
7. 2 mL blood collection clot activator tubes.

2.4 Determining White Blood Cell Count

1. Blood collected either in a tube coated with K_3EDTA or lithium heparin.
2. 3% acetic acid solution.
3. Hemocytometer (Neubauer counting chamber).
4. Light microscope.

2.5 Titration of PBMCs

1. Flat bottom 96-well plate for tissue culture.
2. Vero-dogSLAMtag cells stably expressing canine SLAM [17].
3. DMEM without supplements.
4. DMEM-5%: DMEM supplemented with 5% FBS, 1% L-glutamine, and 1% pen-strep.
5. Blood collected in a tube coated with K_3EDTA.
6. 15 mL centrifuge tubes.
7. 1.5 mL centrifuge tubes.
8. 1× PBS (140 mM NaCl; 2.7 mM KCl; 1.8 mM KH_2PO_4, pH 7.0–7.4).
9. 1× RBC lysis buffer (155 mM NH_4Cl, 10 mM Tris, pH 7.5).
10. Centrifuge suited for Falcon tubes.
11. RPMI-10%: RPMI supplemented with 10% FBS, 1% L-glutamine, and 1% pen-strep.
12. Trypan blue solution.
13. Hemocytometer (Neubauer counting chamber).
14. Tissue culture suited incubator at 37 °C.

2.6 PBMC Proliferation Assay

1. Blood collected in a tube coated with lithium heparin.
2. 15 mL centrifuge tubes.
3. 1× PBS (140 mM NaCl; 2.7 mM KCl; 1.8 mM KH_2PO_4, pH 7.0–7.4).
4. 1× RBC lysis buffer (155 mM NH_4Cl, 10 mM Tris, pH 7.5).
5. Ficoll or Histopaque 1077.
6. Centrifuge suited for Falcon tubes.
7. RPMI-10%: RPMI supplemented with 10% FBS, 1% L-glutamine, and 1% pen-strep.
8. 10 mg/mL Phytohemagglutinin, M-form (PHA-M) stock solution.
9. Hemocytometer (Neubauer counting chamber).
10. Tissue culture suited incubator at 37 °C.
11. U bottom 96-well plate for tissue culture.
12. Black flat bottom 96-well plate for tissue culture.

13. BrdU Cell Proliferation Chemiluminescent Assay Kit (containing 1000× BrdU labeling reagent; FixDenat buffer; anti-BrdU-POD, stabilized; antibody dilution solution; 10× washing buffer; substrate component A; substrate component B).
14. Chemiluminescence plate reader.

3 Methods

3.1 Regular Monitoring of Animals

1. Evaluate activity and general appearance of animals (*see* **Note 1**).
2. Measure rectal temperature (*see* **Note 2**).
3. Weigh animals.
4. Evaluate clinical signs as the occurrence of rash, conjunctivitis, dehydration, hyperkeratosis, and possibly neurological deficits (*see* **Note 3**).

3.2 Infecting Ferrets with CDV

1. Before infection, animals should be monitored as described above.
2. Anesthetize animals via intramuscular injection of 0.1 mL medetomidine hydrochloride solution (1 mg/mL) and 0.1 mL ketamine solution (10%).
3. When animals are anesthetized, infect intranasally by adding the virus inoculum drop by drop into the nostrils (*see* **Note 4**).
4. Reverse the Medetomidine sedation by subcutaneous injection of 0.15 mL atipamezole hydrochloride solution (5 mg/mL).

3.3 Blood Draw

1. Before blood draw, animals should be monitored as described above.
2. Anesthetize animals via intramuscular injection of 0.2 mL Medetomidine hydrochloride solution (1 mg/mL).
3. Draw blood samples from the vena cava. For this purpose, put animals on the back, and insert blood collection needle just above the sternum in an approximate angel of 30° to the skin. As soon as skin layers are pierced, change angle and move the needle horizontally under the sternum/rib cage in the direction toward the heart. When the needle is in the appropriate position, put on the required blood collection tube (*see* **Note 5**). Collect blood (*see* **Note 6**). After blood collection, remove blood collection tube and needle, and stop possible bleeding by application of soft pressure to the puncture site.
4. Reverse the Medetomidine sedation by subcutaneous injection of 0.15 mL atipamezole hydrochloride solution (5 mg/mL).

3.4 Determining White Blood Cell Count

1. Use blood collected in anticoagulant coated collection tube.
2. Add 50 μL full blood to 450 μL 3% acetic acid solution (*see* **Note 7**).
3. Incubate for 1–2 min at room temperature until red blood cells have lysed. The acetic acid-blood suspension should be of brown to black color.
4. Load 10 μL of acetic acid-blood suspension into a hemocytometer (Neubauer counting chamber), and count white blood cells under a light microscope:

$$c = \frac{n}{4} \times 10^4 \times d$$

with c = number of PBMC per 1 mL; n = cells counted in all four big squares; and d = dilution factor (in this case, 10; *see* **step 2**).

3.5 Titration of PBMCs

1. Prepare a cell suspension of 10^5 Vero-dogSLAMtag cells per mL in DMEM-5%, and seed into a 96-well plate with 100 μL per well 1 day prior to blood collection.
2. Use blood collected in anticoagulant coted collection tube (*see* **Note 8**).
3. Transfer blood to 15 mL tubes.
4. Spin down the sample for 5 min at 2000 rpm.
5. Collect plasma in 1.5 mL tubes and store at −20 °C (can be used for immunological assays).
6. Resuspend the remaining red and white blood cells in PBS to an approximate total volume of 2 mL.
7. Add 10 mL of RBC lysis buffer.
8. Incubate at room temperature until the suspension becomes clearer with a cherry red color.
9. Spin down sample for 5 min at 2000 rpm.
10. Remove supernatant.
11. Wash cell pellet in 10 mL PBS.
12. Spin down sample for 5 min at 2000 rpm.
13. Remove supernatant and repeat washing step.
14. Resuspend cell pellet in 1 mL RPMI-10%.
15. Dilute 10 μL of cell suspension in 90 μL Trypan blue solution.
16. Load 10 μL of Trypan blue cell suspension into a hemocytometer (Neubauer counting chamber), and count white blood cells under a light microscope:

$$c = \frac{n}{4} \times 10^4 \times d$$

with c = number of PBMC per 1 mL; n = cells counted in all four big squares; d = dilution factor (in this case, 10; *see* **step 15**).

17. Prepare tenfold dilution series: prepare eight tubes/wells with 450 μL DMEM without supplements, respectively, dilute 50 μL of cell suspension in the first 450 μL DMEM and mix vigorously, dilute 50 μL of the previous dilution step in the next 450 μL DMEM, and repeat until initial sample is diluted to 10^{-8}.
18. Transfer 100 μL of each dilution step in quadruplicates into a 96-well plate seeded with Vero-dogSLAMtag cells (*see* **step 1**).
19. Incubate plate at 37 °C for 3–5 days.
20. Calculate $TCID_{50}/10^6$ PBMCs according to the method described by Spearman and Kärber (*see* **Note 9**).

3.6 PBMC Proliferation Assay

1. Transfer blood to 15 mL tubes.
2. Spin down sample for 5 min at 2000 rpm.
3. Collect plasma in 1.5 mL tubes and store at −20 °C (can be used for immunological assays).
4. Resuspend the remaining red and white blood cells in PBS to an approximate total volume of 4 mL.
5. Gently layer the PBS/blood suspension onto 3 mL Ficoll/Histopaque 1077 in 15 mL Falcon tube.
6. Spin down sample for 40 min at 400× *g* with minimal deceleration to avoid disturbance.
7. Collect the whole Ficoll/Histopaque 1077 layer containing the PBMCs, and transfer into 10 mL PBS in 15 mL Falcon tube.
8. Spin down sample for 5 min at 2000 rpm.
9. Remove supernatant and wash cell pellet in 10 mL PBS.
10. Spin down sample for 5 min at 2000 rpm.
11. Repeat wash step.
12. Resuspend PBMC pellet in 1 mL RPMI-10%.
13. Dilute 10 μL of cell suspension in 90 μL RBC lysis buffer.
14. Load 10 μL of cell suspension into a hemocytometer (Neubauer counting chamber), and count white blood cells under a light microscope.
15. For each sample, seed four wells with 2.5×10^5 PBMC/well in 100 μL RPMI-10%, with two wells supplemented with 1 mg/mL phytohemagglutinin (*see* **Note 10**).
16. Incubate at 37 °C for 24 h.

17. Add 10 μL 100 μM BrdU solution (1:100 dilution of BrdU labeling reagent, 1000× conc. From the Cell Proliferation ELISA, BrdU (chemiluminescent) kit in RPMI-10%) to each well.
18. Incubate at 37 °C for 24 h.
19. Transfer samples into a black 96-well plate.
20. Spin down sample for 5 min at 2000 rpm.
21. Wash cells with 150 μL RBC lysis buffer per well.
22. Spin down sample for 5 min at 2000 rpm.
23. Repeat wash step twice with 150 μL PBS each.
24. Remove PBS and bake cells at 65 °C for 1 h.
25. Store at 4 °C or proceed with staining.
26. For labeling, add 200 μL FixDenat solution from the Cell Proliferation ELISA, BrdU (chemiluminescent) kit, per well, and incubate for 30 min at room temperature.
27. Resolve the stabilized anti-BrdU-POD in 1.1 mL water to prepare the anti-BrdU-POD stock solution.
28. Dilute 100 μL in 10 mL of the anti-BrdU-POD stock solution in antibody dilution solution for one 96-well plate, and add 100 μL per well.
29. Wash three times with 200 μL 1× washing buffer from the Cell Proliferation ELISA, BrdU (chemiluminescent) kit, per well.
30. Add 100 μL substrate component B into 10 mL substrate component A for one 96-well plate; keep under agitation for 15–20 min and add 100 μL per well.
31. Measure chemiluminescence in a microplate luminometer with photomultiplier technology.

3.7 Euthanasia of Moribund Animals (See Note 11)

1. Anesthetize animals via intramuscular injection of 0.1 mL medetomidine hydrochloride solution (1 mg/mL), and 0.1 mL ketamine solution (10%).
2. Bleed out animal by heart punctuation (*see* **Note 12**).
3. Alternatively, euthanasia by intracardiac injection of 1 mL T61® euthanasia solution (*see* **Note 13**).

4 Notes

1. Every animal is individual in terms of their activity. Usually, the animals are very playful and active when they meet their handlers, but this can vary from animal to animal. Thus, evaluation of activity is difficult, but basic hallmarks of behavior include their curiosity toward their handlers, appetite, and thirst.

2. Grab the skin on the neck of the animal to control the head to prevent any movement toward the hand of the handler. Fix the hip of the animal either with a second handler or with the other hand between the palm and middle finger, ring finger, and pinkie while inserting the thermometer with the index finger and thumb.
3. Evaluation of rash development is easier if parts of the body are shaved. We typically shave the throat area and the belly to ease evaluation. The earliest signs of rash often appear around the mouth, especially in the mouth corners.
4. For infection, we suggest a volume for the inoculum of about 200 μL and 2×10^5 $TCID_{50}$.
5. We usually draw blood from the vena cava. Alternatively, the blood draw is also possible from the vena jugularis. However, in our hands, the vena cava presents a more reliable and easier blood collection site for volumes around 2 mL.
6. We usually use blood collected in tubes coated with K_3EDTA or lithium heparin.
7. As an alternative for counting PBMCs in whole blood, the use of veterinary blood counters suitable for ferret blood may be feasible.
8. We usually use blood collected in tubes coated with K_3EDTA.
9. Viremia (infectious CDV in PBMC) is usually already detectable around 3 days after infection, which is before the onset of other clinical symptoms. Typically, peak viremia occurs around 7 days postinfection. Thereafter, the viral load in the blood is decreasing (Fig. 1f).
10. As a result of morbillivirus-induced leucopenia, the amount of PBMCs which can be obtained from the volume of blood drawn may be reduced to numbers lower than 10^6 cells needed for the assay. Thus, the proliferation capacity is calculated as a relative value with the ratio between stimulated and unstimulated cells.
11. The humane endpoint of a study is usually defined by several parameters including general severity of clinical symptoms (or severity of the disease). The cutoff criteria are usually defined to avoid unjustified harm to the animals without gain of knowledge. Importantly, those cutoff criteria need to be defined prior to the animal trial, and authorization by the local legal authorities is eminent.
12. The method of euthanasia of moribund animals depends on whether specimen other than blood are harvested from the animal. If necropsy follows after the pathogenesis study or large blood volumes are required for further analysis, heart bleeding under deep anesthesia is the method of choice.

13. If analysis of other specimen is not desired, T61® euthanasia solution containing embutramide is used for euthanasia. Alternatively, drugs authorized for small animal euthanasia (e.g., pentobarbital) may be used.

References

1. Minta AA, Ferrari M, Antoni S et al (2022) Progress toward regional measles elimination – worldwide, 2000-2021. MMWR Morb Mortal Wkly Rep 71:1489–1495. https://doi.org/10.15585/mmwr.mm7147a1
2. Pfeffermann K, Dörr M, Zirkel F et al (2018) Morbillivirus pathogenesis and virus-host interactions. Adv Virus Res 100:75–98. https://doi.org/10.1016/bs.aivir.2017.12.003
3. Arruda B, Shen H, Zheng Y et al (2021) Novel morbillivirus as putative cause of fetal death and encephalitis among swine. Emerg Infect Dis 27:1858–1866. https://doi.org/10.3201/eid2707.203971
4. Lee B, Ikegame S, Carmichael J et al. (2021) Zoonotic potential of a novel bat morbillivirus. Res Sq. rs.3.rs-926789. https://doi.org/10.21203/rs.3.rs-926789/v1
5. Tatsuo H, Ono N, Tanaka K et al (2000) SLAM (CDw150) is a cellular receptor for measles virus. Nature 406:893–897. https://doi.org/10.1038/35022579
6. Tatsuo H, Ono N, Yanagi Y (2001) Morbilliviruses use signaling lymphocyte activation molecules (CD150) as cellular receptors. J Virol 75:5842–5850. https://doi.org/10.1128/JVI.75.13.5842-5850.2001
7. Mühlebach MD, Mateo M, Sinn PL et al (2011) Adherens junction protein nectin-4 is the epithelial receptor for measles virus. Nature 480:530–533. https://doi.org/10.1038/nature10639
8. Noyce RS, Bondre DG, Ha MN et al (2011) Tumor cell marker PVRL4 (nectin 4) is an epithelial cell receptor for measles virus. PLoS Pathog 7:e1002240. https://doi.org/10.1371/journal.ppat.1002240
9. Albrecht P, Lorenz D, Klutch MJ et al (1980) Fatal measles infection in marmosets pathogenesis and prophylaxis. Infect Immun 27:969–978. https://doi.org/10.1128/iai.27.3.969-978.1980
10. Lorenz D, Albrecht P (1980) Susceptibility of tamarins (Saguinus) to measles virus. Lab Anim Sci 30:661–665
11. El Mubarak HS, Yüksel S, van Amerongen G et al (2007) Infection of cynomolgus macaques (Macaca fascicularis) and rhesus macaques (Macaca mulatta) with different wild-type measles viruses. J Gen Virol 88:2028–2034. https://doi.org/10.1099/vir.0.82804-0
12. Delpeut S, Sawatsky B, Wong X-X et al (2017) Nectin-4 interactions govern measles virus virulence in a new model of pathogenesis, the squirrel monkey (Saimiri sciureus). J Virol 91 (11):e02490-16. https://doi.org/10.1128/JVI.02490-16
13. Dunkin GW, Laidlaw PP (1926) Studies in dog-distemper. J Comp Pathol Ther 39:213–221. https://doi.org/10.1016/S0368-1742(26)80021-9
14. Green RG, Ziegler NR, Green BB et al (1930) Epizootic Fox encephalitis. I. General description*. Am J Epidemiol 12:109–129. https://doi.org/10.1093/oxfordjournals.aje.a115259
15. Roscoe DE (1993) Epizootiology of canine distemper in New Jersey raccoons. J Wildl Dis 29:390–395. https://doi.org/10.7589/0090-3558-29.3.390
16. Roelofs D, Schmitz KS, van Amerongen G et al (2023) Inoculation of raccoons with a wild-type-based recombinant canine distemper virus results in viremia, lymphopenia, fever, and widespread histological lesions. mSphere 8:e0014423. https://doi.org/10.1128/msphere.00144-23
17. von Messling V, Springfeld C, Devaux P et al (2003) A ferret model of canine distemper virus virulence and immunosuppression. J Virol 77:12579–12591. https://doi.org/10.1128/jvi.77.23.12579-12591.2003
18. Perpiñán D, Ramis A, Tomás A et al (2008) Outbreak of canine distemper in domestic ferrets (Mustela putorius furo). Vet Rec 163:246–250. https://doi.org/10.1136/vr.163.8.246
19. Origgi FC, Plattet P, Sattler U et al (2012) Emergence of canine distemper virus strains with modified molecular signature and enhanced neuronal tropism leading to high mortality in wild carnivores. Vet Pathol 49:913–929. https://doi.org/10.1177/0300985812436743
20. Hur K, Bae JS, Choi JH et al (1999) Canine distemper virus infection in binturongs (Arctictis binturong). J Comp Pathol 121:295–299. https://doi.org/10.1053/jcpa.1999.0322

21. Daoust P-Y, McBurney SR, Godson DL et al (2009) Canine distemper virus-associated encephalitis in free-living lynx (Lynx canadensis) and bobcats (Lynx Rufus) of eastern Canada. J Wildl Dis 45:611–624. https://doi.org/10.7589/0090-3558-45.3.611
22. Qiu W, Zheng Y, Zhang S et al (2011) Canine distemper outbreak in rhesus monkeys, China. Emerg Infect Dis 17:1541–1543. https://doi.org/10.3201/eid1708.101153
23. Seimon TA, Miquelle DG, Chang TY et al (2013) Canine distemper virus: an emerging disease in wild endangered Amur tigers (Panthera tigris altaica). mBio 4(4): e00410-13. https://doi.org/10.1128/mBio.00410-13
24. Carré H (1905) Sur la maladie des jeunes chiens. C R Hebd Seances Acad Sci:689–690
25. Dunkin GW, Laidlaw PP (1926) Studies in dog-distemper. J Comp Pathol Ther 39:201–212. https://doi.org/10.1016/S0368-1742(26)80020-7
26. Sidhu MS, Husar W, Cook SD et al (1993) Canine distemper terminal and intergenic non-protein coding nucleotide sequences: completion of the entire CDV genome sequence. Virology 193:66–72. https://doi.org/10.1006/viro.1993.1103
27. Radecke F, Spielhofer P, Schneider H et al (1995) Rescue of measles viruses from cloned DNA. EMBO J 14:5773–5784. https://doi.org/10.1002/j.1460-2075.1995.tb00266.x
28. Gassen U, Collins FM, Duprex WP et al (2000) Establishment of a rescue system for canine distemper virus. J Virol 74:10737–10744. https://doi.org/10.1128/jvi.74.22.10737-10744.2000
29. von Messling V, Milosevic D, Cattaneo R (2004) Tropism illuminated: lymphocyte-based pathways blazed by lethal morbillivirus through the host immune system. Proc Natl Acad Sci USA 101:14216–14221. https://doi.org/10.1073/pnas.0403597101
30. Ludlow M, Nguyen DT, Silin D et al (2012) Recombinant canine distemper virus strain Snyder Hill expressing green or red fluorescent proteins causes meningoencephalitis in the ferret. J Virol 86:7508–7519. https://doi.org/10.1128/JVI.06725-11
31. von Messling V, Zimmer G, Herrler G et al (2001) The hemagglutinin of canine distemper virus determines tropism and cytopathogenicity. J Virol 75:6418–6427. https://doi.org/10.1128/JVI.75.14.6418-6427.2001
32. von Messling V, Svitek N, Cattaneo R (2006) Receptor (SLAM CD150) recognition and the V protein sustain swift lymphocyte-based invasion of mucosal tissue and lymphatic organs by a morbillivirus. J Virol 80:6084–6092. https://doi.org/10.1128/JVI.00357-06
33. Sawatsky B, Wong X-X, Hinkelmann S et al (2012) Canine distemper virus epithelial cell infection is required for clinical disease but not for immunosuppression. J Virol 86:3658–3666. https://doi.org/10.1128/JVI.06414-11
34. Sawatsky B, Cattaneo R, von Messling V (2018) Canine distemper virus spread and transmission to naive ferrets: selective pressure on signaling lymphocyte activation molecule-dependent entry. J Virol 92(15):e00669-18. https://doi.org/10.1128/JVI.00669-18
35. Siering O, Sawatsky B, Pfaller CK (2021) C protein is essential for canine distemper virus virulence and pathogenicity in ferrets. J Virol 95(4):e01840-20. https://doi.org/10.1128/JVI.01840-20
36. Krumm SA, Yan D, Hovingh ES et al (2014) An orally available, small-molecule polymerase inhibitor shows efficacy against a lethal morbillivirus infection in a large animal model. Sci Transl Med 6:232ra52. https://doi.org/10.1126/scitranslmed.3008517
37. Tilston-Lunel NL, Welch SR, Nambulli S et al (2021) Sustained replication of synthetic canine distemper virus defective genomes in vitro and in vivo. mSphere 6:e0053721. https://doi.org/10.1128/mSphere.00537-21
38. Summers BA, Greisen HA, Appel MJ (1984) Canine distemper encephalomyelitis: variation with virus strain. J Comp Pathol 94:65–75. https://doi.org/10.1016/0021-9975(84)90009-4

Chapter 16

Measles Foci Reduction Neutralization Test (FRNT)

Sabine Santibanez and Annette Mankertz

Abstract

The plaque reduction neutralization test (PRNT) and the enzyme-linked immunosorbent assay (ELISA) are both widely used to assess immunity to infectious diseases such as measles, but they use two different measurement principles: ELISA measures the ability of antibodies to bind to virus components, while the PRNT detects the aptitude of antibodies to prevent the infection of a susceptible cell. As a result, detection of measles virus (MV) neutralizing antibodies is the gold standard for assessing immunity to measles. However, the assay is laborious and requires experience and excellent technical skills. In addition, the result is only available after several days. Therefore, the classical PRNT is not suitable for high-throughput testing. By using an immunocolorimetric assay (ICA) to detect MV-infected cells, the standard PRNT has been developed into a focus reduction neutralization test (FRNT). This assay is faster and has improved specificity. The FRNT described here is extremely useful when immunity to measles virus needs to be assessed in patients with a specific medical condition, such as immunocompromised individuals in whom presumed residual immunity needs to be assessed. The FRNT is not generally recommended for use with large numbers of specimens, such as in a seroprevalence study.

Key words Measles virus, Diagnostic test, Immunity, Antibodies, Neutralization, ELISA

1 Introduction

Measles virus (MV) is often transmitted in situations where many people meet or live in close proximity. It is therefore important to assess immunity to MV when refugees are admitted to a migrant shelter, children are enrolled in school, or medical personnel are recruited [1]. In many cases, immunity to MV can be assessed by reviewing the vaccination record: If two doses of MMR vaccine are documented in a valid vaccination card, immunity can be assumed without further action or testing [2]. Unfortunately, these documents are often missing. In these cases, a blood

The original version of the chapter has been revised. A correction to this chapter can be found at https://doi.org/10.1007/978-1-0716-3870-5_19

Dzwokai Z. Ma and Christian K. Pfaller (eds.), *Measles and Related Morbilliviruses: Methods and Protocols*,
Methods in Molecular Biology, vol. 2808, https://doi.org/10.1007/978-1-0716-3870-5_16,

sample can be taken to measure the anti-measles IgG titer as a correlate of immunity [3]. A positive test result for anti-measles IgG indicates immunity. A negative test result indicates the need to vaccinate the presumably unprotected individual. However, the use of a life-attenuated vaccine is not indicated for, e.g., pregnant women, newborns, and young children or immunocompromised patients. If protection needs to be assessed in patients, e.g., with an underlying medical condition who are IgG-negative in a standard ELISA, the FRNT is the gold standard for demonstrating a protective correlate against measles disease [4]. This test should be used for individual patients with a critical clinical question, but it is not recommended for population studies.

Infectious MV particles in a suspension can be quantified by performing an assay for virus titration that counts either visible plaques in a stained cell layer or stained foci of MV infection in a cell layer. The foci are formed by MV-infected cells that can be detected using an immunocolorimetric assay (ICA) with an MV-specific monoclonal antibody. Virus quantification using an ICA is highly specific, saves time, and avoids the use of a toxic substance such as methylene blue. Both principles of virus quantification are applied to MV neutralization assays. Plaque reduction by MV neutralizing antibodies is measured in a conventional PRNT, whereas the more recently developed FRNT measures the reduction of foci of MV infection at an earlier point in the incubation time.

The FRNT is used here to quantify the titer of neutralizing antibodies against MV, but the technique has been adapted for other viruses [5, 6]. The FRNT employs incubation of a patient's serum in serial dilution with virus particles and subsequent inoculation of this mixture onto permissive cells.

MV vaccine strains use the ubiquitous cell surface molecule CD46 as a receptor, whereas wild-type MV strains do not bind to CD46. Unlike CD46, the human SLAM receptor (signaling lymphocyte activation molecule; also known as CDw150) is a cellular receptor for all types of MV regardless of whether these strains are attenuated or not. Transfection with a plasmid expressing human SLAM enables a non-susceptible cell line such as Vero to bind MV, support its replication, and develop cytopathic effects. Vero/hSLAM cells have thus become a useful tool for isolating and culturing MV [7].

Infection of cultured cells with a cytopathogenic virus leads to cell death throughout the culture. In contrast, incubation of infected cells under a semi-solid overlay medium results in the formation of small areas of infected and later on damaged cells (plaques) due to the restriction of the virus particle to spread in the culture dish. The amount of infectious virus particles is reflected by the number of foci or plaques formed and can be expressed as focus or plaque forming units (FFU or PFU). The presence of neutralizing antibodies prevents infection and subsequent foci and plaque formation. To test the neutralizing potency of serum antibodies by FRNT, serum dilutions are mixed with a defined number of virus particles and incubated and inoculated onto a monolayer of cultured Vero/hSLAM cells. The cells are then covered with the overlay and incubated for 3–4 days. After foci formation, the cells are incubated with a MV-specific monoclonal antibody and an enzyme conjugated secondary antibody. Subsequently added substrate reacts with the enzyme and produces a visible color change. A stained spot indicates a focus of MV infection in the cell monolayer and corresponds to a single infectious virus particle in the inoculum. The neutralizing antibody titer is defined as the reciprocal serum dilution that causes a 50% reduction in the number of foci. The 50% endpoint titer is referred to as ND_{50} (neutralizing dose).

When a neutralization assay is used to obtain a comparable result (e.g., to assess immunity with regard to protection or in clinical studies), it should be performed in a standardized format. A standardized PRNT which allows to transform the obtained ND_{50} values into IU/mL has been developed for use in immunogenicity studies of aerosol vaccination [5]. The original Vero cells, the low-passage Edmonston wild-type MV as challenge virus, and the second International Standard (second IS) for anti-measles serum/plasma as reference serum were used in this study. An automatable FRNT using Vero cells, a high-passage Edmonston MV, and the second IS has also been developed [8]. When Vero cells are used, neutralizing antibodies that prevent binding of MV to the hSLAM receptor, which is selectively expressed on lymphoid tissues, cannot be detected. We have therefore modified the previously described assays by establishing Vero/hSLAM cells for use in a standardized FRNT involving the use of the low-passage Edmonston wild-type MV as the challenge virus and the third International Standard (third IS) for anti-measles serum/plasma containing 3 IU/mL [9, 10] as the reference serum (*see* **Notes 1–3**). The standardized FRNT allows the ND_{50} value (expressed as titre) to be converted to IU/mL and thus provides a comparable correlate of measles immunity. A Vero/hSLAM cell-based FRNT also allows the use of a wild-type MV of any genotype (e.g., the currently circulating genotypes B3 and D8) as a challenge virus. In this case, the assay cannot be considered standardized, and consequently, the ND_{50} value obtained cannot be converted to IU/mL, but the values can be expressed, for example, as relative units/mL.

In the following chapter you will learn how to carry out a reliable FRNT.

2 Materials

1. Anti-measles-IgG enzyme-linked immunosorbent assay (ELISA).

2. 30% bovine serum albumin (BSA) solution (in water).
3. Carboxymethyl-cellulose (CMC).
4. Dulbecco's modified Eagle's medium (DMEM), high glucose.
5. Growth medium: DMEM supplemented with FCS, penicillin/streptomycin, geneticin, L-glutamine, and sodium hydrogen carbonate.
6. Fetal calf serum (FCS).
7. 50 mg/mL Geneticin.
8. Goat anti-mouse IgG (H + L) horseradish peroxidase (HRP) conjugate (secondary antibody).
9. 200 mM L-glutamine.
10. MV-specific primary antibody.
11. Methanol (>99.5%).
12. 10× Minimum Essential Medium (MEM).
13. 16% paraformaldehyde.
14. Penicillin/streptomycin mixture (10,000 U/mL penicillin, 10,000 μg/mL streptomycin).
15. Phosphate buffered saline (PBS).
16. 7.5% sodium hydrogen carbonate solution.
17. Standard serum.
18. TMB substrate solution.
19. 0.05% Trypsin-EDTA solution.
20. Tween-20.
21. Vero/hSLAM cells.
22. 48-well cell culture plates.

3 Methods

3.1 Cell Cultivation

1. Obtain Vero/hSLAM cells (*see* **Note 1**) or transfect Vero cells with a plasmid expressing the hSLAM receptor (WHO 2021).
2. Prepare the growth medium for Vero/hSLAM cells. Use one bottle of Dulbecco's modified Eagle's medium (DMEM), high glucose, and supplement it with 10% fetal calf serum (FCS), 1% penicillin/streptomycin (P/S; 10,000 U/mL, 10,000 μg/mL), 0,5% geneticin (50 mg/mL, final concentration 250 μg/mL), 1% L-glutamine (200 mM), and 3 mL 7.5% sodium hydrogen carbonate.
3. Prepare stocks of the cells for storage in liquid nitrogen (*see* **Note 4**).

4. Grow Vero/hSLAM cells at all times in growth medium, and use for 15 passages (*see* **Notes 5** and **6**). Thaw fresh cells when passage 15 is reached.
5. Use commercially available trypsin-EDTA 0.05% for passaging.

3.2 Selection and Handling of Sera

1. Sera should not contain supplements, coagulants, or preservatives; these substances may interfere with the test.
2. All serum specimens should be considered potentially positive for infectious agents including HIV and hepatitis B virus. Universal precautions should be followed: Sera must be handled under a safety cabinet. Protective gloves, a laboratory coat, and goggles or a face shield should be worn when handling sera outside the laminar flow. All work surfaces must be wiped with virucidal disinfectant after the work is completed.
3. Serum samples are inactivated in a thermoblock or water bath at 56 °C for 30 min prior to use in the FRNT.
4. Sera can be pretested by a standard anti-measles-IgG ELISA (*see* **Note 7**).
5. Patient sera should be stored at −20 °C (freezer) if it is not used immediately.

3.3 Preparation of Solutions

1. Aliquot the 30% bovine serum albumin (BSA) solution (in water) into 50 mL Falcon tubes, and store at −20 °C.
2. Prepare blocking buffer by adding 5 mL FCS and 0.5 mL Tween 20 and 10 mL 30% BSA solution to 500 mL PBS. The blocking buffer is used to saturate and block non-specific binding sites.
3. Prepare the semi-solid overlay stock solution: Add 3 g carboxymethyl-cellulose (CMC), and fill up to 300 mL with aqua bidest; dissolve and autoclave (1% CMC) and store at 4 °C.
4. Prepare CMC solution for 48-well plates freshly: Add 25 mL 1% CMC to a 50 mL graduated cylinder. Add 3 mL 10× MEM, 1.5 mL FCS, 0.5 mL P/S (10,000 U/mL, 10,000 μg/mL), and 1.1 mL 7.5% NaHCO3. Add DMEM to a volume of 50 mL. Mix by swirling between the measuring cylinder and a beaker glass. The CMC overlay is used to prevent diffusion of the virus particles.
5. Prepare the MV-specific primary antibody (*see* **Note 8**).
6. Preparation of the horseradish peroxidase (HRP) conjugated secondary antibody: Dissolve 1 mg lyophilized goat anti-mouse IgG peroxidase conjugate in 1 mL PBS. Dissolve and make 30 μL aliquots which should be stored at −20 °C.

7. Preparation of 2% paraformaldehyde: Transfer 210 mL PBS to a 250 mL graduated cylinder, and add 30 mL 16% paraformaldehyde solution and store at 4 °C.

3.4 Seeding of Cells

Work under a biosafety cabinet. Work quickly and do not allow the cells to dry out.

Vero/hSLAM cells are passaged 1:10 and reseeded into 48-well cell culture plates. A 25 cm^2 flask confluent with Vero/hSLAM cells is sufficient for four 48-well cell culture plates.

1. Aspirate the medium, rinse the cells with 5 mL trypsin-EDTA 0.05% solution, and remove it.
2. Add 5 mL trypsin-EDTA 0.05% solution, and incubate the cells for 2 min at room temperature; remove 4 mL from the solution; incubate for 8 min at 37 °C in the CO_2 incubator to allow trypsin to work; and collect the detached cells in 5 mL of growth medium.
3. Transfer the cells to a 75 cm^2 flask containing 95 mL growth medium and gently mix.
4. Pipet 500 μL of the cell suspension into each well of a 48-well plate.
5. Incubate the cells for 24 h in a CO_2 incubator at 37 °C.

3.5 Setup of Virus Suspension

The concentration of infectious MV particles in the stock suspension must be determined in advance by virus titration [11] and is expressed as foci forming units per ml (FFU/mL). However, the amount of infectious MV particles in the suspension may decrease slightly over time. Therefore, the factor used to dilute the virus stock suspension for use in FRNT must be adjusted just prior to the start of the experiment. For a valid FRNT run, the average FFU count determined for the virus control wells (*see* Subheading 3.4) must be between 25 and 60 infectious MV particles (FFU) per well, while the resulting average concentration (42.5 FFU/well) is considered the target value c_t. Dilution of the virus stock suspension by the adjusted factor d_a should ensure that the virus control is within the range required for a valid FRNT.

1. Prepare a 48-well cell culture plate as described in Subheading 3.1.
2. Calculate the factor d required to dilute the virus stock suspension to the given virus concentration c_s (expressed in FFU/mL) using the following formula:

$$d = c_s / 850\ \text{FFU/mL}$$

850 FFU/mL correspond to 42.5 FFU in 50 μL.

3. Dilute an aliquot of the virus stock suspension by factor d with DMEM medium without additives to obtain a virus suspension

with an assumed concentration of 850 FFU/mL; stepwise dilution is recommended, final volume ≥ 1 mL.

4. Prepare the inoculum by adding the same volume of DMEM medium without additives to the diluted virus suspension (e.g., 1 mL DMEM medium +1 mL diluted stock virus suspension).
5. Add 100 μL per well of the prepared inoculum to the 80% confluent Vero/hSLAM cell layer; using a minimum of 12 parallel wells.
6. Incubate the MV-infected cells as described in Subheading 3.6 for the virus control.
7. Fix the cells and perform an ICA to stain the foci of MV infection as described in Subheadings 3.7 and 3.8.
8. Count the number of blue-stained foci indicating MV infection (FFU) per well (c_1, c_2, ... c_n, expressed as FFU/well); only the wells with an intact cell layer should be used.
9. Calculate the arithmetic mean of the FFU counts per well to obtain the actual virus concentration c_a in the suspension (expressed as FFU/well) using the following formula:

$$c_a = (c_1 + c_2 + c_3 + \ldots + c_{n-1} + c_n)/n$$

n = number of wells used for FFU counting.

10. Calculate the adjusted dilution factor d_a using the formula

$$d_a = d \times c_a / c_t$$

c_t = target virus concentration = 42.5 FFU/well, when the FRNT is performed in the recommended 48-well format.

$$d_a = d \times c_a / 42.5 \text{ FFU/well}.$$

3.6 Serial Dilution of the Serum

The patient serum sample should be tested at least in duplicate. The quantities given are based on a duplicate determination of the ND_{50}.

1. Prepare six 1.4 mL blank tubes for the serial dilution of the patient serum (1:4, 1:8, 1:16, 1:32, 1:64, 1:128).
2. Pipette 225 μL DMEM medium without additives into the first tube, and add 75 μL serum, to make a 1:4 dilution.
3. Pipette 150 μL DMEM medium without additives into tubes 2 to 6.
4. Transfer 150 μL from tube 1 (1:4 dilution) to tube 2; mix by pipetting up and down 10 times.
5. Transfer 150 μL from tube 2 to tube 3, and continue until tube 6 is prepared.
6. Remove 150 μL from tube 6.

3.7 Use and Preparation of Controls and Reference Sera

A cell control, a virus control, and a reference serum should be used to assess whether an assay run can be considered valid.

- *Cell control*: The cell control uses medium; no virus is added to the cells. This control must have an intact cell layer. If it does not, a contamination has affected the result of the assay. The cell control should be performed in a minimum of four wells.
- *Virus control*: The virus control uses the virus suspension that is adjusted to the required number of 25–60 FFU in 50 μL medium without additives per well, but no serum is added. Consequently, the formation of foci of MV infection is not inhibited. Parallel testing in at least six wells should result in an average focus count within the required range. If not, the virus was inactivated earlier, or the titer was not been determined correctly.
- *Reference serum:* An International Standard (IS) serum (IS) approved by the WHO for use in FRNT or, alternatively, an appropriate in-house serum calibrated against an IS may be used as the reference serum. In an assay run, the reference serum should give an ND_{50} value within a pre-determined acceptable range that can be established by using ND_{50} values obtained from 10 to 20 valid assay runs [5]. When a standardized assay is performed, the ND_{50} value obtained for the reference serum can be used to convert the ND_{50} value of the tested specimen to IU/mL. A reference serum should also be used in a FRNT of a non-standardized format (e.g., when a wild-type MV of any genotype is used as a challenge virus), as it allows evaluation of the validity of the assay run. The reference serum should be tested at least in duplicate.

3.7.1 Preparation of the Cell Control

The given quantity refers to performance in four wells.

1. Prepare one 1.4 mL blank tube.
2. Pipet 600 μL DMEM medium without additives into this tube.

3.7.2 Preparation of the Virus Control

The given quantities refer to performance in six wells.

1. Prepare two 1.4 mL blank tubes.
2. Pipet into each tube 300 μL DMEM medium without additives, and add 300 μL from the adjusted virus suspension.

3.7.3 Preparation of the Reference Serum

The given quantities refer to performance in duplicate.

1. Prepare six 1.4 mL blank tubes for the serial dilution of the reference serum (1:100, 1:200, 1:400, 1:800, 1:1600, 1: 3200). The reference serum is pre-diluted by 1:50 in DMEM medium and can be stored at −20 °C.

2. Pipette 150 μL DMEM medium without additives into tubes 1 to 6.
3. Pipette 150 μL of the pre-diluted reference serum into tube 1, to make a 1:100 serum dilution.
4. Transfer 150 μL from tube 1 into tube 2; mix by pipetting up and down 10 times.
5. Transfer 150 μL from tube 2 to tube 3, and continue until tube 6 is prepared.
6. Remove 150 μL from tube 6.
7. Pipette 150 μL of the virus suspension into each dilution of reference serum.

3.8 Incubation of Diluted Sera with Virus Suspension

1. Pipette 150 μL of the virus suspension into each dilution of the patient serum and the reference serum.
2. Perform the incubation also for the virus control and the cell control.
3. Seal all the tubes with aluminum foil.
4. Incubate 1 h at 37 °C in a water bath.

3.9 Incubation of Cells with the Serum/Virus Mixture

1. Pre-warm the PBS and blocking buffer to room temperature. Pre cool the methanol to −20 °C.
2. Cells should be 80% confluent.
3. Remove the medium from the 48-well plate.
4. Pipette 100 μL of the serum/virus suspension into each well of the prepared cell culture plate except for the cell control.
5. Pipette 100 μL dilution of each serum/virus suspension into two wells.
6. Pipette 100 μL of each reference serum dilution into two wells.
7. Pipette 100 μL of the virus control into six wells.
8. Pipette 100 μL of the cell control into four wells.
9. Incubate the plate 1 h at 36 °C in a CO_2 incubator.
10. Add 0.5 mL CMC overlay to each well.
11. Incubate at 36 °C for 4 days in a CO_2 incubator.
12. On day 4, check that the cell layer is still intact and begin fixation (see below).

3.10 Fixation of the Cells

1. Remove the medium from the 48-well cell culture plate.
2. Wash each well with 0.5 mL PBS and aspirate.
3. Add 200 μL cold PFA (2% paraformaldehyde, 4 °C) per well.
4. Incubate 30 min at 4–8 °C in a refrigerator.
5. Aspirate PFA.

6. Wash each well 1× with 0.5 mL PBS.
7. Add 200 μL cold ≥99.5% methanol per well.
8. Incubate the cell culture plate for 10 min in a freezer at −20 °C.
9. Aspirate methanol.
10. Wash 1× with 0.5 mL PBS.
11. Aspirate PBS.
12. Add 200 μL blocking buffer per well.
13. Incubate for 30 min at RT.

The assay may be stopped at this point. Cell culture plates containing 200 μL blocking buffer can be stored in the refrigerator at 4 °C for up to 4–8 weeks.

3.11 Immunocolorimetric Staining of Foci of MV Infection by Application of the First and Second Antibody

1. Aspirate blocking buffer.
2. Add 100 μL of the MV-N protein-specific monoclonal antibody to 5 mL blocking buffer (1:50 dilution).
3. Pipette 100 μL of the diluted primary antibody per well.
4. Incubate for 30 min at RT.
5. Wash 2× with 0.5 mL blocking buffer per well.
6. Add 5 μL HRP conjugated secondary antibody diluted in 5 mL blocking buffer (dilution 1:1000) using the previously prepared aliquot.
7. Pipette 100 μL of the diluted secondary antibody per well.
8. Incubate at RT for 30 min.
9. Wash 2× with 0.5 mL blocking buffer per well.
10. Pipette 100 μL TMB substrate solution per well.
11. Incubate for 5–10 min at RT; foci will appear blue-purple.
12. Stop the reaction with aqua dest.
13. Count the foci on a light table using a magnifying glass; see example in Fig. 1.

3.12 Calculation of the MV Neutralizing Antibody Titer

The 50% endpoint titer (ND_{50}) corresponds to the titer given as the reciprocal serum dilution and can be calculated by using the Kaerber formula [12]. This method takes into account the number of foci of MV infection obtained for the whole series of serum dilutions. The term "serum dilution" refers to the total serum dilution (addition of an equal volume of adjusted challenge virus suspension to a volume of a 1:100 diluted serum results in a total serum dilution of 1:200). It is recommended that multiple tests be set up, ideally in parallel rows on the same plate.

3.12.1 Kaerber Formula

$$\log_{10} ND_{50} = d_m - d_f\,(S - 0.5)$$

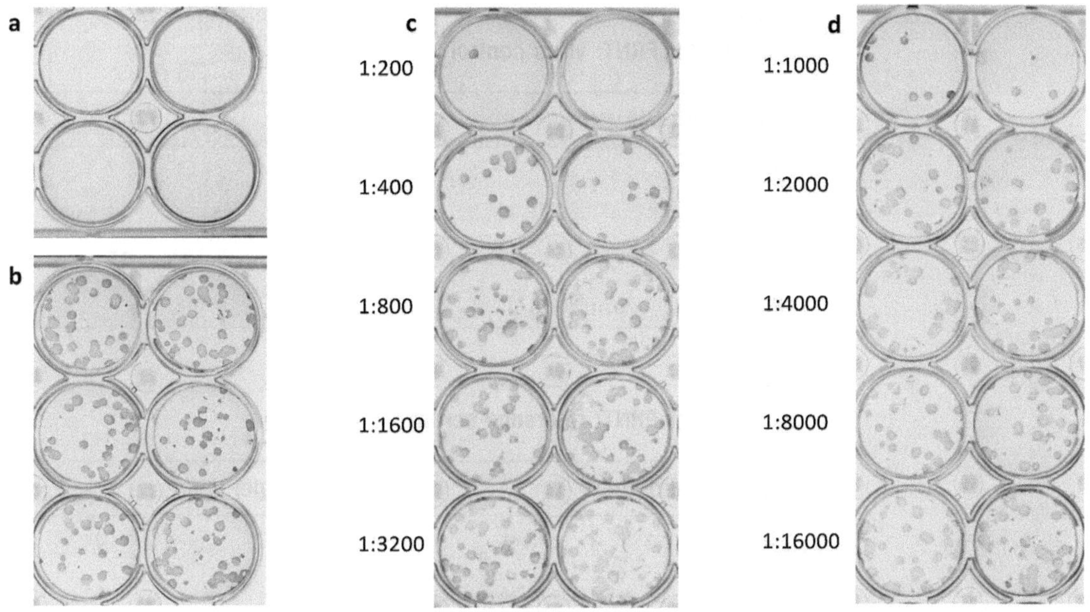

Fig. 1 Result of the FRNT assay run described in the example (*see* Subheading 3.14). The assay run was performed in a 48-well plate. The wells for the cell control (**a**), the virus control (**b**), the reference serum (**c**), and the tested plasma sample (**d**) are shown. The reference serum and the sample were tested in duplicate in parallel rows. The dilutions used for the reference serum and the sample are given to the left of the image. The corresponding foci counts are given in Tables 1a, 1b, and 1c

d_m = decadic logarithm ($\log_{10}$) of the reciprocal of the highest serum dilution (the lowest serum dilution corresponding to a foci count equal to x_{vc} (see below) is referred to as highest serum dilution; the reciprocal of a 1:200 dilution is 200).

$d_f = \log_{10}$ of the serum dilution factor ($\log_{10} 2 = 0.3010$).

$$S = (x_1 + x_2 + x_3 + \ldots + x_{n-1} + x_n)/x_{vc}$$

x_1 = number of foci for the well with the lowest serum dilution, x_2 = number of foci for the well with the second lowest serum dilution, . . ., x_n = foci count for the well with the highest serum dilution.

x_{vc} = the average number of foci count (arithmetic mean) in the wells for the virus control.

3.12.2 Calculating the ND50

1. Calculate the value for x_{vc} (virus control) as given above.
2. For the first row of serum sample 1, count the foci per well along the increasing serum dilution until the value of x_{vc} is reached ($x_n = x_{vc}$).
3. Calculate S (number of foci per well divided by x_{vc}) using the formula above; when the value for x_n exceeds the value for x_{vc} ($x_n > x_{vc}$), x_n is considered as equal to x_{vc} (use for x_n the value of x_{vc}).

Table 1a
Example for FRNT: virus control

Foci count per well	
34	37
28	24
25	35
Average foci count: 30.5	

Table 1b
Example for FRNT: reference serum

Reciprocal dilution	Foci count per well	
200	1	0
400	16	11
800	29	31
1600	31	
ND50 per row:	397.3	440.8
Final ND50 (GMT):	418.5	

Table 1c
Example for FRNT: sample (human plasma)

Reciprocal dilution	Foci count per well	
1000	8	3
2000	24	17
4000	26	25
8000	28	31
16,000	36	
ND50 per row	1602.5	2028.1
Final ND50 (GMT)	1802.8	

4. Calculate the value for d_f as given above.
5. Determine the highest serum dilution, convert to the reciprocal dilution, and calculate the decadic logarithm of the reciprocal dilution to obtain the value for d_m.
6. Insert the values obtained for d_m, d_f, and S into the Kaerber formula.

7. Calculate the value for $\log_{10}$ ND_{50} according to the Kaerber formula.
8. Use the result to calculate the ND_{50}.
9. Use the procedure described above (1–8) to calculate the ND_{50} of the second (parallel) row of sample 1.
10. Calculate the geometric mean of the ND_{50} values obtained for the first and the second row in order to obtain the final ND_{50} value for sample 1.
11. Use the procedure described above (1–10) to calculate the ND_{50} for the other serum samples and the reference serum.

3.13 Use of Validation Criteria

The assay run is considered valid if

- The cell control wells show an intact cell monolayer.
- The average number of foci for the virus control is within the required range of 25–60 FFU per well.
- The rows for the reference serum and for the specimen show a continuous increase in the number of foci per well with increasing serum dilutions until the value of the virus control is reached.
- The ND_{50} value obtained for the reference serum (IS or calibrated in-house serum) is within the previously accepted range.

3.14 Interpreting the Results

A ND_{50} (expressed as titer) of ≥ 8 (1:≥8) is considered positive [5], indicating the presence of MV neutralizing antibodies. The level of MV neutralizing antibodies that correlates with protection against clinical disease may be estimated to be 120 mIU/mL when the conversion of the ND_{50} titer value to IU/mL is based on the use of the second IS containing 5 IU/mL [5]. Published data comparing the second and the third IS using the PRNT (World Health Organization 2006, 2007) suggest that the protective level may also be estimated to be 120 mIU/mL when using the third IS containing 3 IU/mL.

3.15 Practical Example

The potency of a human plasma sample to neutralize an isolated wild-type MV of genotype B3 was determined using a FRNT based on Vero/hSLAM cells. The plasma sample and the reference serum were tested in duplicate in parallel rows. The cell control wells showed an intact cell monolayer (Fig. 1a); the virus control wells yielded an average count of 30.5 FFU/well (Fig. 1b, Table 1a) which was within the required range of 25–60 FFU/well. The reference serum was tested at twofold dilutions ranging from 1: 200 to 1:3200 (the term "dilution" refers to the total dilution; *see* Subheading 3.11). The ND_{50} was calculated for each row using the Kaerber formula, and the geometric mean resulted in a final ND_{50} of 418, which was within the previously established acceptable range of 300–1500 (Fig. 1c, Table 1b). The plasma sample was tested at twofold dilutions ranging from 1:1000 to 1:16,000. The assay run was considered valid, and a final ND_{50} of 1803 (1:1803) was calculated for the plasma sample (Fig. 1d, Table 1c). The use of a genotype B3 MV as a challenge virus does not allow the ND_{50} value obtained to be converted to IU/mL.

4 Notes

1. Cell stocks of the Vero/hSLAM cell line, developed by Dr. Yusuke Yanagi and colleagues at Kyushu University, Japan, have been provided to the GMRLN laboratories by the National Institute of Infectious Diseases, Tokyo, Japan, and are available at regional cell repositories. Vero/hSLAM cells are available to GMRLN laboratories on request and free of charge from WHO. Any publication of work using the Vero/hSLAM cell line should acknowledge the original publication [7]. Alternatively, Vero/hSLAM cells can be obtained from the European Collection of Authenticated Cell Cultures (ECACC), UK. It is essential that geneticin (also known as G418) is added to the culture medium as it is required to maintain the plasmid expressing the hSLAM receptor.
2. The low-passage Edmonston virus (MVi/Maryland/0.54) was provided by Dr. J. Beeler, Center for Biologics Evaluation and Research (CBER), Food and Drug Administration, USA.
3. The third IS (third International Standard for Anti-measles Serum, 97/648) was purchased from the National Institute for Biological Standards and Control (NIBSC), UK.
4. Preparation of cell stocks for liquid nitrogen storage: Cells to be used to prepare stocks for internal use or distribution to other laboratories should be passaged in medium containing 250 μg/mL geneticin. Cells are passaged forward (typically 2 times), increasing the number of flasks to provide a sufficient amount of cells to prepare 20–50 vials of cell stock for liquid nitrogen storage.
5. Passage of Vero/hSLAM cells for routine virus isolation: To prepare cells for virus isolation procedures, Vero/hSLAM cells should be recovered from liquid nitrogen and can be passaged up to 15 times in medium without Geneticin. Geneticin is omitted when a FRNT is performed.
6. Note that Vero/hSLAM cells are classified as a class 1 genetically modified organism (GMO1).
7. Normally, ELISA and FRNT results correlate well [4]. If a serum has given a positive result for anti-measles IgG in

the ELISA, it is possible to start in the FRNT with the 1:8 total serum dilution. On the other hand, it is more likely that equivocal sera will be tested in the FRNT. Therefore, in order to get a good feel for the correlation of the results of your ELISA and FRNT results, start with the 1:4 dilution in the first run and document both results.

8. We use the supernatant obtained from cultured hybridoma cells that secrete an antibody that binds to the MV-N protein. Alternatively, a commercially available MV-specific monoclonal antibody can be used.

Acknowledgments

The authors would like to thank Ms. Cornelia Walter for her excellent technical assistance.

References

1. Hahne SJ, Nic Lochlainn LM, van Burgel ND, Kerkhof J, Sane J, Yap KB, Van Binnendijk RS (2016) Measles outbreak among previously immunized healthcare workers, The Netherlands, 2014. J Infect Dis 214(12):1980–1986. https://doi.org/10.1093/infdis/jiw480
2. McLean HQ, Fiebelkorn AP, Temte JL, Wallace GS, Centers for Disease C, Prevention (2013) Prevention of measles, rubella, congenital rubella syndrome, and mumps, 2013: summary recommendations of the Advisory Committee on Immunization Practices (ACIP). MMWR Recomm Rep 62(RR-04): 1–34
3. Bolotin S, Hughes SL, Gul N, Khan S, Rota PA, Severini A, Hahne S, Tricco A, Moss WJ, Orenstein W, Turner N, Durrheim D, Heffernan JM, Crowcroft N (2020) What is the evidence to support a correlate of protection for measles? A systematic review. J Infect Dis 221(10):1576–1583. https://doi.org/10.1093/infdis/jiz380
4. Lutz CS, Hasan AZ, Bolotin S, Crowcroft NS, Cutts FT, Joh E, Loisate S, Moss WJ, Osman S, Hayford K (2023) Comparison of measles IgG enzyme immunoassays (EIA) versus plaque reduction neutralization test (PRNT) for measuring measles serostatus: a systematic review of head-to-head analyses of measles IgG EIA and PRNT. BMC Infect Dis 23(1):367. https://doi.org/10.1186/s12879-023-08199-8
5. Cohen BJ, Audet S, Andrews N, Beeler J, WHO working group on measles plaque reduction neutralization test (2007) Plaque reduction neutralization test for measles antibodies: description of a standardised laboratory method for use in immunogenicity studies of aerosol vaccination. Vaccine 26 (1):59–66. doi: https://doi.org/10.1016/j.vaccine.2007.10.046
6. Chen MH, Zhu Z, Zhang Y, Favors S, Xu WB, Featherstone DA, Icenogle JP (2007) An indirect immunocolorimetric assay to detect rubella virus infected cells. J Virol Methods 146(1–2):414–418. https://doi.org/10.1016/j.jviromet.2007.08.021
7. Ono N, Tatsuo H, Hidaka Y, Aoki T, Minagawa H, Yanagi Y (2001) Measles viruses on throat swabs from measles patients use signaling lymphocytic activation molecule (CDw150) but not CD46 as a cellular receptor. J Virol 75(9):4399–4401. https://doi.org/10.1128/JVI.75.9.4399-4401.2001
8. Terletskaia-Ladwig E, Enders G, Meier S, Dietz K, Enders M (2011) Development and evaluation of an automatable focus reduction neutralisation test for the detection of measles virus antibodies using imaging analysis. J Virol Methods 178(1–2):124–128. https://doi.org/10.1016/j.jviromet.2011.08.026
9. WHO Expert Committee on Biological Standardization (2006) Report of a collaborative study to assess the suitability of a replacement

for the 2nd international standard for anti-measles serum. WHO/BS/06.2031

10. WHO Expert Committee on Biological Standardization (2007) Report on a collaborative study to investigate the relationship between the 1st IRP and the 2nd and 3rd international standards for anti-measles serum/plasma, in both ELISA and PRNT. WHO/BS/07.2076
11. Dulbecco R, Vogt M (1954) Plaque formation and isolation of pure lines with poliomyelitis viruses. J Exp Med 99(2):167–182. https://doi.org/10.1084/jem.99.2.167
12. Kaerber G (1931) Beitrag zur kollektiven Behandlung pharmakologischer Reihenversuche. Naunyn Schmiedebergs Arch Exp Pathol Pharmakol 162:480–483

Chapter 17

Multiplex Bead Assay for the Serological Surveillance of Measles and Rubella

Melissa M. Coughlin, Gaby Smits, Zachary Matson, Rob van Binnendijk, and Bettina Bankamp

Abstract

There is increasing interest in evaluating antibody responses to multiple antigen targets in a single assay. Immunity to measles and rubella are often evaluated together because immunity is provided through combined vaccines and because routine immunization efforts and surveillance for measles and rubella pathogens are combined in many countries. The multiplex bead assay (MBA) also known as the multiplex immunoassay (MIA) described here combines the measurement of measles- and rubella-specific IgG antibodies in serum quantitatively according to international serum standards and has been successfully utilized in integrated serological surveillance.

Key words Measles, Rubella, Serosurveillance, Multiplex bead assay, Multiplex immunoassay

1 Introduction

Evaluation and documentation of population immunity continues to be important in the public health response to infectious diseases, including vaccine-preventable diseases (VPDs). Serologic surveillance is especially useful for VPDs to evaluate routine immunization systems, inform supplemental immunization activities, and identify possible gaps in immunity that were not identified by other methods such as administrative coverage estimates and post-campaign surveys [1–4]. Serologic data provide support documentation of elimination for measles and rubella [4–6]. The expense of serosurveys has generated growing interest in combining surveillance for multiple diseases onto a single test platform to leverage the utility of serological surveillance data across multiple diseases of public health importance [7].

The multiplex bead assay (MBA), also referred to as the multiplex immunoassay (MIA), uses Luminex®-based technology to evaluate the serological response to multiple target antigens in a

Dzwokai Z. Ma and Christian K. Pfaller (eds.), *Measles and Related Morbilliviruses: Methods and Protocols*,
Methods in Molecular Biology, vol. 2808, https://doi.org/10.1007/978-1-0716-3870-5_17,

single reaction using 5 μL or less of sample and has been used in multiple countries [8–13]. The multiplexing capacity is achieved through the differential internal fluorescent labeling of polystyrene magnetic or nonmagnetic microspheres or "beads." The beads are internally labeled with different concentrations of red and infrared fluorescent dyes, creating a unique signature emission profile for each bead type following excitation [14]. This allows categorization of individual bead types based on their unique emission spectra into a bead "region." Depending on the instrument, the capacity to read different bead regions ranges from 50 to 500 regions, allowing multiplexing of multiple unique targets [14].

Measurement of the beads and reporter fluorophore, which measures reactivity to the antigen target, occurs using a Luminex® instrument equipped with a red and a green light source. The red light excites the fluorescent internal bead dyes leading to the unique emission signature to identify the bead. The green light source excites the phycoerythrin (PE) reporter specific to the target being detected. In addition to the unique bead region read capacity, the light sources and fluidics also differ depending on the instrument [14]. The Luminex 200®, FlexMAP 3D®, and xMAP Intelliflex® systems use laser light sources and a fluidics stream, similar to a flow cytometer, that allows a single bead to be assessed by each laser at a time and captures the emission spectra through a photomultiplier tube [15]. The MAGPIX® instrument uses red and green LED lights and does not use a flow-based sample acquisition. Instead, the sample is passed through a flow cell into an imaging chamber where the collection of beads is bound within the magnetic chamber and imaged with a camera following sequential illumination with the LED lights [16]. All instruments assess multiple bead events, allowing the user to obtain quantitative repeat measurements of the same bead/antigen combination, generating a more precise quantitative measurement when compared to single quantitative assessments by, e.g., plate-bound EIA assays. To this end, the instruments capture median fluorescent intensity (MFI), rather than mean, to also reduce any effect of carryover from subsequent wells, especially if moving from a high signal well to a lower signal well.

Bead target specificity is attained when the surface of the beads is covalently bound with specific antigens of interest through available amine groups using the carbodiimide chemical process. The carbodiimide chemical reaction occurs in two steps: beads are first activated to create a semi-stable intermediary followed by the addition of specific antigen of interest and covalent amide bond formation [14].

The measles and rubella MBA is an indirect immunoassay requiring two independent bead regions. The target antigens bound to the surface of different bead regions are measles and rubella whole virus lysates [17, 18]. Beads coupled with either

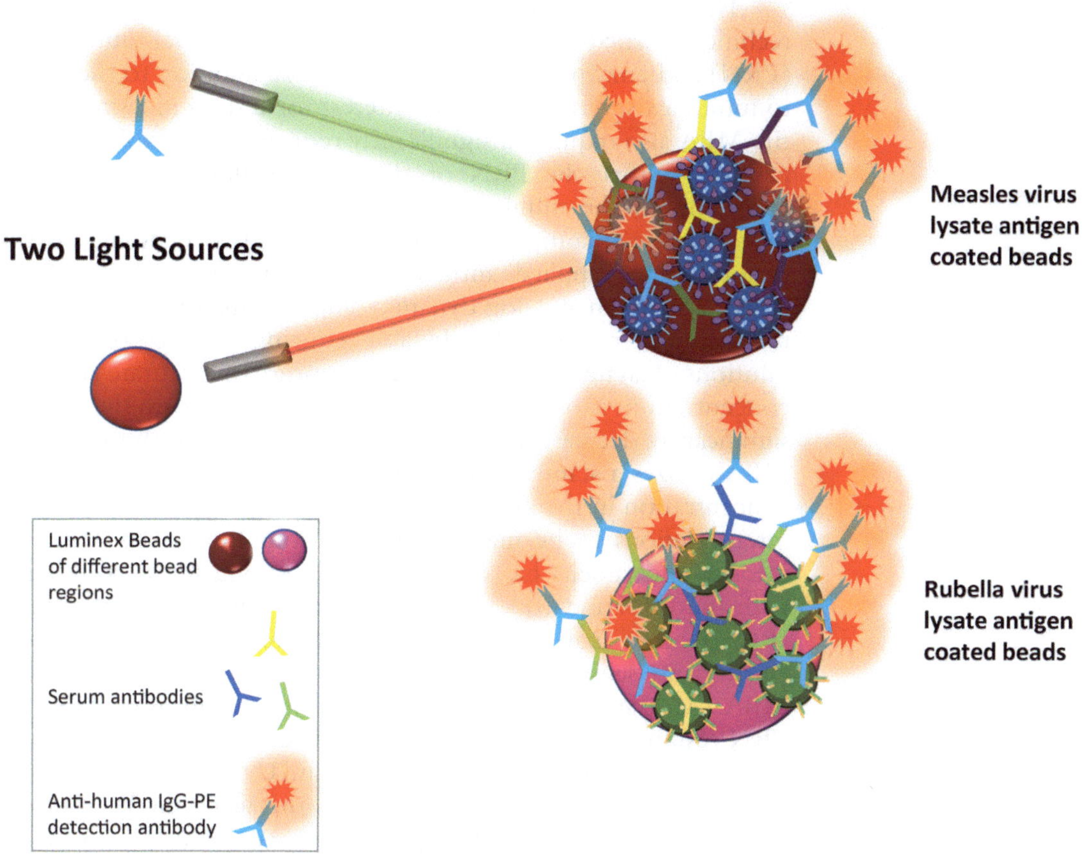

Fig. 1 Graphical depiction of measles and rubella MBA. Luminex instruments contain two light sources. The red light source excites the internal fluorescent dyes of the beads, and the green light source excites the phycoerythrin-conjugated fluorophore on the secondary antibody. Beads of different regions are coupled with measles or rubella virus whole virus lysate, and the concentrations of antibodies specific to viral targets are measured with PE conjugated anti-human IgG secondary and quantitated using the standard curve

measles or rubella virus lysate are mixed into a single well to measure antigen-reactive antibodies in test serum. Test serum is incubated with the antigen-coated beads. Antibodies in the serum recognizing measles and rubella antigens bind to the respective antigen-coupled beads. To detect measles- and rubella-specific IgG, a directly conjugated anti-human IgG phycoerythrin (PE)-labeled secondary antibody is used. The PE fluorophore on the secondary antibody is then detected by the green laser or LED light, and the instrument quantitates the amount of reporter signal on each bead type (Fig. 1). The instrument reports median fluorescent intensity of reporter signal, per bead region, per well. This is quantitated to determine antibody concentration with a National Institute for Biological Standards and Control (NIBSC) standard curve based on the current WHO serum standards for measles and rubella and can be reported by the instrument in mIU/mL or IU/mL depending on the input standard.

The MBA has several advantages over standard serological techniques such as the plaque reduction neutralization assay (PRN) and enzyme immunoassay (EIA). In addition to the ability to multiplex several antigens of interest for testing in a single well, the MBA also requires a significantly reduced serum volume compared to either the PRN or the EIA. Multiple measurements according to the number of beads counted per antigen allows more precise quantitative measurement and, depending on the quality of the applied antigen and instrument type (FlexMAP 3D®), a more than 4-log dynamic range of quantitative measurements can be reached, thereby also reducing the need of multiple dilutions of an unknown specimen to capture the accurate antibody concentration. Most substantially, the time to perform the MBA is significantly decreased compared to the PRN. Like the PRN, the MBA utilizes a standard and can report the quantitative antibody concentration in the sample and therefore allows for similar result interpretations to be made as results from PRN. The MBA for measles and rubella has demonstrated a strong linear correlation with the PRN and EIA [17, 18]. Receiver operator characteristic (ROC) curve analysis was used to determine the seroprotection cutoffs and sensitivity and specificity of the measles and rubella MBA. Measles ROC analysis comparing the measles MBA to PRN resulted in a seroprotection cutoff of 153 mIU/mL, which is close to the PRN based value of 120 mIU/mL using the WHO second measles standard and 200 mIU/mL using the WHO third measles standard, above which persons were previously shown to be protected from developing clinical symptoms upon exposure to measles virus, with a sensitivity of 98% and a specificity of 83% (Fig. 2). Rubella ROC analysis comparing the rubella MBA to quantitative EIA yielded a cutoff of 9.36 IU/mL, similar to the accepted cutoff of 10 IU/mL for seroprotection with a sensitivity of 99% and a specificity of 100% (Fig. 2) [17].

2 Materials

The procedure below outlines the MBA procedure for conducting the assay. The assay should be performed with validated lots of beads coupled with measles and rubella antigen (*see* **Note 1**). The current assay utilizes beads coupled with whole virus lysates for measles and rubella. Follow all recommended biosafety and chemical safety procedures, and follow all waste disposal regulations (*see* **Note 2**). Sodium azide is used as a preservative in the bead storage buffer when preparing coupled beads. Use proper safety guidelines when working with sodium azide (*see* **Note 2**). Quality control should be performed on each run and evaluated across studies to ensure proper performance of the assay and controls (*see* **Note 3**).

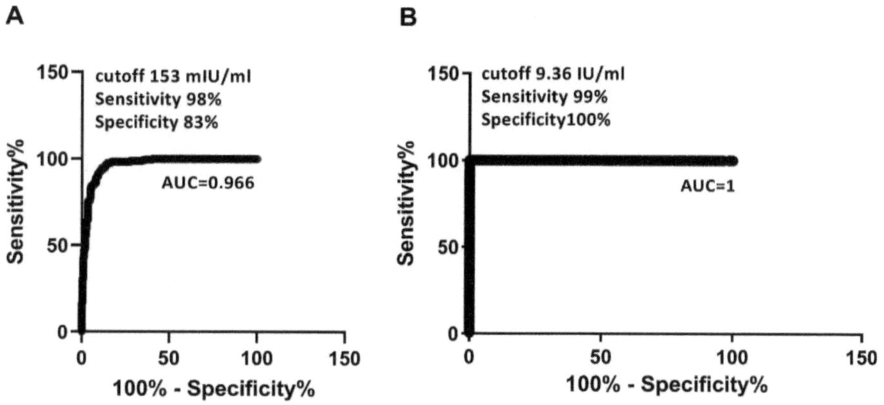

Fig. 2 Measles and rubella MBA receiver operator characteristic (ROC) curve analysis. (**a**) ROC curve for measles MBA using a PRN cutoff of 120 IU/mL as the classifier for positivity. (**b**) ROC curve for rubella MBA using quantitative ELISA cutoff of 10 IU/mL as the classifier for positivity. (Reproduced from ref. 17 with permission from ASM journals)

2.1 Equipment and Software

1. Luminex® instrument: This procedure may be performed on any Luminex® instrument including the BioPlex200®, FlexMAP3D®, or MAGPIX® obtained through a certified Luminex® vendor.
2. Software: The xPONENT® 4.2 software is provided with Luminex® instruments and is required for instrument control, data acquisition, and data reduction. Other software is available from Luminex® vendors that may provide extended options for data analysis and support, such as Bio-Plex Manager® 6.2. Data from xPONENT® is exported as a .csv file, and other vendor software may export in multiple formats for further evaluation in Excel. Data merging of multiple plates may be conducted in R. Neither xPONENT® or Bio-Plex Manager® perform quality control evaluation based on plate run performance. Quality control may be performed externally using plate controls and Levey-Jennings plots [19].
3. Pipets: multichannel and/or repeater pipet. Regularly calibrated multichannel pipets of appropriate volumes are helpful to make serum dilutions; add bead master mix to plates and add secondary antibody to plates. Alternatively, a regularly calibrated repeater pipet can be used to add the bead master mix or secondary antibody as preferred.
4. Vortex.
5. Small sonicating water bath.
6. Vacuum manifold for 96-well plates.
7. Variable speed controlled orbital shaker.
8. Variable speed controlled tube rotator with microcentrifuge tube holder racks.

9. Microcentrifuge if using nonmagnetic beads.
10. A magnetic 96-well plate holder.
11. A hemocytometer or an automated cell counter may be used for counting beads following coupling to determine the number of beads/μL as well as to observe reduction of bead clumping over time following coupling (*see* **Note 4**).
12. Magnetic microcentrifuge separation rack such as DynaMag™-2 Magnet (Thermo Fisher) or Luminex® Magnetic Tube separator (*see* **Note 5**).
13. Parafilm.

2.2 Laboratory Consumables and Reagents

1. Sample needle probe. Probe from the instrument should be regularly cleaned and maintained according to manufacturer recommendations and may be replaced as needed.
2. Off-plate reagent block. This is supplied with the instrument and is used for start-up, cleaning, and shutdown procedures.
3. Sample probe height adjustment kit. This is used for the instrument probe height adjustment function. Appropriate probe height adjustment is required for proper reading of assay plates. The probe height adjustment kit includes metal discs and spheres used as spacers or a probe height adjustment plate with specific distance requirements depending on the type of plate used.
4. 96-well deep well plate. Dilutions of sera can be prepared in a 96-well deep well plate with a sealing mat or tape prior to use (*see* **Note 6**).
5. Assay plates: Filter plates, mylar bottom, and hardbottom opaque well plates compatible with Luminex® instruments are recommended for the procedure. Non-filter plates may only be used with magnetic beads. Example plates used in the procedure outlined above include Millipore Sigma (cat# MABVN1250), Bio-Rad (cat# 171025001), Greiner BioOne® (cat# 655209), and Costar U-bottom (cat #3789) (*see* **Note** 7).
6. Opaque plate covers. Black opaque 96-well lids or aluminum sealing tape may be used to protect reagents from light during incubation steps. When removing sealing tape, remove cautiously to prevent well to well contamination. Alternatively, a clear 96-well plate lid may be used if the plate is wrapped in aluminum foil during the incubation steps.
7. Beads: Luminex® beads of any bead region; this includes Luminex MicroPlex® or MagPlex® beads. Plate type used in the assay may be limited based on bead type; MicroPlex® beads may only be used on filter bottom plates. Bead master mix preparations may be wrapped in aluminum foil to protect from light while working.

8. Antigens: measles virus whole cell viral lysate and rubella virus whole cell viral lysate (*see* **Note 8**). Used for coupling separate bead regions for use in the MBA.
9. Bovine serum albumin (BSA) (*see* **Note 9**).
10. EDC (1-ethyl-3-(3-dimethylaminopropyl)carbodiimide hydrochloride) (*see* **Note 9**).
11. Sulfo-NHS (N-hydroxysulfosuccinimide) (*see* **Note 9**).
12. 5% sodium azide stock solution (*see* **Note 2**).
13. USA Scientific low-binding microcentrifuge tubes (cat #1415-2500 and cat #1420-9700) (*see* **Note 10**).
14. 15 and 50 mL conical tubes.
15. Phosphate buffered saline (0.01 M, pH 7.2) sterile or filter sterilized.
16. MES (2-(N-morpholino)ethanesulfonic acid hydrate, 4-morpholineethanesulfonic acid) buffer (50 mM, pH 5.5).
17. R-phycoerythrin-conjugated goat anti-human IgG secondary antibody (cat # 09-116-098).

2.3 Instrument Maintenance Reagents

1. Deionized H_2O for preparation of instrument maintenance reagents (*see* **Notes 11** and **12**).
2. Bleach (10–20%) for instrument cleaning/sanitization and shutdown procedures (*see* **Note 13**). Fresh dilutions of bleach are recommended to be prepared weekly.
3. NaOH (0.1 N) for instrument cleaning to prevent protein buildup within the fluidics system and is recommended to remove clogs. The dilution of NaOH should be stored in a plastic bottle.
4. Ethanol (70%) is used in instrument cleaning and between plate washing functions.
5. Calibration kits (Luminex® or other approved vendor) used weekly to perform instrument calibration.
6. Verification kits (Luminex® or other approved vendor) used daily prior to instrument use for verification that the instrument is functioning within current calibration settings.
7. Sheath fluid: Only approved sheath fluid should be used in instruments based on model as recommended by the manufacturer.

2.4 Buffer Preparation

1. Working stock solution of Tween-20: 10% Tween-20 solutions should be prepared by pipetting 10 mL of Tween-20 into 90 mL PBS. Pipetting technique for working with viscous solutions should be used when pipetting Tween-20. Pipet Tween-20 solution by holding bottle, and pipet at angle and

pipetting slowly. Ensure no bubbles form in the pipet tip. The working stock may be stored at room temperature (~20–25 °C) and should be monitored for any particulate formation. If particulates are noted within the solution, prepare a fresh working stock.

2. PBS wash buffer: 1× PBS (0.01 M pH 7.2) sterile filtered through 0.22 μm filter. This buffer is used for washing beads during coupling procedure.
3. Bead storage buffer: 1% BSA, 0.05% sodium azide in PBS. To prepare 100 mL in 1× PBS (0.01 M pH 7.2), add 1 g BSA and 1 mL 5% sodium azide stock solution. 2 mL of storage buffer is needed per tube of coupled beads prepared.
4. Bead activation buffer: 5 mg/mL EDC in 1× PBS or 50 mM MES pH 5.5 and 5 mg/mL sulfo-NHS in 1× PBS or 50 mM MES pH 5.5, prepared separately. Just prior to use, mix equal amounts of EDC solution and sulfo-NHS solution; use immediately for bead activation.
5. Serum dilution buffer (SDB): 0.5% BSA in PBS. To prepare 100 mL in 1× PBS (0.01 M pH 7.2), add 0.5 g BSA and 1 mL 10% Tween-20 solution.
6. PBS-Tween wash buffer: PBS + 0.05% Tween-20 (PBST). To prepare 1 L in 1× PBS (0.01 M pH 7.2), add 500 μL Tween-20. This buffer is used for washing beads during assay procedure.
7. WHO standard: A WHO serum standard for measles and rubella or an inhouse serum standard quantitated to a WHO serum standard should be used to quantitate antibody concentration in samples. Current available WHO standards from NIBSC include the NIBSC WHO measles third standard 97/648, the WHO rubella standards 67/182, and RUBI-1-94. The NIBSC standard RUBI-1-94 was used in the procedure (*see* **Note 14**).
8. Secondary antibody: A 1:200 or 1:400 dilution of secondary antibody should be prepared in PBST (*see* **Note 15**).

3 Methods

3.1 Coupling of Luminex® Beads for Assay Procedure

1. Uncoupled beads are stored at 2–8 °C; remove from storage and let the beads come to room temperature (~20–25 °C) (approximately 30 min) prior to starting the coupling procedure. When mixing uncoupled beads in their original container, do so by gently inverting the tube several times, and do not vortex at this step. When ready, transfer a maximum of 1 mL of beads to a USA Scientific low-binding microcentrifuge tube.

2. Remove the manufacturer bead storage buffer from the beads according to the use of magnetic or nonmagnetic beads. All washing procedures will follow the same method for removal of supernatant according to bead type.
   (a) If using nonmagnetic beads: Centrifuge for 4 min at 13,400 rpm in a microcentrifuge. Then remove the supernatant using a pipet carefully as to not disturb the bead pellet.
   (b) If using MagPlex® beads: Place tube into the magnetic microcentrifuge single tube holder or rack, and allow separation to occur for 30–60 s. With the tube still positioned in the magnetic microcentrifuge tube rack, remove the supernatant using a pipet. Take care not to disturb the microspheres which will be adhered to the side of the tube toward the magnet.
3. To activate the beads for the coupling procedure, prepare the bead activation buffer either in PBS or MES immediately prior to use by combining the EDC and sulfo-NHS solutions at a 1:1 ratio. Remove the beads from the magnetic rack or tube holder, and add 1 mL of freshly prepared bead activation buffer. Resuspend the beads in the bead activation buffer by vortexing vigorously for 30–60 s. It is best to seal the top of the tube using parafilm and wrap the tube in aluminum foil to protect beads from light and then incubate for 20 min at room temperature (~20–25 °C) under constant rotation at 10 rpm.
4. Remove the bead activation buffer and wash the beads with PBS wash buffer according to bead type. Repeat the wash step two times (for a total of three washes), removing the supernatant each time.
   (a) If using nonmagnetic beads: Centrifuge for 4 min at 13,400 rpm in a microcentrifuge. Then remove the supernatant using a pipet carefully as to not disturb the bead pellet. Resuspend the beads in 1 mL of wash buffer. Sonicate and vortex the beads for approximately 20 s each.
   (b) If using MagPlex® beads: Place tube into the magnetic microcentrifuge single tube holder or rack, and allow separation to occur for 30–60 s. With the tube still positioned in the magnetic microcentrifuge tube rack, remove the supernatant using a pipet. Take care not to disturb the microspheres which will be adhered to the side of the tube toward the magnet. Carefully remove the supernatant, and resuspend the microspheres in 1 mL of wash buffer. Vortex briefly and sonicate for 20 s.
5. Meanwhile, prepare the antigen solutions for bead coupling individually with a total volume of each antigen mixture to be 550 μL according to below (*see* **Note 8**):

(a) For measles virus whole virus lysate antigen, prepare an antigen stock with a concentration of 0.9 μg/μL. The final concentration of antigen for 12.5 × 10^6 beads is 150 μg (*see* **Note 8**).

(b) For rubella virus whole virus lysate antigen, prepare an antigen stock with a concentration of 0.5 μg/μL. The final concentration for 12.5 × 10^6 beads is 15–30 μg (*see* **Note 8**).

(c) Vero uninfected lysate antigen stock with a concentration of 0.9 μg/μL. The final concentration of antigen for 12.5 × 10^6 beads is 150 μg.

6. Remove final wash supernatant from the beads. Add 500 μL of antigen mixture to the microspheres and vortex briefly. Seal the top of the tube using parafilm, and wrap the tube in aluminum foil. Incubate for 2 h under constant rotation at 10 rpm at room temperature (~20–25 °C).

7. Remove the antigen coupling buffer, and wash the beads with PBS wash buffer according to bead type. Repeat the wash step two times, for a total of three washes for rubella and Vero uninfected lysate coupled beads, and three times, for a total of four washes, for measles coupled beads, removing the supernatant each time.

   (a) If using nonmagnetic beads: Centrifuge for 4 min at 13,400 rpm in a microcentrifuge. Then remove the supernatant using a pipet carefully as to not disturb the bead pellet. Resuspend the beads in 1 mL of wash buffer. Sonicate and vortex the beads for approximately 20 s each.

   (b) If using MagPlex® beads: Place tube into the magnetic microcentrifuge single tube holder or rack, and allow separation to occur for 30–60 s. With the tube still positioned in the magnetic microcentrifuge tube rack, remove the supernatant using a pipet. Take care not to disturb the microspheres which will be adhered to the side of the tube toward the magnet. Carefully remove the supernatant and resuspend the microspheres in 1 mL of wash buffer. Vortex briefly and sonicate for 20 s.

8. *Optional*: For the measles coupled beads, an additional wash/rotation step can be helpful to reduce bead aggregation. To do this for the measles antigen-coupled beads, aspirate the supernatant of the final washing step after separation is complete. Resuspend the microspheres in 1 mL of bead storage solution without azide. Seal the top of the tube with parafilm, and wrap in aluminum foil, and incubate for 30 min under constant rotation at 10 rpm at room temperature (~20–25 °C). Remove supernatant according to bead type as described in washing steps.

9. After removing the last wash or the optional additional rotation wash for measles coupled beads, resuspend the beads in bead storage buffer by adding 1 mL of bead storage buffer, vortex beads for 30 s, and sonicate beads for 30 s.
10. If preparing multiple 1 mL concentrations of any individual antigen type, then the 1 mL individual bead suspensions should be pooled and aliquoted into fresh 1 mL portions or stored in larger volume USA Scientific tubes (*see* **Note 16**).
11. Coupled beads should be stored at 2–8 °C protected from light. Beads should be counted following any necessary "resting" time (*see* **Notes 1** and **4**).

3.2 Preparation of Luminex® System for Data Acquisition by Performing the System Initialization Routines (See Note 17)

1. Perform a daily start-up routine each day when using the instrument (*see* **Note 18**).
2. An instrument calibration using the appropriate calibration kit should be performed weekly or as needed such as following probe replacement.
3. An instrument verification should be performed before each day's runs using the appropriate verification kit.
4. A weekly fluidics maintenance is recommended following probe cleaning, sonication, and replacement.

3.3 Preparation of Samples, Controls, and Standard

1. Dilute serum to desired testing concentration (e.g., 1:200, 1: 400, 1:4000) (*see* **Notes 19–22**). Alternatively, dried blood spot (DBS) punches may also be used and eluted into serum dilution buffer (*see* **Note 23**) [9, 20]. Thoroughly mix serum dilution or dry blood spot elution prior to using.
2. Prepare the NIBSC reference standard in serum dilution buffer (*see* **Note 14**). A 1:400 initial dilution of standard is used in this procedure followed by either 1:3 or 1:2 serial dilutions to generate a 10- or 11-point standard curve.
3. Samples, controls, and standards may be prepared in serum dilution buffer up to 12 weeks prior to running the assay and stored at 2–8 °C (*see* **Note 21**).

3.4 Preparation of Beads for Assay Procedure

1. The internal bead dyes are light sensitive, and therefore beads should be protected from exposure to light at all times. Tubes used for bead dilutions and assay plates may be wrapped in aluminum foil; additionally, black opaque 96-well plate lids may be used to protect the assay plate from light during incubation steps.
2. Conjugated beads tend to form clumps and must be resuspended into a single bead suspension before use. To achieve this, vortex the microspheres for 30–60 s, sonicate in a small

$$\left(\frac{2{,}500\ beads}{0.025mL}\right)(V_{total}) = \left(\frac{Stock\ bead\ count}{1\mu L}\right)(V_{stock})$$

$$\left(\frac{2{,}500\ beads}{0.025mL}\right)(5.6mL) = \left(\frac{12{,}500\ beads}{1\mu L}\right)X$$

$$560{,}000\ beads = \left(\frac{12{,}500\ beads}{1\mu L}\right)(X)$$

$$X = \frac{560{,}000\ beads}{\left(\frac{12{,}500\ beads}{1\mu L}\right)} = 44.8uL\ bead\ stoc$$

Fig. 3 Example bead dilution calculation. Example calculated volume of beads for 5.5 mL; this is enough volume to conduct the MBA in two 96-well plates, at 2500 beads/25 μL using a 12,500 beads/μL stock. Answer is in μL because mL cancels out and bead count is in beads/μL

ultrasonic bath for 2–3 min, and then vortex again for 30–60 s prior to pipetting from tube. A homogenous suspension of beads can be confirmed visually.

3. Bead solutions can be prepared with a concentration range of 1250–4000 beads per well depending on the quality of the beads; this should be sufficient to reach the minimally required amount of 50 beads to be analyzed within an acceptable timeframe. Prepare a microsphere solution with the desired concentration of beads/region/25 μL/well in serum dilution buffer. Add beads for all regions to the same master bead mix. Sample calculation is depicted in Fig. 3 (*see* **Note 24**). For one plate, it is recommended to make a master mix with a total final volume of 2.8 mL.
4. Vortex the bead solution for 30–60 s, and add 25 μL to each well with a repeater pipet or multichannel pipet. Holding the pipet tip at a 45° angle allows addition of bead mixture without touching the bottom of the well.

3.5 MBA Procedure

1. If samples, controls, and standards are prepared and stored at 2–8 °C prior to the day of the assay procedure, then allow diluted samples, controls, and standards to reach room temperature (~20–25 °C) for at least 30 min before beginning the MBA procedure.
2. It is recommended to include blank wells and control sera on every plate. Blank wells consist of wells that have beads, but with no sera added during the assay run; instead, serum dilution buffer should be added to these wells. See an example plate layout in Fig. 4.

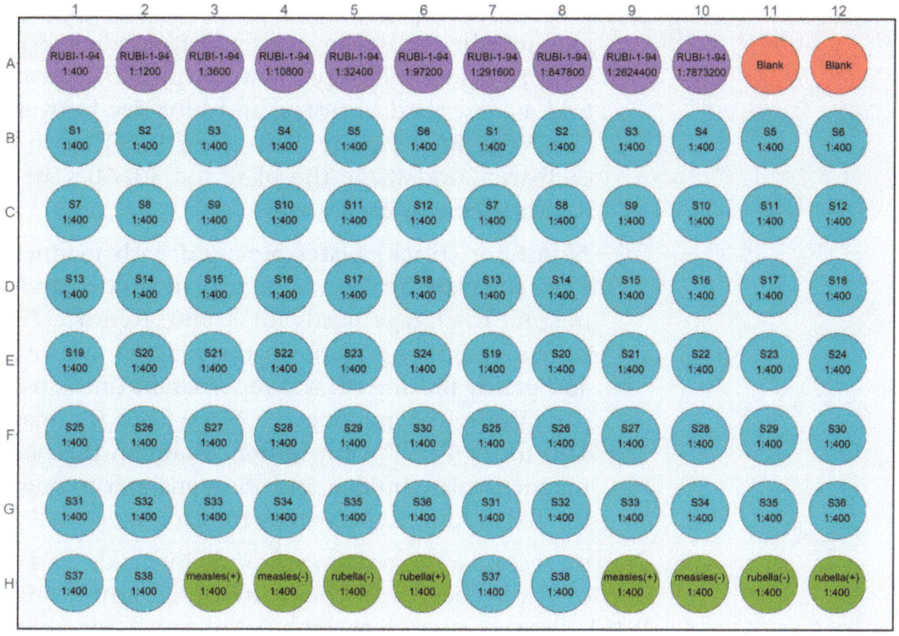

Fig. 4 Example MBA plate layout. An example plate layout for duplicate sera sample is provided. This layout is optimal for bead lot validation with a standard curve and duplicate sample measurements on replicate plates with the comparison bead lots on separate plates. Alternative plate options for serosurveillance that do not include the standard curve on each plate and that allow the option to run samples in singlet for large studies may also be used. Minimally, each plate should include two blank wells and appropriate controls as depicted in the example layout

3. Following addition of beads, add an equal volume (25 μL) of blank (serum dilution buffer), standard (vortex for 10–15 s before adding), control sera and samples to the designated wells. If sample dilutions were prepared in a 96-well deep well plate, they may be added using a multichannel pipet. Holding the pipet tip at a 45° angle allows addition of sera without touching the bottom of the well. Standard samples may be included on the plate for within plate direct quantitation of target antibodies in sera; *see* **Note 25** below for further explanation of when within plate standards may be less efficient and recommendations for quantitation in those cases.
4. Cover the plate with aluminum sealing foil or opaque plate lid, and secure plate on plate shaker. Shake for 30–60 s at 800 rpm, and then reduce speed to 600 rpm, and continue shaking for 45 min at room temperature (~20–25 °C).
5. After incubation, wash the plate three times with 100 μL PBST wash buffer according to plate type used; *see* **step (a)** or **(b)** based on plate type used. For one plate, approximately 10 mL of PBST wash buffer is used per wash.

(a) Filter bottom plates may be used for magnetic and non-magnetic beads. If using a filter bottom plate, do not allow wells to dry out. Empty the wells using the vacuum manifold as described in notes, and blot dry bottom of plate (*see* **Note 26**). Pipette 100 μL of PBST wash buffer to each well, and shake the plate for 30–60 s at 800 rpm. Repeat steps for each wash.

(b) Non-filter (black) plates are used with magnetic beads. Place the plate on the magnetic plate holder, and allow the magnet to engage beads for 3 min. Remove PBST wash buffer by holding plate on the magnet and overturning and gently flicking off the supernatant onto an absorbent pad. Without removing the plate from the magnet, add 100 μL of PBST wash buffer. Maintain the plate on the magnetic plate holder, and allow magnet to engage beads for 1–1.5 min. Repeat steps for each wash.

6. Prepare secondary detection antibody by making a 1:200 dilution of R-phycoerythrin-conjugated goat anti-human IgG in PBST. Prepare enough to apply 50 μL/well; for one plate, this is 25 μL into a total volume of 5 mL of PBST. Minimize exposure of the detection solution to light by covering the tube with aluminum foil.

7. Remove final PBST wash buffer from plate, as described above. If using magnetic plate washer, remove plate from magnet after removing the last PBST wash buffer.

8. Add 50 μL detection antibody to each well with a multichannel pipet.

9. Cover the plate with opaque lid or aluminum sealing foil, secure, and place on shaker. Shake for 30–60 s at 800 rpm on a shaker, and then reduce speed to 600 rpm, and continue shaking for 30 min at room temperature (~20–25 °C) (*see* **Note 27**).

10. Wash the plate three times as described above, remove PBST wash buffer, and add 100 μL PBS or PBST to each well (*see* **Note 28**).

11. Shake the plate covered with aluminum sealing foil or opaque plate lid for 5 min at 600 rpm at room temperature (~20–25 ° C). If reading more than one plate, reduce shaking speed to reduce unnecessary shear stress on or aggregation of bead/antibody complexes while the first plate reads, increasing to 600 rpm 5 min before reading on instrument.

12. To read the plate on a Luminex® instrument, follow recommended instrument start-up procedures. Plates should only be tested on instruments that have passed weekly calibration and daily validation procedures. Prior to running on the instrument, remove the plate lid. It is recommended to record a minimum of 50 beads/region.

13. If necessary, a plate can be stored at 2–8 °C and measured the next day after bringing plate to room temperature (~20–25 °C) for at least 30 min, changing buffer to fresh PBST and shaking for 5 min on plate shaker at 600 rpm (*see* **Note 29**).

3.6 MBA Analysis

1. Verify plate quality control evaluating that the blank wells on the tested plate yield MFI values within background range and no well contamination is suspected (*see* **Note 30**).
2. Verify other quality control parameters including well bead count, MFI patterns, and serum controls on each tested plate (*see* **Note 3**).
3. Use the standard curve to interpolate antibody concentrations in each sample accounting for the sample dilution (*see* **Notes 25** and **30**).

4 Notes

1. Bead lot validation should be performed using a defined panel of sera consisting of at least 32 sera with a range of antibody concentrations. A new lot of beads should be directly compared to a previously coupled and validated lot. Recommended criteria for validation of bead lots include sample and control observed concentrations yielding a Pearson's correlation coefficient of a minimum of 0.97 and average percent coefficient of variance (CV) of <20% between bead lots. Additionally, the standard curves should be evaluated across bead lots to ensure that parallel curves yield similar peak curve values.
2. It is recommended to consult with the institutional biosafety officer before beginning any work. Some safety precautions to note: Laboratory personnel should be trained in handling potentially infectious material. All serum samples and DBS should be treated as infectious and handled using universal precautions. Personal protective equipment (PPE) includes gloves, lab coat, safety glasses, and use of a certified biological safety cabinet when diluting concentrated serum samples and working with sodium azide. Sodium azide should be handled with extreme caution. Double nitrile gloves should be worn and changed immediately following dilution of the stock or working stock into bead storage buffer. Dilutions of serum should be performed in a certified biological safety cabinet (BSC). All waste from the Luminex® instrument should be collected in an appropriate waste container. All potentially infectious material should be treated with 10% bleach prior to disposal according to institutional biohazardous waste disposal policy.

3. Quality control should be performed on every plate tested and across plates during the course of a study. Plate quality control should evaluate bead count per well. The bead count entered into the instrument protocol for collection for each bead region is 50 beads/region. It is ideal if a well reaches this bead count for the region; however, data may be acceptable with lower bead counts. A bead count of 35 is the minimum recommended by Luminex®. Bead counts as low as 20 for the measles and rubella assay have been verified to yield acceptable data but should be evaluated carefully. Quality control evaluation should also include reviewing the MFI across plate wells and antigens to identify any patterns of background or below background MFI that would indicate pipetting error. Controls should include a measles and rubella positive and a negative serum that have been previously characterized and have sufficient volume to be used repeatedly for an entire study and ideally over several studies. Controls may be positive or negative for both measles and rubella in a single sample, or multiple controls may be used in order to meet the criteria to have a positive and a negative for each antigen. Controls should fall within the range of expected values for each control based on repeated measurements in the assay. Control MFI values may be tracked across multiple plate runs using Levey-Jennings plots [19]. Control values confirm that the assay is performing as expected. When tracked across multiple plates within a study, this can help identify plates that are falling outside of expected ranges and may warrant repeat testing. Uninfected control beads conjugated with Vero cell lysate are useful to identify samples yielding high signal to background results due to cellular lysate components present in the viral lysate used to conjugate beads.
4. After coupling of beads, beads should be counted using a hemacytometer or automatic cell counter. Note that not all automatic cell counters can accurately capture beads or accurately count small groupings of beads. Any automatic cell counter used should be verified for counting accuracy by comparison with manual counting using a hemacytometer prior to regular use. Accurate counting of beads may require the beads to be "rested" for several weeks following coupling at 2–8 °C with periodic sonication and vortexing to break up clumps. Some antigens such as the measles virus whole lysate are particularly susceptible to bead aggregation following bead coupling. This aggregation improves over time, and the beads achieve a uniform suspension. Bead stability has been evaluated at 1 year.
5. The magnetic tube separator from Luminex holds a single tube and works for single tube production of beads. If multiple bead

sets are being coupled simultaneously, then a separation rack such as the Thermo Fisher option can hold multiple tubes at a time.

6. Serum dilutions may be made in deep well plates. The size of wells should be chosen based on total dilution volume. Formats are available in 96-well plate layouts sufficient for most dilution applications of the MBA; for larger dilution volumes, 48-well plates may be required.
7. Some mylar bottom plates can result in a hook effect resulting from steric hinderance caused by nonspecific protein sticking. This can result in a reduction in attributable signal particularly in samples or standard dilutions with high antibody concentrations. This can be addressed through the use of filter plates or hard bottom plates or additional dilution of sample.
8. There are several commercially available sources of whole virus lysate for measles and rubella; however, not all function with the same sensitivity and specificity [17]. The above procedure has been validated using measles virus lysate from Zeptometrix (cat #0810025) and rubella virus lysate from Meridian (cat #6123) or GenWay Bio (cat #GWB-HG4389) and in accordance with previously established bead-based assays validated with a noncommercially but similarly produced whole measles virus lysate, which has also been applied in large population serosurveys [11, 18]. Additionally, even for commercial sources of measles and rubella lysates, the starting viral titer appears to influence assay sensitivity even for different antigen lot preparations from the same vendor. Antigen preparations from lower titer viral stock result in lower viral protein concentrations within the total lysate protein concentration and yield beads that will produce lower signals. The measles whole virus lysate antigen from Zeptometrix must be dialyzed against PBS to remove buffer components incompatible with coupling prior to the coupling procedure. Protein concentration should be reconfirmed using BCA or other similar procedure. For rubella whole virus antigen, a final concentration of 15 μg has been published for the GenWay Bio obtained antigen, and 15 and 30 μg has been used for the Meridian produced antigen; however, the 30 μg concentration has demonstrated better bead lot to lot consistency.
9. The above procedure has been validated with EDC from Thermo Fisher (cat# PI22980), sulfo-NHS from Thermo Fisher (cat# PI24510), and bovine serum albumin (BSA) from Sigma (cat #A3803). These are critical reagents and therefore should be validated within the assay procedure.
10. The USA Scientific tubes are a critical reagent, and other tube types could result in adherence of the beads to the side of the

tube resulting in bead aggregation, insufficient coupling, and bead loss. Other larger tubes are also available from USA Scientific that are low bind and may be used for storage of larger volumes of beads. The 1.5 mL microcentrifuge flip top tubes are best used for the coupling procedure, while the screw top tubes are best used for storage of beads following coupling.

11. Regular maintenance of instruments is essential. This includes daily running of unclog procedures with 0.1 N NaOH in addition to instrument shutdown procedures with 10–20% bleach. If running multiple plates with several bead regions, it may be necessary to perform more frequent or enhanced cleaning procedures, such as running unclog procedures after every two plates.
12. Performing an unclog procedure without the drain function is beneficial to prevent probe clogging.
13. Bleach dilutions of 10–20% may be used for instrument maintenance procedures. More stringent cleaning procedures when instrument clogging has been noted should be carried out using 20% bleach. Ideally bleach solutions used should have a final concentration of 1–2% active sodium hypochlorite.
14. Multiple NIBSC reference standards are available for measles and rubella. The NIBSC WHO rubella standard RUBI-1-94 was used in this assay. RUBI 1-94 may be used as a reference standard for both measles and rubella. Rubella antibody concentrations are provided by NIBSC. Measles antibody concentrations were quantitated using WHO measles second international standard for comparison [18]. The measles third international standard has not been calibrated for ELISA-based or MBA based assays, and further evaluation of this standard in the MBA is necessary to obtain accurate cutoff values to determine sample positivity.
15. The above procedure has been validated with secondary antibody from Jackson Immunoresearch 1 (cat # 09-116-098). Secondary antibody is a critical reagent and therefore should be validated within the assay procedure.
16. When preparing multiple individual 1 mL bead coupling reactions of a single antigen, each 1 mL coupling aliquot may be validated prior to pooling followed by a final validation of the pooled aliquot. Because each 1 mL bead portion undergoes a separate coupling reaction procedure, this step can ensure that any coupled beads that don't meet validation requirements are not included in the final pool.
17. Instrument start-up procedures of the manufacturer should be followed. Typically, a daily start-up procedure is conducted to initialize fluidics. If using a laser-based instrument, the laser warm-up time is required prior to running any calibration or validation beads.

18. The daily start-up routine preloaded with the xPONENT® software includes the laser warm-up step, fluidics initialization, calibration, and verification routines. The start-up procedures may be modified to add a routine that performs the fluidics steps of the Luminex®-provided daily start-up routine separately from the laser warm-up and calibration/verification routines to allow more flexibility. The laser warm-up begins when the instrument is turned on, the warmed-up laser is not required for the fluidics start-up but is required to be completed prior to any calibration and verification procedures. Separating the steps in the daily start-up routine allows required calibrations and verifications to be run as indicated by procedures for the day of run.
19. Recommended storage conditions or specimens prior to testing are as follows: Serum specimens should ideally be frozen and stored at −20 °C or lower but may be stored at 2–8 °C for up to 14 days prior to testing [21]. Prior to testing, ideally DBS should be frozen at −20 °C or lower with desiccant present in storage bags; however, DBS may be stored at room temperature (20–25 °C) for up to 14 days and stored at 2–8 °C for up to 3 months [22].
20. Serum may be diluted to concentrations determined to be appropriate. After 1:400 dilution, most serum samples will be within range of the standard curve for measles and rubella. Alternatively, dual serum dilution technique may be used in which samples are diluted both at 1:200 and a higher dilution such as 1:4000 to obtain results with at least one dilution point within the linear range of the standard curve. Note that on average, approximately 30% of samples will have a single dilution point within range.
21. Sera, controls, and standards may be diluted and stored at 2–8 ° C prior to performing assay. Diluted standards, controls, and sera are stable at 2–8 °C up to 12 weeks; however, serum stability should be confirmed in the operating laboratory. If storing samples prior to use in the MBA, sera, controls, and standards should be allowed to come to room temperature (~20–25 °C) (approximately 30 min) before using in the assay.
22. When preparing standard, control, and serum dilutions, the minimum volume recommended to pipet from serum is 2 μL to allow accurate pipetting.
23. For DBS, a 6 mm or 3 mm punch may be taken from the filter paper using a core punch tool. A 6 mm punch is equivalent to approximately 10 μL of blood and 5 μL of serum. A 3 mm punch is equivalent to approximately 1.25 μL of serum. The puncher should be wiped with a 70% ethanol dampened wipe and dried prior to punching the next sample. Residual ethanol on the puncher will inhibit blood elution.

24. When preparing bead dilutions, always add extra volume to the total required volume to account for pipetting loss. It is recommended to add a minimum of 300 μL when using a repeat pipet.
25. For large serosurveys targeting multiple antigens that also have standard curves, running standards for each antigen on each plate becomes inefficient. Therefore, standards can be run separately from samples at regular intervals throughout the study to generate an average standard curve, which may be defined using a 5-PL curve fit in statistical software. Cutoffs for positivity may be determined by extrapolation of an MFI cutoff based on the cutoffs defined above or the equation used to calculate concentrations of samples.
26. Filter plates may be used with a vacuum manifold for removal of liquid during washing steps. If using filter plates, pre-wet the wells of the plates by adding 100 μL of 0.05% PBST to each well. Perform this step first, and then prepare bead master mix. Apply gentle vacuum to the filter plate (max 5 mmHg for 2–5 s) to remove PBST. Gently blot the bottom of the plate by pressing onto an absorbent pad or towel to remove excess liquid. Do not allow the wells to dry out.
27. Orbital shaker speeds may vary between equipment. Shaking speed should be vigorous enough to allow bead suspension and mixing during incubations, but not so vigorous to result in spillover between wells.
28. If bead aggregation or instrument clogging is an issue while running a plate in the instrument, PBST may help to make instrument runs more efficient and alleviate instrument clogging. If this is not occurring, PBS is recommended for running in the instrument.
29. For large serosurveys, if plates will be tested over multiple days, consider standardizing storage times for prepared dilutions as much as possible.
30. Background well MFI range will vary depending on instrument used. For the FlexMAP, background wells generally yield an MFI less than 150, with an average MFI of 66 for measles and 88 for rubella beads, and for the MagPix, background wells generally yield an MFI less than 25, with an average MFI of 18 for measles and 20 for rubella beads. The BioPlex manager software is helpful when quantifying samples using a standard curve within the plate. The software generates a 5-parameter logistic (5-PL) regression curve and applies the equation from the standard curve to automatically calculate the sample antibody concentration.

Acknowledgments

We thank Logan Melot for contributing to the images of measles and rubella viruses in Fig. 1 using BioRender. We would like to thank Dr. Diana Martin and Dr. Gretchen Cooley for their continued collaboration and expert advice on integrated serosurveillance by MBA and use of DBS.

Funding

The findings and conclusions in this report are those of the authors (Melissa M. Coughlin, Zachary Matson, and Bettina Bankamp) and do not necessarily represent the official position of the Centers for Disease Control and Prevention, US Department of Health and Human Services.

References

1. Choisy M, Trinh ST, Nguyen TND, Nguyen TH, Mai QL, Pham QT, Tran ND, Dang DA, Horby PW, Boni MF, Bryant J, Lewycka SO, Nadjm B, Van Doorn HR, Wertheim HFL (2019) Sero-prevalence surveillance to predict vaccine-preventable disease outbreaks; A lesson from the 2014 measles epidemic in Northern Vietnam. Open Forum Infect Dis 6(3):ofz030. https://doi.org/10.1093/ofid/ofz030
2. Hagan JE, Takashima Y, Sarankhuu A, Dashpagma O, Jantsansengee B, Pastore R, Nyamaa G, Yadamsuren B, Mulders MN, Wannemuehler KA, Anderson R, Bankamp B, Rota P, Goodson JL (2017) Risk factors for measles virus infection among adults during a large outbreak in postelimination era in Mongolia, 2015. J Infect Dis 216(10):1187–1195. https://doi.org/10.1093/infdis/jix449
3. Hales CM, Johnson E, Helgenberger L, Papania MJ, Larzelere M, Gopalani SV, Lebo E, Wallace G, Moturi E, Hickman CJ, Rota PA, Alexander HS, Marin M (2016) Measles outbreak associated with low vaccine effectiveness among adults in Pohnpei State, Federated States of Micronesia, 2014. Open Forum Infect Dis 3(2):ofw064. https://doi.org/10.1093/ofid/ofw064
4. Cutts FT, Claquin P, Danovaro-Holliday MC, Rhoda DA (2016) Monitoring vaccination coverage: defining the role of surveys. Vaccine 34(35):4103–4109. https://doi.org/10.1016/j.vaccine.2016.06.053
5. Dabbagh A, Patel MK, Dumolard L, Gacic-Dobo M, Mulders MN, Okwo-Bele JM, Kretsinger K, Papania MJ, Rota PA, Goodson JL (2017) Progress toward regional measles elimination – worldwide, 2000-2016. MMWR Morb Mortal Wkly Rep 66(42): 1148–1153. https://doi.org/10.15585/mmwr.mm6642a6
6. Durrheim DN, Orenstein WA, Schluter WW (2018) Assessing population immunity for measles elimination – the promise and peril of serosurveys. Vaccine 36(28):4001–4003. https://doi.org/10.1016/j.vaccine.2018.04.036
7. Arnold BF, Scobie HM, Priest JW, Lammie PJ (2018) Integrated serologic surveillance of population immunity and disease transmission. Emerg Infect Dis 24(7):1188–1194. https://doi.org/10.3201/eid2407.171928
8. Breakwell L, Anga J, Cooley G, Ropiti L, Gwyn S, Wannemuehler K, Woodring J, Ogaoga D, Martin D, Patel M, Tohme RA (2020) Seroprevalence of chronic hepatitis B virus infection and immunity to measles, rubella, tetanus and diphtheria among schoolchildren aged 6–7 years old in the Solomon Islands, 2016. Vaccine 38(30):4679–4686. https://doi.org/10.1016/j.vaccine.2020.05.029
9. Cooley GM, Mitja O, Goodhew B, Pillay A, Lammie PJ, Castro A, Moses P, Chen C, Ye T, Ballard R, Martin DL (2016) Evaluation of multiplex-based antibody testing for use in large-scale surveillance for yaws: a comparative study. J Clin Microbiol 54(5):1321–1325. https://doi.org/10.1128/JCM.02572-15

10. Feldstein LR, Bennett SD, Estivariz CF, Cooley GM, Weil L, Billah MM, Uzzaman MS, Bohara R, Vandenent M, Adhikari JM, Leidman E, Hasan M, Akhtar S, Hasman A, Conklin L, Ehlman D, Alamgir A, Flora MS (2020) Vaccination coverage survey and seroprevalence among forcibly displaced Rohingya children, Cox's Bazar, Bangladesh, 2018: a cross-sectional study. PLoS Med 17(3): e1003071. https://doi.org/10.1371/journal.pmed.1003071

11. Smits G, Mollema L, Hahne S, de Melker H, Tcherniaeva I, van der Klis F, Berbers G (2014) Seroprevalence of rubella antibodies in The Netherlands after 32 years of high vaccination coverage. Vaccine 32(16):1890–1895. https://doi.org/10.1016/j.vaccine.2014.01.066

12. Vos RA, Mollema L, Kerkhof J, van den Kerkhof J, Gerstenbluth I, Janga-Jansen AVA, Stienstra Y, de Melker HE, van der Klis FRM (2019) Risk of measles and diphtheria introduction and transmission on Bonaire, Caribbean Netherlands, 2018. Am J Trop Med Hyg 101(1):237–241. https://doi.org/10.4269/ajtmh.18-0824

13. Vos RA, Mollema L, van Binnendijk R, Veldhuijzen IK, Smits G, Janga-Jansen AVA, Baboe-Kalpoe S, Hulshof K, van der Klis FRM, Melker HE (2019) Seroepidemiology of measles, mumps and rubella on Bonaire, St. Eustatius and Saba: the first population-based serosurveillance study in Caribbean Netherlands. Vaccines (Basel) 7(4):137. https://doi.org/10.3390/vaccines7040137

14. Angeloni S, Das S, De Jager W, Dunbar S (2022) xMAP Cookbook, 5th edn. Luminex

15. Luminex (2017) Luminex FlexMAP 3D hardware user manual. Luminex. https://info.luminexcorp.com

16. Luminex (2017) MAGPIX hardware installation and user manual. Luminex. https://info.luminexcorp.com/

17. Coughlin MM, Matson Z, Sowers SB, Priest JW, Smits GP, van der Klis FRM, Mitchell A, Hickman CJ, Scobie HM, Goodson JL, Alexander JP Jr, Rota PA, Bankamp B (2021) Development of a measles and rubella multiplex bead serological assay for assessing population immunity. J Clin Microbiol 59(6): e02716-20. https://doi.org/10.1128/JCM.02716-20

18. Smits GP, van Gageldonk PG, Schouls LM, van der Klis FR, Berbers GA (2012) Development of a bead-based multiplex immunoassay for simultaneous quantitative detection of IgG serum antibodies against measles, mumps, rubella, and varicella-zoster virus. Clin Vaccine Immunol 19(3):396–400. https://doi.org/10.1128/CVI.05537-11

19. Matson Z, Cooley G, Simon NPA, Bankamp B, Coughlin MM (2022) shinyMBA: a novel R-shiny application for quality control of the multiplex bead assay for large scale serosurveillance studies. *Manuscript accepted in Scientific Reports.*

20. Vos RA, Mollema L, van Boven M, van Lier A, Smits G, Janga-Jansen AVA, Baboe-Kalpoe S, Hulshof K, Stienstra Y, van der Klis FRM, de Melker HE (2020) High varicella zoster virus susceptibility in Caribbean island populations: implications for vaccination. Int J Infect Dis 94:16–24. https://doi.org/10.1016/j.ijid.2020.02.047

21. Crooke SN, Coughlin MM, Perelygina LM (2023) Measles and rubella viruses. In: Manual of clinical microbiology, 13th edn. ASM Press, Washington, DC

22. Gwyn S, Aragie S, Wittberg DM, Melo JS, Dagnew A, Hailu D, Tadesse Z, Haile M, Zeru T, Nash SD, Arnold BF, Martin DL, Keenan JD (2021) Precision of serologic testing from dried blood spots using a multiplex bead assay. Am J Trop Med Hyg 105(3): 822–827. https://doi.org/10.4269/ajtmh.21-0140

Chapter 18

Measles IgG Avidity Assay

Sara Mercader and Stephen Crooke

Abstract

Measles IgG avidity assays determine the overall strength of molecular binding between measles-specific IgG antibodies and measles virus antigens. Avidity results can distinguish recent from distant measles virus infections. Individuals who are immunologically naïve to measles virus develop low-avidity antibodies upon measles virus infection or first-time vaccination. Within 4–6 months, antibodies mature to high avidity. Measles avidity assays are most useful in the context of measles elimination. In such settings, avidity and epidemiological and clinical information are used to classify measles breakthrough infections for control and surveillance purposes and to assist in case confirmation when other laboratory results are inconclusive or nonexistent. We present a highly accurate end-titer measles avidity assay that delivers results based on IgG quality (avidity) that are independent of IgG concentration.

Key words Immunoglobulin G, Antibodies, Antigen, Avidity, Immunoassay, Measles, Elimination, Vaccine failure classification, Serum, Serology

1 Introduction

Avidity describes the overall binding strength of multivalent antibodies to multivalent antigens. In individuals who are immunologically naïve to measles virus antigens, low-avidity antibodies are synthesized upon measles virus infection or first-time vaccination during the host primary immune response. Over time, IgG antibodies gradually mature to high-avidity during a process known as somatic hypermutation [1–3]; measles-specific IgG maturation may take 4–6 months [4–6]. Avidity is used as an indicator of the quality of IgG antibodies. Avidity enzyme-linked immunosorbent assays (ELISAs) are designed to differentiate between primary antibody responses with detection of immature low-avidity IgG antibodies and secondary (or distant) antibody responses with detection of mature high-avidity IgG. Avidity ELISAs rely on a denaturant wash step added after serum incubation, when

Dzwokai Z. Ma and Christian K. Pfaller (eds.), *Measles and Related Morbilliviruses: Methods and Protocols*, Methods in Molecular Biology, vol. 2808, https://doi.org/10.1007/978-1-0716-3870-5_18,

measles-specific IgG antibodies have interacted with measles virus antigen coated on the plate. For an avidity ELISA to perform properly, denaturant solutions need to be optimized so that the weaker interactions of low-avidity antibodies are disrupted by the denaturant solution and IgG antibodies are washed away, while the stronger interactions of high-avidity antibodies remain undisrupted. Diethylamine, ammonium thiocyanate, and urea are denaturants that can be used in this critical step. Avidity results are generally calculated as the ratio of absorbance values measured in wells treated with the denaturant wash over absorbance values measured in wells treated with wash buffer and classified into low-, high-, and, depending on the assay, intermediate- or equivocal avidity results based on established avidity thresholds [5, 7–12]. Alternatively, avidity results are calculated as the concentration of ammonium thiocyanate required to elute 50% of bound IgG antibodies [13].

In measles elimination settings, avidity results are used to classify confirmed measles cases with documented vaccination records (breakthrough measles cases) into those who never responded to the vaccine (primary vaccine failure) and those who responded to the vaccine and later became susceptible to measles virus infection due to antibody waning (secondary vaccine failure) [14, 15]. Vaccine failure classification is informative in measles control and surveillance in areas where endemic measles virus circulation has been interrupted [16–21]. Avidity results are especially useful in elimination settings for providing support to confirm cases that do not have an epidemiological link (sporadic cases), and their IgM and RT-PCR results are unavailable or inconclusive [5, 6, 12]. In this specific context, low-avidity results can confirm unvaccinated acute measles cases; and high-avidity results with detection of elevated plaque reduction neutralization titers ($\geq$40,000 mIU/mL) can identify breakthrough cases [21–25]. However, high-avidity results alone cannot rule out nor rule in a measles case. Additionally, the quality of measles IgG has been studied in the context of novel measles vaccine development and in persons with immunodeficiencies or with underlying diseases [13, 26–29].

The measles IgG avidity assay method described in this chapter was developed by modifying the protocol of a commercially available measles-specific IgG enzyme immunoassay to include three 5-min washes with the denaturant diethylamine (60 mM, pH 10.25) following serum incubation [5]. Serum is serially diluted in this assay, and results are expressed as the end-titer avidity index [30]. This approach is considered optimal since results are independent of IgG concentration in the serum specimen [9]. Limitations of this assay can be found in [5].

2 Test Systems and Kits

2.1 ZEUS ELISA Measles IgG Test System (Zeus Scientific, Branchburg, NJ) (Or Equivalent), Which Contains the Following Components

1. Microtiter plate with 96 wells configured in twelve 8-well strips coated with measles virus antigen. Strips are shipped in a strip holder and inside a sealed envelope with a desiccant.
2. Serum diluent (SD).
3. Calibrator (human serum) (CAL).
4. Positive control (human serum) (PC).
5. Negative control (human serum) (NC).
6. Conjugated (horseradish peroxidase) goat antihuman IgG (ready for use).
7. 3,3′,5,5’-Tetramethylbenzidine (TMB): chromogen substrate for horseradish peroxidase (ready for use).
8. Stop solution: 1 N H_2SO_4 (ready for use).
9. 10× Wash buffer concentrate.

2.2 Captia™ Measles IgG Kit (Trinity Biotech, Jamestown, NY) (See Note 1), Which Contains the Following Components

1. Microtiter plate with 96 wells configured in twelve 8-well strips coated with measles virus antigen. Strips are shipped in a strip holder and inside a sealed envelope with a desiccant. Two strip holders are needed per run.
2. Serum diluent type 1 (SD).
3. Calibrator (human serum) (CAL).
4. Positive control (human serum) (PC).
5. Negative control (human serum) (NC).
6. Conjugated (horseradish peroxidase) goat antihuman IgG (ready for use).
7. 3,3′,5,5’-Tetramethylbenzidine (TMB): chromogen substrate for horseradish peroxidase (ready for use).
8. Stop Solution: 1 N H_2SO_4 (ready for use).
9. 20×Wash buffer type 1 concentrate.

2.3 In-House Avidity Control Sera (See Note 2)

1. High-avidity control: Select measles IgG positive serum samples from healthy individuals with known history of measles virus infection or vaccination more than a year before sample collection. Test specimen to confirm that it contains high-avidity IgG and with a mean end-titer avidity index % (etAI%) near the midpoint between 70% and 100%.
2. Intermediate-avidity control: Select measles IgG positive samples collected within 2–4 months after measles rash onset in an unvaccinated individual or after first-time vaccination from measles IgM-positive individuals aged 9 months and older (to avoid maternal antibody interference). Alternatively,

intermediate-avidity samples might be found among some healthy individuals that have been exposed to measles virus or vaccination in the distant past (more than a year). Test specimen to confirm that it contains intermediate-avidity IgG and with a mean etAI% near the midpoint between 30% and 70%.

3. Low-avidity controls: Select measles IgG positive samples collected within 2 months after measles rash onset or first-time vaccination from measles IgM-positive individuals aged 9 months and older (to avoid maternal antibody interference). Test specimen to confirm that it contains low-avidity IgG and with a mean etAI% near the midpoint between 0% and 30%.
4. In-house measles IgG positive control to have an internal control that is independent from the kit's controls.

2.4 Liquid Reagents

1. 60 mM diethylamine (DEA) (pH 10.25 (±0.05)). Preparation for this reagent is explained in Subheading 3.2. DEA is used in Subheading 3.5 as denaturant wash.
2. 1 M hydrochloric acid (HCl) (*see* **Note 3**) is used in Subheading 3.2 to lower the pH of the DEA.

2.5 Equipment

1. pH meter with pH calibrator solutions (pH 4, 7, and 10) and electrode storage solution.
2. Motorized serological pipette controller.
3. 50 mL conical tube and appropriate rack.
4. 1 L graduated cylinder.
5. 250 mL Erlenmeyer flask.
6. Parafilm.
7. 50 ml plastic pipettes.
8. 5 mL plastic pipettes.
9. 1 mL single micropipette and adequate tips.
10. 10 μL single micropipette and adequate tips.
11. 20 μL single micropipette and adequate tips.
12. 200 μL single micropipette and adequate tips.
13. 300 μL multichannel pipette (12 channels) and adequate tips.
14. Repeater eight-channel pipettor with 100 μL capability and adequate tips (optional).
15. Rack for serum tubes and 1 mL microtiter tubes for serum dilutions.
16. Calibrated timer: preset at (a) 20 min, (b) 5 min, (c) 5 min, (d) 10 min.
17. Paper towels.

18. Two distinct troughs for washes: one labeled "WB" for kit's wash buffer (WB solution) and the other labeled "DEA" for diethylamine wash (DEA solution).
19. Four troughs for reagents: one each for serum diluent, conjugate, TMB, and stopping solution.
20. Two basins to separately discard DEA and WB solutions.
21. Marker and tape.
22. Plate reader with filters for 450 nm and 630 nm.
23. 1 L bottle labeled "Diethylamine to discard <60 mM."
24. 1 L (or 0.5 L) bottle to store working-strength Captia wash at 2–8 °C.
25. Personal protective equipment: laboratory coat, disposable gloves, laboratory safety glasses.

2.6 Dilution Hardware

Nine empty boxes to hold microtiter tubes for serum dilutions, or similar. Label as follows:

1. Box 1: "Kit controls"—place eight tubes on column 1.
2. Box 2: "TOP, DEA only 1:10"—place *n* tubes on row A (*n* is number of samples and avidity controls to test at the top half of plate).
3. Box 3: "TOP, BOTH 1:10^2 to 1:10^4"—place *n* tubes on rows C, D, and F.
4. Box 4: "TOP, WB only 1:10^5"—place *n* tubes on row H.
5. Box 5: "BOTTOM, DEA only 1:10"—place *n* tubes on row A.
6. Box 6: "BOTTOM, BOTH 1:10^2 to 1:10^4"—place *n* tubes on rows C, D, and F.
7. Box 7: "BOTTOM, WB only 1:10^5"—place *n* tubes on row H.
8. Box 8 "TOP BOTH"—place *n* tubes on row A, F, G, and H.
9. Box 9: "BOTTOM BOTH"—place *n* tubes on row A, F, G, and H.

3 Methods

3.1 Assay Preparation

1. Read prior to use the manufacturer's instructions of the Captia™ Measles IgG kit and ZEUS ELISA Measles IgG Test System (or equivalent), including described reagent storage and stability, biosafety precautions, and specimen collection and storage.
2. Determine measles IgG content of specimens to be tested for measles IgG avidity using the Zeus IgG measles kit (or equivalent) according to manufacturer's instructions. Only specimens that are positive for measles IgG can be tested

Table 1
Selection of dilution scheme based on IgG content as measured with the ZEUS ELISA Measles IgG Test System (Zeus Scientific, Branchburg, NJ) (*see* Notes 1 and 12)

IgG level	Zeus ISR	Suggested dilution scheme	Required serum volume
Extra low IgG	<1.2	Likely IgG N on avidity	10 μL to test IgG content
Low IgG	1.2–4	Both curves start at 1:10	At least 60 μL
Mid-IgG	4–6	DEA curve starts at 1:10 and wash buffer curve at 1:100	At least 40 μL
High IgG	>6	Both curves start at 1:100	At least 10 μL

for measles IgG avidity. Record the ISR (or equivalent value) for each IgG positive sample. This value is used to determine the dilution scheme used in the measles avidity assay (*see* Table 1 and **Note 4**).

3. Remove the Captia™ Measles IgG kit from refrigerator, calculate reagent volumes needed for each run, and place reagents at room temperature (21–25 °C) (*see* **Note 5**). Decide which of the three dilution schemes each sample will be run. This decision is based on sample IgG level (*see* Table 1). The dilution indicated below is the starting dilution for a four-point series dilution curve:
4. For samples with mid-IgG levels, prepare a 1:10 DEA curve and a 1:100 WB curve (*see* **steps 5–10** in Subheading 3.3). Use boxes 1 through 7 for this dilution scheme.
5. For samples with low IgG levels, prepare 1:10 DEA curve and 1:10 WB curve (*see* **steps 12–16** in Subheading 3.3). Use boxes 1 through 9.
6. For samples with high IgG levels, prepare 1:100 DEA curve and 1:100 WB curve (*see* **steps 18–22** in Subheading 3.3). Use boxes 1 through 9.
7. Calculate the number of well strips required by counting the number of specimens to test in duplicate and adding one extra strip for the kit controls. Attach half of the strips plus the additional strip to one strip holder starting on column 1. Tape down strips. Label this plate "WB." This plate will be washed with WB provided by the kit. Attach the remaining strips to the other strip holder beginning on column 2 and leaving column 1 empty. Tape down strips. Label this plate DEA. This plate will be washed with 60 mM diethylamine (DEA) at pH 10.25 (±0.05). Place both strip holders inside the envelope with desiccant, and seal with tape.

3.2 Preparation of Denaturant Wash (60 mM Diethylamine, pH 10.25)

1. Review Material Safety Data Sheets for diethylamine, which can be found online.
2. Calibrate the pH meter according to manufacturer's instructions.
3. Prepare wash buffer type 1 (WB) per kit's instructions.
4. Prepare 60 mM DEA wash at pH 10.25 (±0.05) (*see* **steps 5–9** in Subheading 3.2).
5. Measure 120 mL of WB and add to a 250 mL Erlenmeyer flask.
6. Measure pH of the WB to confirm pH is 7.2–7.4. Discard solution and prepare new one if pH is not in this range.
7. Under a chemical fume hood, add 0.75 mL stock DEA to 120 mL WB by using a 1 mL single-channel micropipette (*see* **Note 6**).
8. Measure pH of the DEA solution to confirm pH is at least 11.6. Discard solution and prepare new one if pH is below 11.6.
9. Lower pH of DEA solution to 10.25 (±0.05) by adding 1 M HCl; usually a volume between 5 mL and 6 mL is needed. Start by adding about 4 mL and then slowly adding 1 M HCl to get to 10.25. If final pH falls below 10.20, discard DEA solution and start over.

3.3 Preparation of Serum Dilutions (See Note 7)

1. *See* Fig. 1 for control and sample distribution.
2. In microtiter tubes of box 1, dilute kit controls, calibrators, and in-house IgG control to a 1:21 dilution by combining and mixing 200 μL of serum diluent and 10 μL of serum. Two more samples can be added if testing samples with extra low IgG content.
3. *See* **step 2** in Subheading 3.1 to select appropriate dilution scheme. To control for the effect of the denaturant wash on the assay, high-, intermediate-, and low-avidity controls are always included on each run, and they are diluted per the 1:10 DEA curve and 1:100 WB curve dilution scheme; these controls also control for preparation of the 1:10 DEA curve and 1:100 WB curve dilution scheme. All runs will include the 1:10 DEA curve and 1:100 WB curve dilution scheme to run the high-, intermediate-, and low-avidity controls. If a different dilution scheme is needed (both curves starting at 1:10 or 1:100), then the high-, intermediate-, and low-avidity controls are run as described above plus an additional high-avidity control that is run at the selected scheme (either starting both curves at 1:10 or both at 1:100).

	1	2	3	4	5	6	7	8	9	10	11	12
	Kit controls	A1	A2	A3	S1	S2	S3	S4	S5	S6	S7	S8
A	S.D.	High	Intermediate	Low	Test -1	Test -2	Test -3	Test -4	Test -5	Test -6	Test -7	Test -8
B	N.C.											
C	CAL 1											
D	CAL 2											
E	P.C.	High	Intermediate	Low	Test -1	Test -2	Test -3	Test -4	Test -5	Test -6	Test -7	Test -8
F	I.H. IgG											
G	S.D.											
H	S.D.											

	1	2	3	4	5	6	7	8	9	10	11	12
		A1	A2	A3	S1	S2	S3	S4	S5	S6	S7	S8
A	No strip	High	Intermediate	Low	Test -1	Test -2	Test -3	Test -4	Test -5	Test -6	Test -7	Test -8
B												
C												
D												
E		High	Intermediate	Low	Test -2	Test -3	Test -4	Test -5	Test -6	Test -7	Test -8	Test -8
F												
G												
H												

Fig. 1 Sample distribution for plate to be washed with (**a**) wash buffer or (**b**) diethylamine wash. Cells in yellow indicate location of kit's controls and calibrators. Cell in orange indicates location of in-house IgG control. Columns in green indicate location of avidity control samples. Columns in blue indicate location of clinical samples to test. Column in gray indicates a space without a strip. A1 through A3 are avidity control samples, S1 through S8 are clinical samples to test, SD is serum diluent, NC is negative control, CAL 1 and CAL 2 are calibrator test 1 and calibrator test 2, PC is positive control from measles IgG kit, and IH IgG is in-house IgG control

4. For 1:10 DEA curve and 1:100 WB curve dilution scheme, prepare two serially diluted curves for each sample with four-point dilutions (*see* **Note 7** and **steps 5–10** in Subheading 3.3):

   1:10 → 135 μL serum diluent (kit) + 15 μL serum stock

   $1{:}10^2$ → 270 μL serum diluent (kit) + 30 μL previous dilution

   $1{:}10^3$ → 270 μL serum diluent (kit) + 30 μL previous dilution

   $1{:}10^4$ → 270 μL serum diluent (kit) + 30 μL previous dilution

   $1{:}10^5$ → 270 μL serum diluent (kit) + 30 μL previous dilution

Fig. 2 A

	1	2	3	4	5	6	7	8	9	10	11	12
		A1	A2	A3	S1	S2	S3	S4	S5	S6	S7	S8
A	No strip	1:10	1:10	1:10	1:10	1:10	1:10	1:10	1:10	1:10	1:10	1:10
B		1:10^2	1:10^2	1:10^2	1:10^2	1:10^2	1:10^2	1:10^2	1:10^2	1:10^2	1:10^2	1:10^2
C		1:10^3	1:10^3	1:10^3	1:10^3	1:10^3	1:10^3	1:10^3	1:10^3	1:10^3	1:10^3	1:10^3
D		1:10^4	1:10^4	1:10^4	1:10^4	1:10^4	1:10^4	1:10^4	1:10^4	1:10^4	1:10^4	1:10^4
E		1:10	1:10	1:10	1:10	1:10	1:10	1:10	1:10	1:10	1:10	1:10
F		1:10^2	1:10^2	1:10^2	1:10^2	1:10^2	1:10^2	1:10^2	1:10^2	1:10^2	1:10^2	1:10^2
G		1:10^3	1:10^3	1:10^3	1:10^3	1:10^3	1:10^3	1:10^3	1:10^3	1:10^3	1:10^3	1:10^3
H		1:10^4	1:10^4	1:10^4	1:10^4	1:10^4	1:10^4	1:10^4	1:10^4	1:10^4	1:10^4	1:10^4

Fig. 2 B

	1	2	3	4	5	6	7	8	9	10	11	12
		A1	A2	A3	S1	S2	S3	S4	S5	S6	S7	S8
A		1:10^2	1:10^2	1:10^2	1:10^2	1:10^2	1:10^2	1:10^2	1:10^2	1:10^2	1:10^2	1:10^2
B	1:21	1:10^3	1:10^3	1:10^3	1:10^3	1:10^3	1:10^3	1:10^3	1:10^3	1:10^3	1:10^3	1:10^3
C	1:21	1:10^4	1:10^4	1:10^4	1:10^4	1:10^4	1:10^4	1:10^4	1:10^4	1:10^4	1:10^4	1:10^4
D	1:21	1:10^5	1:10^5	1:10^5	1:10^5	1:10^5	1:10^5	1:10^5	1:10^5	1:10^5	1:10^5	1:10^5
E	1:21	1:10^2	1:10^2	1:10^2	1:10^2	1:10^2	1:10^2	1:10^2	1:10^2	1:10^2	1:10^2	1:10^2
F	1:21	1:10^3	1:10^3	1:10^3	1:10^3	1:10^3	1:10^3	1:10^3	1:10^3	1:10^3	1:10^3	1:10^3
G		1:10^4	1:10^4	1:10^4	1:10^4	1:10^4	1:10^4	1:10^4	1:10^4	1:10^4	1:10^4	1:10^4
H		1:10^5	1:10^5	1:10^5	1:10^5	1:10^5	1:10^5	1:10^5	1:10^5	1:10^5	1:10^5	1:10^5

Fig. 2 (**a**): Distribution of serum dilutions on the plate to be washed with diethylamine wash. Strip for column 1 is removed (gray column). Columns 2 through 4 (green columns) are used to test avidity control samples (A1–A3), and columns 5 through 12 (blue columns) are used to test clinical samples (S1–S8). Dilution curves for avidity controls and test samples start at 1:10 and end at 1:10^4. (**b**): Distribution of serum dilutions on the plate to be washed with wash buffer. Column 1 is used to test measles IgG kit controls and calibrators (yellow cells) and the in-house IgG control (orange cell). Kit controls and calibrators are analyzed at a 1:21 dilution. Columns 2 through 4 (green columns) are used to test avidity control samples (A1–A3), and columns 5 through 12 (blue columns) are used to test clinical samples (S1–S8). Dilution curves for avidity controls and test samples start at 1:100 and end at 1:10^5

5. One curve to be washed with DEA starts at a 1:10 dilution and ends at 1:10^4 dilution (*see* Fig. 2a), and the other curve to be washed with WB starts at a 1:10^2 dilution and ends at 1:10^5 dilution on (*see* Fig. 2b). Prepare as follows:
6. In microtiter tubes in box 2, prepare a 1:10 serum dilution by combining and mixing 135 μL serum diluent and 15 μL patient serum or avidity control (high-, intermediate-, and low-avidity) to be tested on top half of the plate (one tube per specimen).

7. In microtiter tubes in boxes 3 and 4, add 270 μL of serum diluent (four additional tubes per specimen).
8. To prepare a $1{:}10^2$ dilution, transfer and mix 30 μL of the 1:10 serum dilution from box 2 row A into microtiter tube with 270 μL of serum diluent in box 3 row C.
9. To prepare $1{:}10^3$ through $1{:}10^5$ dilutions, transfer and mix 30 μL of previous dilution into 270 μL of serum diluent for three additional dilutions using the remaining microtiter tubes in box 3 (rows D and F) and box 4 (row H).
10. Repeat **steps 1** through **4** for specimens to be tested in bottom half of the plate using boxes 5 through 7.
11. For 1:10 DEA curve and 1:10 WB curve dilution scheme, prepare two serially diluted curves for each sample with four-point dilutions (*see* **steps 12–16** in Subheading 3.3):

    1:10 → 225 μL serum diluent (kit) + 25 μL serum stock

    $1{:}10^2$ → 270 μL serum diluent (kit) + 30 μL previous dilution

    $1{:}10^3$ → 270 μL serum diluent (kit) + 30 μL previous dilution

    $1{:}10^4$ → 270 μL serum diluent (kit) + 30 μL previous dilution
12. Both curves start at a 1:10 dilution and end at $1{:}10^4$ dilution. Prepare as follows:
13. In microtiter tubes in row A of box 8 (top half) and box 9 (bottom half), prepare a 1:10 serum dilution by combining and mixing 225 μL serum diluent and 25 μL patient serum or high-avidity control (one tube per specimen).
14. In microtiter tubes in rows F–H of box 8 (top half) and box 9 (bottom half), add 270 μL of serum diluent (three tubes per specimen).
15. To prepare a $1{:}10^2$ dilution, transfer and mix 30 μL of the 1:10 serum dilution from row A into row F with 270 μL of serum diluent.
16. To prepare $1{:}10^3$ and $1{:}10^4$ dilutions, transfer and mix 30 μL of previous dilution into 270 μL of serum diluent for two additional dilutions using the remaining microtiter tubes with 270 μL of serum diluent (rows G and H).
17. For 1:100 DEA curve and 1:100 WB curve dilution scheme, prepare two serially diluted curves for each sample with four-point dilutions (*see* **steps 18–22** in Subheading 3.3):

    $1{:}10^2$ → 300 μL serum diluent (kit) + 3 μL serum stock

    $1{:}10^3$ → 270 μL serum diluent (kit) + 30 μL previous dilution

    $1{:}10^4$ → 270 μL serum diluent (kit) + 30 μL previous dilution

    $1{:}10^5$ → 270 μL serum diluent (kit) + 30 μL previous dilution

18. Both curves start at an approximate 1:100 dilution and end at an approximate $1:10^5$ dilution. Prepare as follows:
19. In microtiter tubes in row A of box 8 (top half) and box 9 (bottom half), prepare a $1:10^2$ serum dilution by combining and mixing 300 μL serum diluent and 3 μL patient serum or avidity control.
20. In microtiter tubes in rows F–H of box 8 (top half) and box 9 (bottom half), add 270 μL of serum diluent (three tubes per specimen).
21. To prepare a $1:10^3$ dilution, transfer and mix 30 μL of the $1:10^2$ serum dilution from row A into row F with 270 μL of serum diluent.
22. To prepare $1:10^4$ and $1:10^5$ dilutions, transfer and mix 30 μL of previous dilution into 270 μL of serum diluent for two additional dilutions using the remaining microtiter tubes (rows G and H).
23. Once all dilutions are prepared, cover boxes and order boxes from box 4 to 2 and box 7 to 5 so that the most diluted samples are available first for pipetting and less diluted samples are available last.

3.4 Incubation of Primary Antibody (Serum Incubation)

1. Remove plates from envelope.
2. Observe serum distribution in Figs. 1 and 2.
3. On plate labeled WB, load controls and specimens from box 1 into column 1.
4. On plate labeled WB, load 100 μL of the $1:10^5$ dilution point from box 7 into row H of columns 2 through *n*.
5. On plates labeled WB and DEA on columns 2 through *n*, load 100 μL of the $1:10^4$ dilution point on WB plate first (row G) and then 100 μL on DEA plate (row H). Continue by loading 100 μL of $1:10^3$ dilution point on WB plate first (row F) and then 100 μL on DEA plate (row G). Follow by loading 100 μL of the $1:10^2$ dilution point on WB plate first (row E) and then 100 μL on DEA plate (row F). Alternate plates to ensure that all dilution points are loaded at approximately the same time. This is best done in one continuous flowing movement (*see* **Note 8**).
6. On plate labeled DEA, load 100 μL of the 1:10 dilution point from box 4 into row E of columns 2 to *n*.
7. Without stopping, continue loading the top half of the plate by following **steps 4** through **6** using boxes 4 through 2. Start on row D of WB plate and end on row A of DEA plate.
8. Incubate at room temperature (21–25 °C) for 20 min as described in kit's insert.

3.5 Washing with Diethylamine (DEA) Solution

1. Add 300 μL of kit wash buffer to each well of the plate labeled WB starting on row H (*see* **Notes 9** and **10**). Start timer for 5 min.
2. Wait about 30 s and add 300 μL of 60 mM DEA wash to each well of the plate labeled DEA starting on row H. Start second timer for 5 min.
3. After 5 min of WB soak, take plate labeled WB, invert and shake out to discard liquid in appropriate container, blot plate on a paper towel with wells facing down, and repeat **step 1** two more times for a total of three washes.
4. After 5 min of DEA soak, discard the contents of the plate labeled WB, blot plate on a paper towel, and repeat **step 2** two more times for a total of three washes.
5. After the three 5-min washes, add 300 μL of kit wash buffer to all wells of the plate labeled WB. Discard the plate contents immediately (no soak), blot plate on a paper towel, and leave plate upside down in a moist area of the towel to prevent plate from drying while washing the DEA plate.
6. After the three 5-min washes, add 300 μL of wash buffer to all wells of the plate labeled DEA. Discard the plate contents immediately (no soak), blot plate on a paper towel, and leave plate upside down in a moist area of the towel to prevent plate from drying while washing the plate labeled WB.
7. Repeat wash **steps 5** and **6** in Subheading 3.4 two more times for a total of three washes by alternating plates (*see* **Note 11**). Perform a final blot on the paper towel for both plates. Ensure that there are no bubbles in the wells before proceeding to the next step.

3.6 Incubation with Secondary Antibody

1. Add 100 μL of horseradish-peroxidase conjugate (ready for use) to each well on both plates.
2. Cover plate and incubate at RT (21–25 °C) for 20 min.
3. Repeat wash **steps 5** and **6** in Subheading 3.5 for a total of three WB washes.
4. Add 100 μL of chromogen/substrate solution (TMB) to each well on both plates. Pay attention to order added to wells.
5. Cover and incubate at room temperature (21–25 °C) for 10 min.
6. Stop reaction immediately by adding 100 μL/well of stop solution and following same loading order as in **step 4** in Subheading 3.6.
7. Read absorbance on a dual wavelength plate reader at 450 nm and 630 nm.

Table 2
Avidity result interpretation

End-titer Avidity Index (etAI%)	Result classification
etAI% ≤ 30%	Low-avidity
etAI% ≥ 70%	High-avidity
30% < etAI% <70%	Equivocal; sample requires retesting. If the results after retest are still between these values, the sample is considered to be of intermediate avidity
If IgG is negative on the Captia™ test, an avidity result cannot be determined	Undetermined
If IgG positive on the WB plate and IgG negative at 1:10 on the DEA plate	Low-avidity

3.7 Calculations and Result Interpretation

Results are based on a ratio of the endpoint titer after the curve is washed with DEA over the endpoint titer after the curve is washed with wash buffer (*see* **Note 13**). This ratio is the end-titer avidity index (etAI%). To calculate etAI%, the formula described by Jenum et al. [30] was modified to account for the dilution scheme used in the measles avidity assay, as follows (Fig. 4):

$$\text{etAI\%} = (\textit{end}-\textit{titer}\ \text{diethylamine curve}/\textit{end}-\textit{titer}\ \text{wash buffer curve}) \times 100\text{, where}$$

$$\textit{end}-\textit{titer} = \text{dilution}_{\text{x}}{}^{-1} \times 10^{a}\text{, where}$$

$$a = \log 10 \times \left(\text{OD}_{\text{x}}\text{ODthresholdvalue}/\text{OD}_{\text{x}} - \text{OD}_{\text{y}}\right)\text{, where}$$

10 is the dilution factor and OD_x is the optical density (OD) value at dilution_x.

OD_y is the optical density value at dilution_y. Dilution_x and dilution_y always bracket the OD threshold value and therefore bracket the end-titer dilution.

OD threshold value is 0.9 × calibrator value as described in kit's insert.

etAI% are then interpretated and classified as indicated on Table 2 [5].

4 Notes

1. This measles IgG avidity assay was developed for use with the Captia™ Measles IgG kit [5]. If another kit is used, then appropriate validation of the alternative kit needs to be performed.

2. An alternative to using in-house serum specimens as avidity controls is to obtain the serum in large quantities from a commercial source to ensure large volumes of control serum. If a commercial source is used, provide exact instructions regarding timing of rash onset in a measles immunologically naïve individual, or timing of first measles vaccination, and timing of sample collection. Test measles IgG positive samples for measles IgG avidity.
3. Prepare 1 M hydrochloric acid (HCl) under a chemical fume hood by measuring 41 mL 37% HCl in 50 mL graduated cylinder, carefully transfer to a 500 mL graduated cylinder containing 400 mL of distilled water, and then add distilled water up to 500 mL.
4. If an alternative measles IgG assay is used to determine IgG content, then it is advisable to establish reference ranges that help determine the appropriate dilution scheme to run the samples on the measles IgG avidity assay (*see* Table 1).
5. When preparing the wash buffer type 1 (WB), it is helpful to fill a 50 mL conical tube to the top with wash buffer concentrate and to bring to room temperature (21–25 °C). Then pipette out the exact 50 mL of wash buffer concentrate to prepare the wash. The opening of the bottle with wash buffer concentrate is too narrow for a 50 mL serological pipette. Alternatively, a 50 mL graduated cylinder can be used.
6. To transfer the 0.75 mL stock DEA to WB solution, do not use polystyrene serological pipettes as they may melt.
7. An easy way to prepare serum dilutions is to load the required amount of serum diluent to all tubes as specified. Then prepare first 1:10 or 1:100 dilution (depending on dilution scheme) and progress to serially dilute samples as described.
8. To gain muscle memory and be able to load the plates efficiently, it helps to practice using water.
9. This step can be performed using the same tips for each 5-min wash as long as the first wash is WB and then DEA. To do so, remove one tip from the end to adapt to the number of columns of the DEA plate. Change tips every time WB is loaded.
10. Figure 3 illustrates a good way to have tips ready for all washes. Using a tip holder, place a tip at each specified location indicated in Fig. 3.
11. In this step, all tips can be reused except one that is removed to perform the wash on the DEA plate. Once the DEA plate is washed, a fresh tip is added to adapt to the number of columns of the WB pate (*see* Fig. 3).

		1	2	3	4	5	6	7	8	9	10	11	12
Washes after conjugate incubation	Tips for R.W.	t											
		t											
		t	t	t	t	t	t	t	t	t	t	t	t
Washes after serum incubation	Tips for R.W.	t											
		t											
		t	t	t	t	t	t	t	t	t	t	t	t
	Tips for DEA and R.W.	t	t	t	t	t	t	t	t	t	t	t	t
		t	t	t	t	t	t	t	t	t	t	t	t
		t	t	t	t	t	t	t	t	t	t	t	t

Fig. 3 Organization of tips in preparation to perform assay washes. Using a tip holder, place a tip (t) at each specified location indicated on the figure. The number of tips used will depend on the number of columns being used on the plate to be washed with wash buffer. RW is regular wash and DEA is diethylamine wash

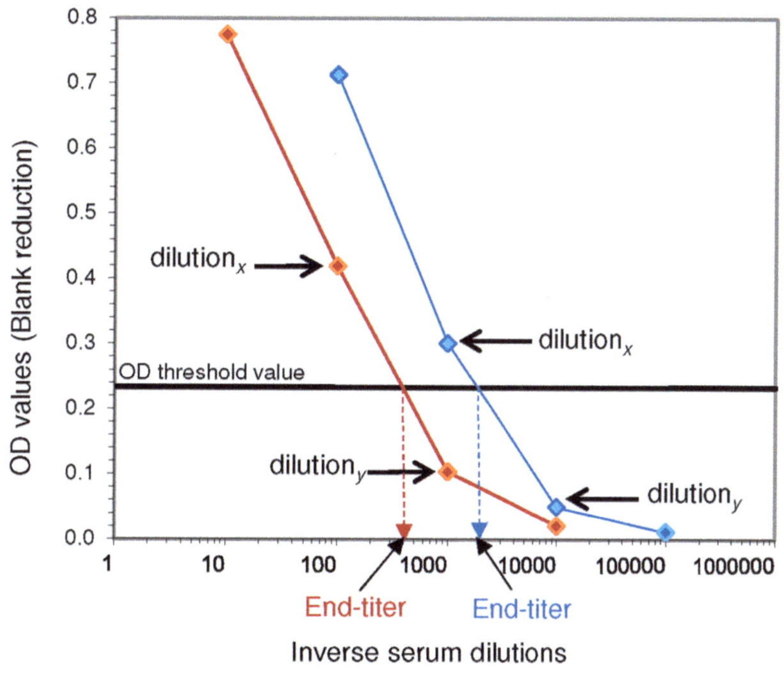

Fig. 4 Graphic representation of dilution series washed either with wash buffer (blue curve) or diethylamine wash (red curve) depicted for reference with elements used in the formula to calculate results for the endpoint titration measles IgG avidity assay [30]. OD is optical density value. Dilution$_x$ is dilution immediately before the OD threshold value and before curve reaches the end-titer. Dilution$_y$ is dilution immediately after the OD threshold value and after curve reaches the end-titer

12. Sometimes using the 1:10 DEA curve and 1:100 WB curve dilution scheme, an avidity result cannot be calculated. This may occur for two reasons:
    1. The regular wash (R.W) curve is IgG negative at a 1:100 dilution. If so, retest and start both curves at 1:10 and make 1:10 dilution curves (four points: 1:10 to $1{:}10^4$).
    2. The DEA wash curve is IgG negative at dilution 1:10 and IgG positive on the RW curve. If so, interpret as low-avidity result. Samples with very low IgG content can be tested at dilution 1:21 in column 1 of the RW plate in wells G and H to confirm samples are IgG positive by the Captia™ IgG assay.
    3. The DEA wash curve is IgG positive in all four dilution points, and an end-titer is not obtained when the dilution series is started at 1:10. If so, retest and start both curves at 1:100 and make 1:10 dilution curves (four points: $1{:}10^2$ to $1{:}10^5$).
13. Use of a program or a simple spreadsheet can assist in the calculation of end-titer avidity results.

Disclaimer

The findings and conclusions in this report are those of the authors and do not necessarily represent the official position of the Centers for Disease Control and Prevention.

References

1. Eisen H, Siskind G (1964) Variation in affinities of antibodies during immune response. Biochemistry 3:996
2. Kuby J (1997) Immunology. W. H. Freeman, New York
3. Siskind GW, Benacerraf B (1969) Cell selection by antigen in the immune response. Adv Immunol 10:1
4. El Mubarak HS, Ibrahim SA, Vos HW et al (2004) Measles virus protein-specific IgM, IgA, and IgG subclass responses during the acute and convalescent phase of infection. J Med Virol 72:290–298. https://doi.org/10.1002/jmv.10553
5. Mercader S, Garcia P, Bellini WJ (2012) Measles virus IgG avidity assay for use in classification of measles vaccine failure in measles elimination settings. Clin Vaccine Immunol 19:1810–1817. https://doi.org/10.1128/CVI.00406-12
6. Tuokko H (1995) Detection of acute measles infections by indirect and mu-capture enzyme immunoassays for immunoglobulin M antibodies and measles immunoglobulin G antibody avidity enzyme immunoassay. J Med Virol 45:306–311. https://doi.org/10.1002/jmv.1890450312
7. De Souza VA, Pannuti CS, Sumita LM et al (1997) Enzyme-linked immunosorbent assay-IgG antibody avidity test for single sample serologic evaluation of measles vaccines. J Med Virol 52:275–279
8. Hedman K, Lappalainen M, Seppaia I et al (1989) Recent primary toxoplasma infection indicated by a low avidity of specific IgG. J Infect Dis 159:736–740. https://doi.org/10.1093/infdis/159.4.736
9. Hedman K, Lappalainen M, Söderlund M et al (1993) Avidity of IgG in serodiagnosis of infectious diseases. Rev Med Microbiol 4:123–129

10. Mercader S, Mcgrew M, Sowers SB et al (2018) Development and use of an endpoint titration assay to characterize mumps IgG avidity following measles, mumps, and rubella vaccination and wild-type mumps infection. mSphere 3(5):e00320-18. https://doi.org/10.1128/msphere.00320-18
11. Narita M, Yamada S, Matsuzono Y et al (1996) Immunoglobulin G avidity testing in serum and cerebrospinal fluid for analysis of measles virus infection. Clin Diagn Lab Immunol 3: 211–215. https://doi.org/10.1128/cdli.3.2.211-215.1996
12. Thomas HI, Barrett E, Hesketh LM et al (1999) Simultaneous IgM reactivity by EIA against more than one virus in measles, parvovirus B19 and rubella infection. J Clin Virol 14: 107–118. https://doi.org/10.1016/S1386-6532(99)00051-7
13. Nair N, Gans H, Lew-Yasukawa L et al (2007) Age-dependent differences in IgG isotype and avidity induced by measles vaccine received during the first year of life. J Infect Dis 196: 1339–1345. https://doi.org/10.1086/522519
14. Pannuti CS, Morello RJ, Moraes JC et al (2004) Identification of primary and secondary measles vaccine failures by measurement of immunoglobulin G avidity in measles cases during the 1997 Sao Paulo epidemic. Clin Diagn Lab Immunol 11:119–122. https://doi.org/10.1128/cdli.11.1.119-122.2004
15. Paunio M, Hedman K, Davidkin I et al (2003) IgG avidity to distinguish secondary from primary measles vaccination failures: prospects for a more effective global measles elimination strategy. Expert Opin Pharmacother 4:1215–1225. https://doi.org/10.1517/14656566.4.8.1215
16. Atrasheuskaya AV, Blatun EM, Neverov AA et al (2005) Measles in Minsk, Belarus, 2001-2003: clinical, virological and serological parameters. J Clin Virol 34:179–185. https://doi.org/10.1016/j.jcv.2004.11.024
17. Bianchi S, Gori M, Fappani C et al (2022) Characterization of vaccine breakthrough cases during measles outbreaks in Milan and surrounding areas, Italy, 2017-2021. Viruses 14(5):1068. https://doi.org/10.3390/v14051068
18. Hamkar R, Mahmoodi M, Nategh R et al (2006) Distinguishing between primary measles infection and vaccine failure reinfection by IgG avidity assay. East Mediterr Health J 12: 775–782
19. Iwamoto M, Hickman CJ, Colley H et al (2021) Measles infection in persons with secondary vaccine failure, New York City, 2018-19. Vaccine 39:5346–5350. https://doi.org/10.1016/j.vaccine.2021.07.078
20. Kurata T, Yamamoto SP, Nishimura H et al (2021) A measles outbreak in Kansai International Airport, Japan, 2016: analysis of the quantitative difference and infectivity of measles virus between patients who are immunologically naive versus those with secondary vaccine failure. J Med Virol 93:3446–3454. https://doi.org/10.1002/jmv.26733
21. Mercader S, Dominguez A, Torner N et al (2021) Classification of measles breakthrough cases in an elimination setting using a comprehensive algorithm of laboratory results: why sensitive and specific IgM assays are important. Int J Infect Dis 112:21–24. https://doi.org/10.1016/j.ijid.2021.09.004
22. Hickman CJ, Hyde TB, Sowers SB et al (2011) Laboratory characterization of measles virus infection in previously vaccinated and unvaccinated individuals. J Infect Dis 204(Suppl 1): S549–S558. https://doi.org/10.1093/infdis/jir106
23. Rota JS, Hickman CJ, Sowers SB et al (2011) Two case studies of modified measles in vaccinated physicians exposed to primary measles cases: high risk of infection but low risk of transmission. J Infect Dis 204(Suppl 1):S559–S563. https://doi.org/10.1093/infdis/jir098
24. Sowers SB, Rota JS, Hickman CJ et al (2016) High concentrations of measles neutralizing antibodies and high-avidity measles IgG accurately identify measles reinfection cases. Clin Vaccine Immunol 23:707–716. https://doi.org/10.1128/CVI.00268-16
25. Venkat H, Briggs G, Brady S et al (2019) Measles outbreak at a privately operated detention facility: Arizona, 2016. Clin Infect Dis 68: 2018–2025. https://doi.org/10.1093/cid/ciy819
26. Brunell PA, Vimal V, Sandu M et al (1995) Abnormalities of measles antibody response in human immunodeficiency virus type 1 (HIV-1) infection. J Acquir Immune Defic Syndr Hum Retrovirol 10:540–548
27. Song MK, Vindurampulle CJ, Capozzo AV et al (2005) Characterization of immune responses induced by intramuscular vaccination with DNA vaccines encoding measles virus hemagglutinin and/or fusion proteins. J Virol 79:9854–9861. https://doi.org/10.1128/jvi.79.15.9854-9861.2005
28. Spoulou V, Giannaki M, Vounatsou M et al (2004) Long-term immunity to measles, mumps and rubella after MMR vaccination among children with bone marrow transplants. Bone Marrow Transplant 33:1187–

1190. https://doi.org/10.1038/sj.bmt.1704476

29. Zlamy M, Zoggeler T, Bachmann M et al (2022) Immunological memory and affinity maturation after vaccination in patients with propionic acidemia. Front Immunol 13: 774503. https://doi.org/10.3389/fimmu.2022.774503

30. Jenum PA, Stray-Pedersen B, Gundersen AG (1997) Improved diagnosis of primary Toxoplasma gondii infection in early pregnancy by determination of antitoxoplasma immunoglobulin G avidity. J Clin Microbiol 35:1972–1977. https://doi.org/10.1128/jcm.35.8.1972-1977.1997

Correction to: Measles Foci Reduction Neutralization Test (FRNT)

Sabine Santibanez and Annette Mankertz

Correction to:
Chapter 16 in: Dzwokai Z. Ma and Christian K. Pfaller (eds.), *Measles and Related Morbilliviruses: Methods and Protocols*, Methods in Molecular Biology, vol. 2808, https://doi.org/10.1007/978-1-0716-3870-5_16

This book was inadvertently published without replacing "1.6" with "1600" in Reciprocal dilution column of table 1b on page 220, which has now been incorporated.

The updated version of this chapter can be found at https://doi.org/10.1007/978-1-0716-3870-5_16

Dzwokai Z. Ma and Christian K. Pfaller (eds.), *Measles and Related Morbilliviruses: Methods and Protocols*,
Methods in Molecular Biology, vol. 2808, https://doi.org/10.1007/978-1-0716-3870-5_19,

INDEX

Dzwokai Z. Ma and Christian K. Pfaller (eds.), *Measles and Related Morbilliviruses: Methods and Protocols*, Methods in Molecular Biology, vol. 2808, https://doi.org/10.1007/978-1-0716-3870-5,

MIX
Papier aus verantwortungsvollen Quellen
Paper from responsible sources
FSC® C105338

If you have any concerns about our products,
you can contact us on
ProductSafety@springernature.com

In case Publisher is established outside the EU,
the EU authorized representative is:
Springer Nature Customer Service Center GmbH
Europaplatz 3, 69115 Heidelberg, Germany

Printed by Libri Plureos GmbH
in Hamburg, Germany